Introduction to Polymer–Clay Nanocomposites

Introduction to Polymer–Clay Nanocomposites

Ahmet Gürses

Pan Stanford Publishing

Published by

Pan Stanford Publishing Pte. Ltd.
Penthouse Level, Suntec Tower 3
8 Temasek Boulevard
Singapore 038988

Email: editorial@panstanford.com
Web: www.panstanford.com

British Library Cataloguing-in-Publication Data
A catalogue record for this book is available from the British Library.

Introduction to Polymer–Clay Nanocomposites

ISBN 978-981-4613-02-6 (Hardcover)
ISBN 978-981-4613-03-3 (eBook)

Printed in the USA

To my daughter and son

Contents

Preface xiii
Acknowledgment xv

1. Polymers and Polymer Synthesis **1**
1.1 Introduction 1
1.1.1 Historical Developments 2
1.1.2 Classifications of Polymers 4
1.1.2.1 Based on origin 5
1.1.2.2 Based on the configuration 6
1.1.3 Structures and Properties of Polymers 7
1.1.3.1 Polyethylene 7
1.1.3.2 Polypropylene 8
1.1.3.3 Poly(methyl methacrylate) 9
1.1.3.4 Polystyrene 9
1.1.3.5 Poly(vinyl chloride) 11
1.1.3.6 Poly(vinyl acetate) 12
1.1.3.7 Poly(vinyl alcohol) 12
1.1.3.8 Engineering polymers 13
1.1.3.9 Nylons 14
1.1.3.10 Epoxy resins 15
1.1.3.11 Phenol formaldehyde resins 16
1.1.3.12 Amino resins 18
1.1.3.13 Polytetrafluoroethylene 19
1.1.3.14 Polyurethanes 19
1.1.3.15 Poly(ether ether ketone) 21
1.1.3.16 Silicones 21
1.1.4 Natural Polymers 22
1.1.4.1 Cellulose 22
1.1.4.2 Starch 23
1.1.4.3 Natural rubber 23
1.1.4.4 Proteins 25
1.1.4.5 Poly-3-hydroxybutyrate 26
1.2 Polymerization Reactions 26
1.2.1 Step-Growth Polymerization 27
1.2.1.1 Typical step-growth polymers 29
1.2.2 Chain Polymerization 40

1.2.2.1 Initiation 41
1.2.2.2 Propagation 43
1.2.2.3 Termination 43
1.2.2.4 Chain transfer 44
1.2.3 Diene Polymerization 45
1.2.4 Practical Methods of Chain Polymerization 47
1.2.4.1 Bulk polymerization 47
1.2.4.2 Solution polymerization 48
1.2.4.3 Suspension polymerization 49
1.2.4.4 Emulsion polymerization 49
1.2.5 Ionic and Coordination Polymerizations 50
1.2.5.1 Cationic polymerization 51
1.2.5.2 Anionic polymerization 52
1.2.5.3 Coordination polymerization 53
1.2.6 Ring-Opening Polymerization 54
1.2.6.1 Poly(propylene oxide) 55
1.2.6.2 Epoxy resins 56
1.2.6.3 Polycaprolactam (nylon-6) 56
1.3 Properties of Polymers 59
1.3.1 Mechanical Properties 59
1.3.1.1 Mechanical tests 61
1.3.2 Thermal Properties 66
1.3.2.1 Glass transition temperature 66
1.4 Polymer Characterization 69
1.4.1 Nuclear Magnetic Resonance 69
1.4.2 Infrared Spectroscopy 70
1.4.3 Thermogravimetric Analysis 70
1.4.4 Differential Scanning Calorimetry 71
1.4.5 X-Ray Diffraction 71
1.4.6 Optical Microscopy 72
1.4.7 Dynamic Mechanical Analysis 73

2. Clay Minerals, Surface Chemistry of Clays and Organoclays, and the Mechanisms of Organoclay Synthesis **87**
2.1 Introduction 87
2.2 Clay Minerals 88
2.2.1 Structure of Clay Minerals 89

2.2.2 Classification of Clay Minerals 91
2.2.2.1 The 1:1 layer type 91
2.2.2.2 The 2:1 layer type 94
2.2.3 Characterization Techniques of Clay Minerals 101
2.2.3.1 X-ray diffraction 101
2.2.3.2 Nuclear magnetic resonance spectroscopy 102
2.2.3.3 Thermal analysis 103
2.2.3.4 Fourier transform infrared spectroscopy 103
2.2.3.5 Electron and scanning probe microscopy 104
2.3 Surface Chemistry of Clay and Organoclays 106
2.3.1 Adsorption at the Clay–Liquid Interface and Chemical Grafting 115
2.3.2 Mechanisms of Adsorption and Aggregation 115
2.3.2.1 Ion exchange 122
2.3.2.2 Ion pairing 125
2.3.2.3 Acid–base interaction and either hydrogen bonding between substrate and adsorbate or Lewis acid–Lewis base reaction 125
2.3.2.4 Adsorption by polarization of π electrons 127
2.3.2.5 Adsorption by dispersion forces 127
2.3.2.6 Hydrophobic bonding 129
2.3.3 Organoclay Structures 144

3. Organoclay Synthesis Methods **191**
3.1 Introduction 191
3.2 Synthesis of Organoclays 195
3.2.1 Cation Exchange 195
3.2.2 Solid-State Interaction 199
3.3 Quaternary Alkylammonium Cations Used to Prepare Organoclays 200
3.4 Applications of Organoclays as Sorbents 222

3.4.1 Aromatic Compounds 222
3.4.1.1 Phenol and its derivatives 223
3.4.1.2 Trichloroethylene 224
3.4.1.3 Pesticides and herbicides 224
3.4.1.4 Anionic contaminants 225

4. Polymer–Clay Nanocomposite Synthesis Methods 239
4.1 Introduction 239
4.2 Types of Polymer–Clay Nanocomposites 241
4.3 Preparation of Polymer–Clay Nanocomposites 244
4.3.1 Template Synthesis (Sol–Gel Technology) 244
4.3.2 Solution Intercalation 246
4.3.3 In situ Intercalative Polymerization 248
4.3.3.1 Thermoplastic nanocomposites 250
4.3.3.2 Thermoset nanocomposites 251
4.3.3.3 Rubber-modified epoxy nanocomposites 252
4.3.4 Melt Intercalation 252
4.3.4.1 Thermoplastic nanocomposites 254
4.3.4.2 Elastomer nanocomposites 254
4.4 Characterization of Polymer–Clay Nanocomposites 256
4.4.1 X-Ray Diffraction 256
4.4.2 Transmission Electron Microscopy and Scanning Electron Microscopy 258
4.4.3 Atomic Force Microscopy 261
4.4.4 Differential Scanning Calorimetry and Thermogravimetric Analysis 261
4.4.5 Fourier Transform Infrared Spectroscopy 262
4.4.6 Nuclear Magnetic Resonance 263
4.4.7 Energy-Dispersive X-Ray Spectroscopy 263
4.4.8 Small-Angle X-Ray Scattering 264
4.4.9 Cone Calorimeter 264
4.4.10 Mass Loss Calorimeter 266

4.5 Properties of Polymer–Clay Nanocomposites 266
4.5.1 Mechanical Properties 267
4.5.2 Thermal Properties 269
4.5.3 Flame Retardancy 270
4.5.4 Barrier Properties 271
4.5.5 Anticorrosive Properties 273
4.5.6 Chemical Properties of Polymer–Silicate Nanocomposites 273

5. Applications of Polymer–Clay Nanocomposites 293
5.1 Introduction 293
5.2 Food Packing 294
5.3 Biomedical Applications 300
5.3.1 Drug Delivery Applications 307
5.3.2 Tissue Engineering and Repair 310
5.3.3 Hydrogels with Biomedical Potential 313
5.4 Wastewater Pretreatment 314
5.5 Electricals/Electronics, Optoelectronics, Sensors, and Automobiles 317
5.6 Other Uses 321
5.6.1 Engineering Plastics 321
5.6.2 Polyester Cloth or Film 321
5.6.3 Reinforcing Plastics 321
5.6.4 Functionalized Plastics and Fibers 322
5.6.5 UHMWPE–MMT Nanocomposite Plastics 323
5.6.6 Polypropylene-Layered Silicate Nanocomposite Films 323
5.6.7 PP–Silica Nanoprecursor Films 324

Index 339

Preface

Nowadays, polymer–clay nanocomposite materials are of great interest in terms of scientific research and industrial applications. This book has been written to present a new practical overview of polymer–clay nanocomposites from a different perspective by offering a comprehensive introduction devoted to the typical characteristics of polymers, clays, and organo clays.

In this context, the first two chapters are devoted to polymers, polymerization mechanisms, and clay and characterization techniques. The second chapter examines predominantly the surface chemistry of the modifications with the main mechanisms of surface modification of the clay. Hence, this chapter starts with concise information describing the interface phenomenon and ends with a comprehensive analysis of interfacial characteristics of organoclay synthesized by considering adsorption and the other modification mechanisms.

Organoclays have critical importance not only in the creation of new materials and applications in materials science but also in the preparation of polymer nanocomposites, which is one of the most developed fields of nanotechnology. Therefore, the third chapter encloses synthesis methods of organoclay and the numerous related research results.

The fourth chapter is predominantly focused on the synthesizing methods of polymer–clay nanocomposites along with the structural, thermal, and mechanical characterization techniques of these composites.

Finally, in the fifth chapter, a wide range of industrial and technological applications of polymer–clay nanocomposite materials are proposed with their numerous practical examples.

To sum up, this book, which focuses on clay-reinforced polymer composites, can be regarded as a detailed review of adsorptive interactions in particular and other interparticle interactions facilitating the synthesis of organoclay, including the interactions between ions and functional groups in the interlayer region.

I believe that this book containing original figures, schemas, and numerous relevant references in each chapter would be an essential reference source for the readers working on the research and development of polymer–clay nanocomposites.

I wish to thank my colleagues, particularly Dr. Kübra Güneş, and Dr. Metin Açıkyıldız, Zafer Eroğlu, and Büşra Kuzey for their support in checking the text, drawing figures and schemas, and creating equations.

Ahmet Gürses
Erzurum, Turkey
Summer, 2015

Acknowledgment

I would like to express my gratitude to my PhD students Dr. Mehtap Ejder Korucu and Dr. Çetin Doğar for their moral support and valuable contribution related to the preliminary literature review.

Chapter 1

Polymers and Polymer Synthesis

1.1 Introduction

A polymer is a macromolecule composed of repeating structural units connected by covalent bonds. DNA, plastics, and proteins are well-known examples of polymers. Polymers in popular usage are referred to as plastics, but actually the term "plastic" corresponds to a large class of natural and synthetic materials in terms of various properties and usage fields. Natural polymer materials such as shellac and amber have been used for centuries. There are also many other natural polymers such as cellulose, which is the main constituent in the content of wood and paper. Biopolymers such as proteins and nucleic acids play crucial roles in biological processes. Bakelite, neoprene, nylon, poly(vinyl chloride) (PVC), polystyrene, polyacrylonitrile, and poly(vinyl butyral) (PVB) are examples of some common synthetic polymers. Polymers have been examined in broad fields such as polymer chemistry, polymer physics, and polymer science. Hermann Staudinger developed the concept of macromolecules during the 1920s. Staudinger's proposal was openly opposed by leading scientists, but the data eventually confirmed the existence of macromolecules (Heeger, 2001).

Nowadays the word "macromolecule" is used almost interchangeably with the word "polymer." But again, these two terms

Introduction to Polymer–Clay Nanocomposites
Ahmet Gürses

ISBN 978-981-4613-02-6 (Hardcover), 978-981-4613-03-3 (eBook)
www.panstanford.com

cannot be claimed to be exactly synonymous. Three main reasons for the difference may be mentioned.

First, many macromolecular compounds in solution behave like colloids. Therefore, it has been assumed that they are identical to inorganic colloids, whereas inorganic colloids are not macromolecular compounds, and they are actually composed of many small molecules aggregated by weak secondary interparticle interactions. German chemist C. D. Harries, who was working on the rubber molecule, pioneered such thinking in the early years of the 20th century (Bernal, 2006).

The second reason for opposition to Staudinger's hypothesis was that it meant the loss of the concept of a single formula for a single compound. Macromolecules had to be written in the form $(CH_2CHX)_n$, where n was a large number. Moreover, no means were available, or indeed are available, for discretely separating molecules where $n = 100$ from those where $n = 101$. Any such attempted fractionation always gives a distribution of values of n, and even if the mean value of a fraction is actually $n = 100$, there are significant numbers of molecules of $n = 99$, $n = 101$, and so on. Now, the concept of "one compound to one formula" using one formula capable of both physical and chemical interpretation had been developed.

Third, polymeric materials were both not well characterized and frequently crystalline. In other words, the reason of opposition was based on the nature of polymeric materials. Extracting crystalline substances of well-characterized physical and chemical properties had become possible for organic chemists.

1.1.1 Historical Developments

Modification of the properties of certain natural polymers created the modern polymer industry. Elastic properties of natural rubber, which was first extracted from a tropical tree in the 19th century, were developed by American chemist Charles Goodyear with a process known as vulcanization. Christian Schoenbein, in 1846, developed cellulose nitrate derived from cellulose (Fenichell, 1996; Heinze and Liebert, 2001). At the beginning, cellulose nitrate was recognized as an explosive material, but then it was noticed that cellulose nitrate was a hard, elastic material that could be molded into different shapes. A plasticized version of cellulose nitrate was used for commercial purposes and it made a development in photography (Vijayaram, 2009). In addition to these, cellophane and viscose rayon fibers were developed with the modification of cellulose at the end of the 19th century.

The first synthetic plastic was developed in the 20th century. Phenol and formaldehyde had been reacted to obtain heat-resistant material (Bakelite) by Leo Baekeland. This polymer was then used for commercial purposes such as household goods and development of electronic devices and in the car industry. The dark color, which was the only disadvantage of Bakelite, was overcome with urea formaldehyde (UF) polymers having a wide range of colors. The commercially important material polyethylene was first developed in the 1930s by chemists in the United Kingdom who were trying to observe the differences in the behavior of ethylene at different temperatures and pressures. On the other hand, nylon was developed deliberately (Kauffman, 1988). World War II led to the development of many low-cost synthetic polymers. After the war, polymers like polyethylene, polystyrene, and PVC had become widely used materials in everyday life (Nkwachukwu et al., 2013).

After the discovery of polymerization catalysis based on various coordination compounds of transition metals by Karl Ziegler in 1955, a striking change in the polymer field was started. With such a catalyst, Ziegler found that polyethylene could be synthesized from ethylene rapidly at ambient temperature and pressure. Furthermore, this polyethylene was almost entirely linear, unlike the low-density branched polyethylene known since 1935, which is produced only in excess elevated pressure and temperature conditions.

Giulio Natta then succeeded with catalysts of this type in polymerizing propylene and discovered the first synthetic stereospecific polymerization (Kulshreshtha and Talapatra, 2000). The Ziegler–Natta catalyst permits the synthesis of stereoregular polymers, in which the monomeric units have either the same or regularly alternating configurations, or the synthesis of isomeric randomly oriented polymers, which have quite different physical properties. Some stereoregular synthetic polymers may occur in a semicrystalline state; the randomly oriented polymers are always amorphous. Natural stereospecific reactions occur in the formation of proteins and other polymers of biologic origin, such as rubber and Gutta-percha (Sinha and Kumar, 2010).

1.1.2 Classifications of Polymers

Generally, polymers can be classified according to their thermal and mechanical properties into thermoplastics, thermosets, and elastomers. Thermoplastics are amorphous or semicrystalline polymers that soften or melt during heating and solidify during cooling. The heating/cooling/heating process can be repeated without perceptible changes in thermal and mechanical properties of thermoplastics. Thermosets undergo chemical changes during heating, and this process is irreversible. Elastomers can be vulcanized (crosslinked under assistance of heat, light, or special chemicals like sulfur and peroxides), which makes them reversibly stretchable for small deformations, but vulcanization is an irreversible process (Gregorova, 2013; Nicholson, 2006).

In 1929, Wallace Carothers proposed a different classification for polymers into two broad groups: condensation and addition polymers. In the case of condensation polymers, the molecular formula of the polymer is different from the formula of the monomer from which it is derived, whereas in the case of addition polymers, the molecular formula of the structural unit is identical to that of the monomer from which the polymer is derived.

Condensation polymers may be formed from monomers bearing two or more reactive groups of such a character that they may condense intermolecularly with the elimination of a by-product, often water. The polyamides and polyesters referred to above afford prime examples of condensation polymers. The formation of polyesters from a suitable diol and carboxylic acid is shown in Reaction 1.1 (Flory, 1953).

$$n[R(OH)_2] + n[R'(COOH)_2] \longrightarrow HO{+}R{-}COO{-}R'{-}COO{+}_n H + (n-1)H_2O \quad (1.1)$$

By contrast, addition polymers are those formed by the addition reaction of an unsaturated monomer, such as those that take place in the polymerization of vinyl chloride (Reaction 1.2).

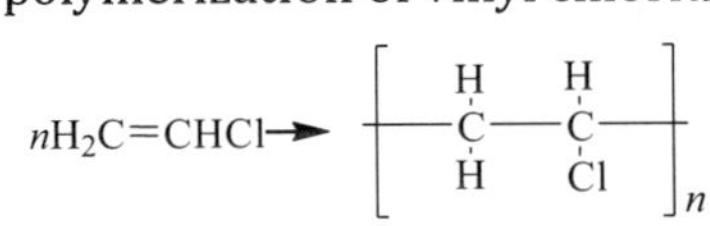

(1.2)

This reaction has been modified slightly by P. J. Flory, considering the mechanisms of the polymerization reactions. He has reclassified the polymerization reactions in the form of step reactions and chain reactions corresponding to the condensation and addition reactions in Carothers's scheme.

A notable exception is the synthesis of polyurethane, which is formed by the reaction of isocyanates with hydroxy compounds, followed by step kinetics without the elimination of a small molecule from the respective units (Reaction 1.3).

$$n\mathrm{R(OCN)_2} + n\mathrm{R'(OH)_2} \longrightarrow \mathrm{OCN}\left[\mathrm{R{-}NHCO_2{-}R'}\right]_n\mathrm{OH} \qquad (1.3)$$

In the first of these, the reaction kinetics reveals a gradual formation of product with high relative molar mass as the reaction proceeds. On the other hand, the chain reactions occur with rapid buildup of a few high-relative-molar-mass molecules, whereas the rest of the monomer remains unreacted. This means that the large molecules with high relative molar mass are formed at the beginning of the polymerization reaction, and also the monomer molecules exist during reaction.

Step reactions can produce macromolecules, having simple no-branched linear structures and also having an intensive crosslinked network. The final structure of the polymer depends on the number of functional groups of the parent monomer, whereas the highly crosslinked network can emerge if a sufficient degree of the branching exists. By contrast, the chain reactions can produce only linear or low-branched polymers. By taking into account the thermoset and thermoplastic classification, it can be said that the chain reactions form thermoplastics, while the step reactions may give either thermoplastics or thermosets. Also there are many classifications based on different perspectives. Some of these classifications are given next.

1.1.2.1 Based on origin

Polymers can be obtained from natural sources and as synthetics. Natural polymers can be originated from proteins and carbohydrates. The typical polymer examples based on proteins and carbohydrates are silk, collagen, keratin, cellulose, starch, and glycogen. Synthetic polymers, which are prepared in the laboratory, are referred to as synthetic polymers or man-made polymers. Some examples of synthetic polymers are polyethylene, polystyrene, Teflon, PVC, synthetic rubber, nylon, Bakelite, Orlon, polyester, and Terylene (Charles and Carraher, 2003).

1.1.2.2 Based on the configuration

A polymer can exhibit three different configurations depending on the functionality of the monomer with which it was synthesized. The main configurations are linear (chain), branched, and crosslinked, or network.

1.1.2.2.1 *Linear (chain) polymers*

The linear polymers that are soluble in a suitable solvent and fusible are made up of one long chain, without any branch, as can be seen in Fig. 1.1.

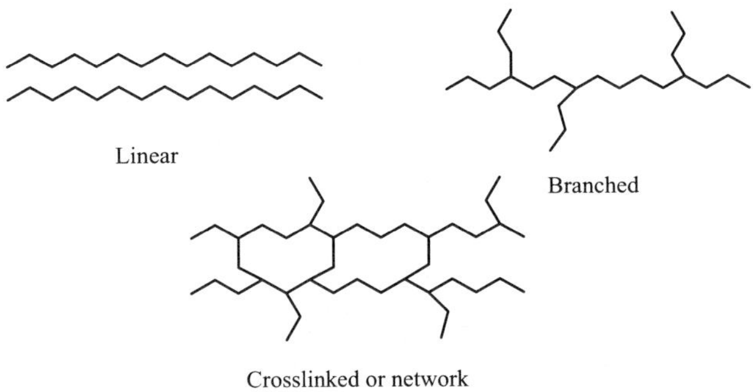

Figure 1.1 Structures of linear branched and crosslinked or network polymers.

1.1.2.2.2 *Branched polymers*

Branched polymers have a backbone to which the smaller-chain fragments in the main chain of molecules are attached (Fig. 1.1). The branched polymers have a lower crystallization tendency and lower densities. Branching emerges as a result of the various uncontrollable side reactions taking place during the process of polymerization. Branched polymers are easily soluble than linear polymers due to the weaker intermolecular interactions between the chains.

1.1.2.2.3 *Crosslinked or network polymers*

Crosslinked polymers have a wide backbone consisting of chains connected to each other by crosslinks to form a three-dimensional network (Fig. 1.1). The crosslinking in the polymers occurs, provided that the separate polymer chains are bonded chemically with each other. Thermoset polymers are crosslinked, with the restrictive structure preventing melting behavior. Some crosslinked polymers may show rubber-like characteristics, and these are known as *elastomers.* Such materials can be extensively stretched, but they rapidly recover their original dimensions (Stuart, 2002).

1.1.3 Structures and Properties of Polymers

In this section, by considering examples of some common polymers, the structural, physical, and chemical properties of polymers will be discussed. Most of the examples selected are of commercial importance. The fact that commercially important polymers are thermoplastics will also be discussed.

1.1.3.1 Polyethylene

Polyethylene, with its simple molecular structure ($[-CH_2CH_2-]_n$), is the most produced plastic material and the first that was used for electrical insulation in 1939. There is difficulty over the nomenclature of this polymer. The name of the polyethylene monomer is recommended as ethene by IUPAC. However, despite this confusion, chemists prefer to use the name "polyethylene" worldwide.

Four different industrial processes for the synthesis of polyethylene—high-pressure process, Ziegler process, Phillips process, and Standard Oil (Indiana) process—are used despite the emergence of products with partly different properties (Whiteley, 2012).

The high-pressure process uses pressures of 1000–3000 atm and temperatures of 80–300°C, together with free-radical initiators, such as benzoyl peroxide. During the 1950s, three research groups working independently discovered three different catalysts that allowed the production of essentially linear polyethylene at low pressure and temperature. These polymers had densities in the region of 960 kg/m^3 and became known as high-density polyethylenes (HDPEs), in contrast to the polymers produced by the extensively commercialized high-pressure process, which were named low-density polyethylenes (LDPEs). These discoveries laid the basis for the coordination catalysis of ethylene polymerization, which has continued to diversify.

Ziegler processes are based on coordination reactions catalyzed by using metal alkyl systems discovered by Ziegler in Germany and developed by Natta in Milan in the early 1950s. A typical Ziegler–Natta catalyst is prepared by using titanium tetrachloride and triethylaluminum and the reaction is carried out at low pressures and temperatures lower than 70°C. Lastly, both Phillips and Standard Oil (Indiana) processes have produced an HDPE using relatively low pressures and temperatures (Whiteley, 2012). This polymer has a wide range of applications, consisting of packing, houseware, appliances, transportation, communications, electric power, agriculture, construction, wire and cable insulation, coatings, and injection-molded products.

1.1.3.2 Polypropylene

Polypropylene, $[-CH_2CH(CH_3)-]_n$, has emerged as a commercial material, following the works of Natta on catalysts and polymerization of ethylene for the preparation of relatively high-molar-mass polymer from alkenes. Following his work on the polymerization of ethylene, Natta revealed the possibility of the synthesis of polymers analogous to propylene in 1957. Polypropylene, which is named polypropene by IUPAC, was first marketed in 1957.

Polypropylene is one of many polymers that exhibit stereoisomerism due to the presence of an asymmetric carbon center in its repeating unit. The polymer chains may be isotactic, where each asymmetric carbon has the same configuration; syndiotactic, where alternating asymmetric carbons have the same configuration; or atactic, where there is no regular configuration of the asymmetric carbon (Fratini, 2006). The first isomer is similar to polyethylene, but it has greater rigidity and hardness, whereas the other isomer has an amorphous structure and little strength.

Commercial polypropylene is usually about 90–95% isotactic and is very similar to polyethylene, but it has a lower density and a higher softening point and thus can be used at higher temperatures.

Polypropylene is not susceptible to the environmental stress cracking and is oxidized more easily as a consequence of the fragmentation of tertiary C–H bonds within the molecule. Another superior property of isotactic polypropylene is that it can be flexed repeatedly without causing weakness (Nicholson, 2006).

1.1.3.3 Poly(methyl methacrylate)

Poly(methyl methacrylate), which is often abbreviated as PMMA, is the most important commercial acrylic polymers. Acrylic polymers like PMMA are generally derived from poly(acrylic acid) (PAA), [-$CH_2CH(COOH)$-]$_n$. In the case of PMMA, this derivation can be achieved by replacing the tertiary hydrogen atom with a methyl group (CH_3) and by etherification between the carboxylic acid group and methanol, CH_3OH (Fig. 1.2).

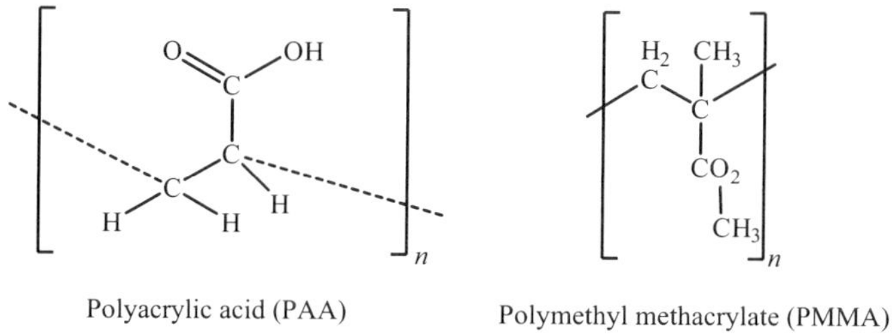

Figure 1.2 Structures of PAA and PMMA.

PMMA is a polymer with a simple carbon backbone containing in its monomer a methyl group (CH_3) and a methacrylate group (COO–CH_3); it is a glassy, transparent, tough material with high resistance to outdoor weathering; and it is used extensively as a replacement for glass in constructions (Koutsos, 2009). PMMA, as all amorphous polymers, has a glass transition temperature (T_g), which is 99°C. Below this temperature, PMMA exists in a stiff and relatively brittle state and above T_g in an elastic or rubber-like state, in which the polymer is soft but does not really flow yet. At a molecular level, rotation about the single bonds in the backbone of the polymer becomes possible above T_g (Carpentier, 2013).

The main applications for PMMA arise from the combination of its transparency and its good outdoor weathering properties. These are coupled with reasonable toughness and rigidity, so PMMA is a useful material in a range of glazing applications.

1.1.3.4 Polystyrene

Polystyrene is a hard, rigid, transparent thermoplastic that emits a characteristic metallic ring when dropped. It is free from odor and taste, burns with a sooty flame, and has a low specific gravity of

1.054 g/mL. Because of its low cost, good moldability, low moisture absorption, good dimensional stability, good electric insulation properties, colorability, and reasonable chemical resistance, it is widely used in injection molding and vacuum forming (Brydson, 1999).

The styrene monomer is vinyl benzene, being a derivative of benzene (Fig. 1.3). The first report of the polymerization reaction came in 1839, when E. Simon described the transformation of what was then called "styrol." He believed he had oxidized the material and called the product styrol oxide. Later, when it was realized that it contained no oxygen, the product became known as metastyrene.

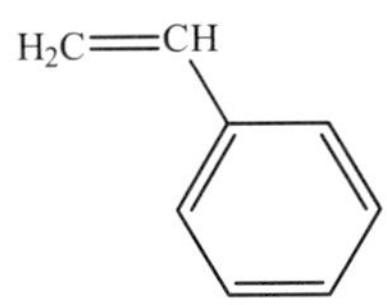

Figure 1.3 Structure of styrene.

This is essentially polyethylene with a phenyl group replacing a hydrogen group bonded to every other carbon atom in the chain (see Fig. 1.3). The effect of the benzene ring, which can be attached in any direction in contrast to the methyl group in polyethylene is to make the polymer stiff and brittle but amorphous, and it has poor chemical resistance. It is used for applications mainly promoted by its low cost. The impact strength of polystyrene can be improved by alloying, progressively with a rubber (styrene butadiene or polybutadiene) or with acrylonitrile to form styrene–acrylonitrile (SAN) and with both butadiene rubber and acronitrile to produce acrylonitrile–butadiene–styrene (ABS) copolymer. The improved toughness greatly increases the applications, and ABS competes with engineering plastics at a significantly lower cost (Wyatt et al., 1994). Also, the polymerization of styrene is used for crosslinking unsaturated polyesters. For this, typically a polyester containing double bonds is produced as a prepolymer. This prepolymer is then mixed with styrene to give a viscous solution, and this solution is then worked into a glass fiber mat.

The resulting product develops its final properties by polymerization of the styrene, which essentially undergoes homopolymerization but also incorporates the occasional double bond from the prepolymer molecule. This gives a large, three-dimensional network intertwined with the glass reinforcement. The composite material has excellent properties and is widely used for boat hulls, shower units, and baths (Nicholson, 2006).

1.1.3.5 Poly(vinyl chloride)

PVC is applied in a variety of products, including packaging, medical devices, food, and also long-life applications such as pipes, window frames, cable insulation, floors coverings, and roofing sheets (Alwaan, 2014). The structural arrangement of PVC has been shown to have the head-to-tail structure.

This arrangement has been experimentally confirmed by showing that it undergoes a Wurtz-type reaction to yield a product containing a small amount of chlorine and no detectable unsaturation when dissolved in dioxan and treated with zinc dust. The alternative possible structure is the head-to-head arrangement, giving a double bond instead of the two eliminated chlorine atoms (Reaction 1.4).

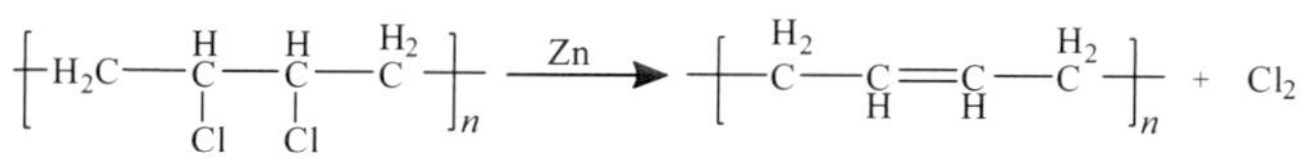

(1.4)

Despite its extensive use and very superior properties, PVC still has some problems due to its poor thermal stability, leading to its degradation by a dehydrochlorination reaction initiated at the labile sites present on the polymeric chains, such as branching, chloroallyl groups, end groups, oxygen-containing groups, and head-to-head structures, in addition to tacticity. This leads to strong discoloration of the polymer and deterioration of its physical and mechanical properties (Abdel-Kader and Mohamed, 2013).

When the materials manufactured from PVC are used in locations where they will be exposed to sunlight, they are deformed, and so it is necessary to add stabilizers, such as barium and cadmium laurates and tin di-iso-octylthioglycollate at the compounding stage. Numerous studies on PVC stabilizers and their mechanisms of action have been made. Stabilizers are used to terminate the unzipping of the chain and to reduce the autocatalytic effect of HCI produced as a result of degradation (Hussain and Mahmood, 1994; Okieimen, 2000; Petre et al., 1999; Taghizadeh and Dasdar Ligvan, 2007).

1.1.3.6 Poly(vinyl acetate)

Poly(vinyl acetate) (PVAc) is the most widely used polymer of vinyl ester (Fig. 1.4). It is not only used as a plastic in the form of emulsion but also used as the precursor for the two polymers, poly(vinyl alcohol) (PVA) and poly(vinyl acetal), which cannot be prepared by direct polymerization. PVC and PVAc are synthetic polymers with the formulae $(CH_2CHCl)_n$ and $(C_4H_6O_2)_n$, respectively.

PVC is widely used in construction because it is cheap, durable, and easy to assemble, while PVAc is a polymer of industrial importance, which includes its application as strengthening agents for cements (Tabhane et al., 2012).

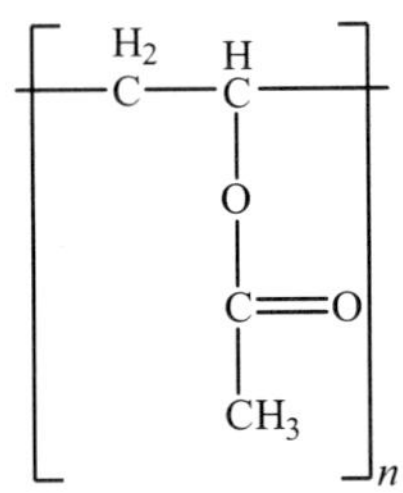

Figure 1.4 Structure of poly(vinyl acetal).

PVAc latex is used in the production of water-based stable and low-cost emulsion paints, adhesives, and textile and paper treatments. PVAc emulsion adhesives are generally used in labeling and packaging and as white glue, which is a popular adhesive.

1.1.3.7 Poly(vinyl alcohol)

PVA, or PVOH, was first prepared by Hermann and Haehnel in 1924 by hydrolyzing PVAc in ethanol with potassium hydroxide (Reaction 1.5). PVA is produced commercially from PVAc, usually by a continuous process (Fig. 1.5) (Singh et al., 2013).

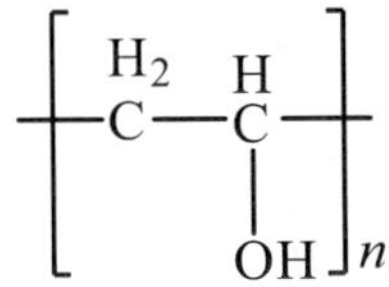

Figure 1.5 Structure of PVA.

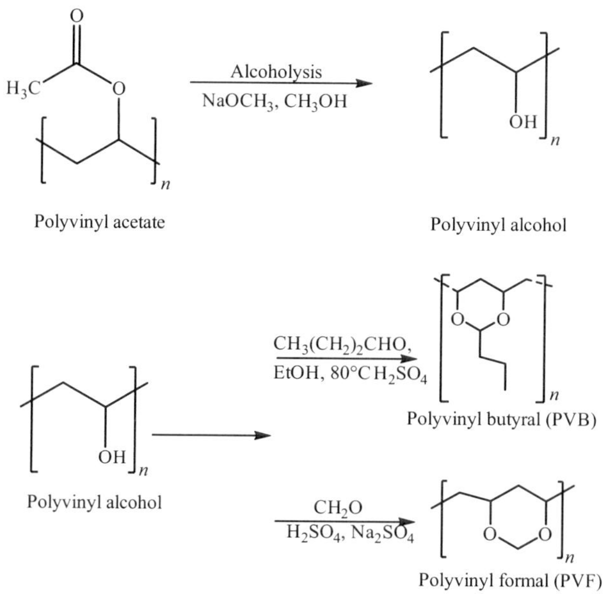

(1.5)

PVA has excellent film-forming, emulsifying, and adhesive properties (Kavitha et al., 2011). The formaldehyde-modified cottony PVA fibers that have too much water absorption capacity are a good alternative to cotton in applications requiring body contact. The fibers have properties such as good dimensional stability and high abrasion resistance, and they are easily washable and quick drying (Ebewele, 2000).

PVA is used also in the manufacture of PVB and poly(vinyl fluoride) (PVF). PVB is a resin usually used for applications that require strong binding, optical clarity, adhesion to many surfaces, toughness, and flexibility. It is prepared from PVA by reaction with butyraldehyde. The major application is laminated safety glass for automobile windshields (Hallensleben, 2000). PVF, which is a thermoplastic fluoropolymer with repeating vinyl fluoride units, has been mainly used in the flammability-lowering coatings of airplane interiors and photovoltaic module backsheets and also in raincoats and metal sheeting.

1.1.3.8 Engineering polymers

Engineering polymers are materials with exceptional mechanical properties such as stiffness, toughness, and low creep, which make them valuable in the manufacture of structural products like gears, bearings, electronic devices, and auto parts. They are polymers capable of withstanding a wide range of temperatures and having relatively high tensile, flexural, and impact strengths. The outstanding properties of engineering plastics are based mainly on their inherently strong intermolecular interactions.

Engineering plastics are an important alternative for metals in the automotive sector and home appliance manufacturing. As engineering polymers continue to take the place of traditional materials in many applications, products with more superior properties compared to the alternatives in many respects are being developed. The superior properties of engineering polymers can be enhanced by the addition of various types of reinforcements, by blending and alloy formation, and through chemical modification such as crosslinking.

1.1.3.9 Nylons

Nylon, which is one of the most commonly used polymers, was first produced in 1935 by Carothers. The term "nylon" is a generic designation for synthetic polymers known generically as aliphatic polyamides, and its other known representatives are nylon-6,6, nylon-6, nylon-6,9, nylon-6,10, nylon-6,12, nylon-11, nylon-12, and nylon-4,6. The first synthetic fibers appeared in the market in 1940 as a result of the work of Carothers, who developed nylon-6,6 (Matar and Hatch, 2000).

Nylon and its representatives are condensation or step polymers, and because of this, they are different from all the other commercially important polymers described so far.

Polyamides are formed by the polymerization of a diamine and a dicarboxylic acid and are commonly known as *nylons* (Khandal, 2006). They are distinguished from each other by a numbering system based on the number of carbon atoms of their precursors. For example, nylon-6,6, which is still the most important commercial nylon, is prepared by a reaction of hexamethylenediamine with adipic acid having six carbon atoms (Reaction 1.6) (Khandal, 2006).

$$n\mathrm{H_2N(CH_2)_6NH_2} + n\mathrm{NHOOC(CH_2)_4COOH} \longrightarrow \mathrm{HN_2}\left[\mathrm{(CH_2)_6NHCO(CH_2)_4CO}\right]_n + (n-1)\mathrm{H_2O} \tag{1.6}$$

Similarly, nylon-6,10 is prepared by using hexamethylenediamine with sebacic acid having 10 carbon atoms. Some nylons prepared by using a single precursor are termed with a single number. Such a starting material or a precursor can be either a long-chain amino acid, such as ω-amino dodecanoic acid with 11 carbon atoms, which will yield nylon-11 by self-condensation, or a closed-ring amide-type compound, known as lactam, which undergoes polymerization by a ring-opening reaction.

Nylon-6, which is used for filament manufacture (polycaproamide), is formed by the hydrolytic polymerization of caprolactam, which is the precursor to nylon-6. Also, nylon-12 can be synthesized by using dodecyl-lactam having 12 carbon atoms by the same method as that of nylon-6 (Datye, 1991). The reaction for nylon-12 has been shown as follows (Reaction 1.7):

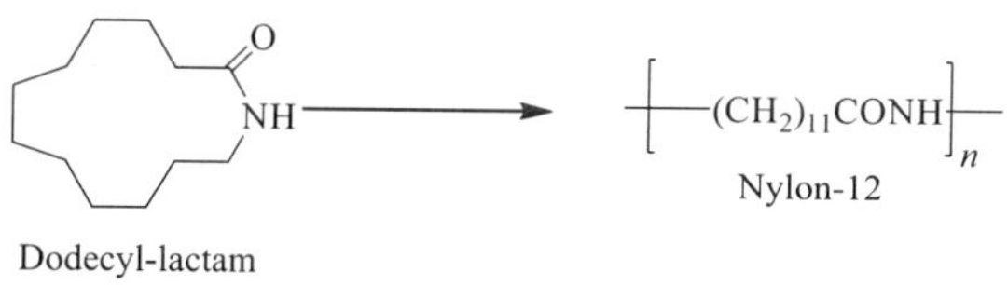

(1.7)

The various nylons tend to have similar physical properties, such as high impact strength, toughness, flexibility, and abrasion resistance. Nylons have very high resistance to organic solvents and also to fuels and oils but not to concentrated mineral acids at room temperature and to alkali solutions at high temperatures (Benhui, 1994). All nylons can be injection-molded, and the products are commonly used in engineering applications, such as bearings and gears (Nicholson, 2006).

1.1.3.10 Epoxy resins

Epoxy polymers that are based on the three-membered heterocyclic matrix as an epoxy or an oxirane ring (Fig. 1.6) are formed from an oligomer containing at least two epoxide groups and a curing agent, usually an amine or a diacid compound (Fink, 2005).

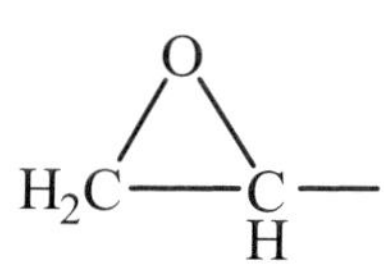

Figure 1.6 Three-membered heterocyclic systems.

Today, the most frequently used epoxy resin is produced from epichlorohydrin and bisphenol A (Rocks, 2004). They are cured through ring-opening crosslinking reactions such that the epoxy group can easily enter them.

Bisphenol A is formed by 2 moles of phenol and acetone and so it is called bisphenol (Reaction 1.8).

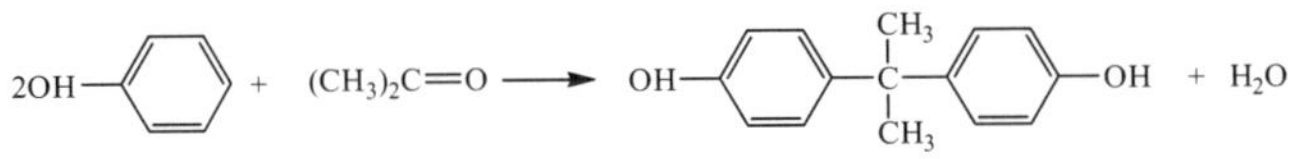

(1.8)

The ring opening is carried out by using a variety of reagents, namely, amines and anhydrides or Lewis acids ($SnCl_4$) and Lewis bases (tertiary amines). The crosslinking reactions, which are given in Reaction 1.8, may show that, in principle, only a trace amount of the curing agent is necessary for curing the epoxy resins. Selection of a curing agent depends on various considerations, such as cost, ease of handling, pot life, cure rate, and the required mechanical, electrical, or thermal properties of the final resin.

In general, epoxies are known for their excellent adhesion, chemical and heat resistance, good-to-excellent mechanical properties, and very good electrical insulating properties. The applications for epoxy-based materials are extensive and include coatings, adhesives, and resin matrices for composite materials such as those using carbon fibers and fiberglass reinforcements (although polyester, vinyl ester, and other thermosetting resins are also used for glass-reinforced plastic) (May, 1987).

1.1.3.11 Phenol formaldehyde resins

Phenol formaldehyde (PF) resins are synthetic polymers obtained by the reaction of phenol or substituted phenol with formaldehyde (Doğar et al., 2014). Phenolic resins are mainly used in the production of circuit boards. They are better known, however, for the production of molded products, including pool balls and laboratory countertops, and as coatings and adhesives. In the form of Bakelite, they are the earliest commercial synthetic polymers (Gardziella et al., 2000).

These polymers were first discovered in the early of 1890s, during the making of artificial dyes by using phenol and formaldehyde. But this discovery did not find an opportunity of becoming a commercial material before it received a patent in 1907 for the materials produced using these two compounds by Leo Baekeland (Pilato, 2010).

PF resins are formed by a step-growth polymerization reaction that can be either acid- or base-catalyzed. Since formaldehyde exists predominantly in solution as a dynamic equilibrium of methylene glycol oligomers, the concentration of the reactive form of formaldehyde depends on temperature and pH. Phenol reacts with formaldehyde at the *o* and *p* sites, allowing up to three units of formaldehyde to attach to the ring. The initial reaction in all cases involves the formation of a hydroxymethyl phenol. The hydroxymethyl group is capable of reacting with either another free *o* or *p* sites or with another hydroxymethyl group.

The first reaction gives a methylene bridge, and the second forms an ether bridge. If the base-catalyzed condensation of phenol with formaldehyde is carefully controlled, an intermediate product can be isolated, such as liquid or solid, depending on the extent of reaction.

At this stage, the material consists of essentially linear chains and is both fusible and soluble in appropriate solvents. When heated under pressure to 150°C, this intermediate is converted to a hard solid product called Bakelite. This second stage is the one at which the three-dimensional crosslinked network matrix develops. The cured phenolic resins are resistant to many chemicals, organic solvents, and water. Novolacs are PF resins with a formaldehyde-to-phenol molar ratio of less than 1. The polymerization is brought to completion using acid catalysis such as oxalic acid, hydrochloric acid, or sulfonate acids. The phenol units are mainly linked by methylene and/or ether groups. Novolacs are commonly used as photoresists. Hexamethylenetetramine, or hexamine, is a hardener added to a crosslinked novolac. At a temperature below 90°C, it forms methylene and dimethylene amino bridges (Charles and Carraher, 2003).

Base-catalyzed PF resins called resoles are made with a formaldehyde-to-phenol ratio of greater than 1 (usually around 1.5). Resoles are referred to as "one step" resins because they cure without a crosslinker, unlike novolacs, called "two step" resins. The typical application examples of resoles used widely for gluing and bonding building materials are exterior plywood, oriented strand boards, and engineered, laminated composite lumber (Parameswaran, 2009).

1.1.3.12 Amino resins

Amino resins are formed by the reaction of amines or amides with aldehydes. Among them, the UF and melamine formaldehyde (MF) resins have significant commercial importance (Bhattacharya and Ray, 2009). These two polymers having an intensively crosslinked network structure are in the thermoset polymer group. The structures of the two parent amino compounds are shown in Fig. 1.7.

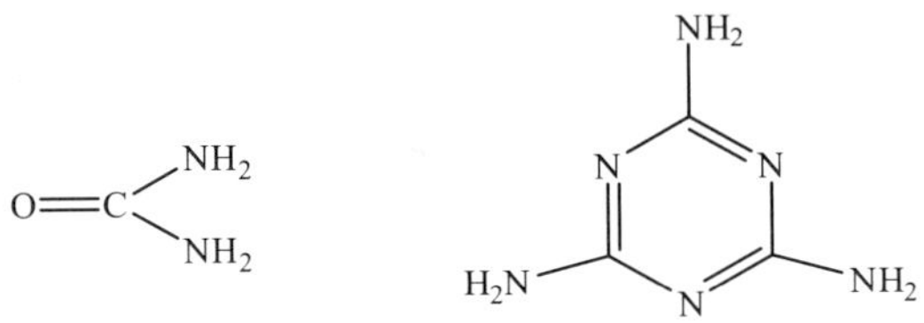

Urea Melamine

Figure 1.7 Structures of urea and melamine.

The initial step of the polymerization process is the reaction of the amine groups with formaldehyde to form methylol units (Reaction 1.9). At higher temperatures, the product can proceed to a different reaction, such as the self-condensation of the methylol groups (Reaction 1.9).

$$\begin{aligned} &-NH_2 + HCHO \longrightarrow -NHCH_2OH \\ &-NHCH_2OH \longrightarrow -NHCH_2OCH_2NH- + H_2O \end{aligned} \qquad (1.9)$$

Alternatively, the methylol groups can proceed to another condensation reaction with free amino groups by removing a mole of water (Reaction 1.10).

$$-NHCH_2OH + H_2N- \longrightarrow -NHCH_2NH- + H_2O \qquad (1.10)$$

Amino resins have been employed for over 60 years as crosslinkers for polyesters, acrylics, and epoxies as they offer good adhesion and excellent chemical and solvent resistance properties. The MF resin is the main constituent of high-pressure laminates, such as Formica, and laminate flooring (Bhattacharya and Ray, 2009). The main commercial products of UF resins are decorative laminates, textiles, paper, foundry sand molds, wrinkle-resistant fabrics, and cotton blends. The UF resin's attributes include high tensile strength, flexural modulus, and heat distortion temperature;

low water absorption and mold shrinkage; and high surface hardness, elongation at break, and volume resistance. Also, the crosslinked amino resins are very resistant toward organic solvents, acidic and alkali solutions.

1.1.3.13 Polytetrafluoroethylene

Polytetrafluoroethylene (PTFE), which is a synthetic fluoropolymer of tetrafluoroethylene, is the completely fluorine-substituted analogue of polyethylene, that is, (-CF_2CF_2-). The best-known brand name of PTFE-based formulas is Teflon, and there are numerous applications (Ebnesajjad, 2000).

PTFE, which is a white solid with a waxy appearance at room temperature and is also very tough and flexible, with good electrical insulation properties, is a thermoplastic polymer. This polymer has a very low surface energy, and the coefficient of friction exhibits excellent nonstick characteristics. Owing to this feature, it has been indispensable for the fabrication of nonstick kitchen utensils.

In industrial applications, owing to its low friction, PTFE is used for applications where sliding action of parts is needed, namely, plain bearings, gears, slide plates, etc. In these applications, it is significantly better than nylon and acetal and it is comparable to ultra-high-molecular-weight polyethylene (UHMWPE). In addition to its unique mechanical properties, PTFE has an almost all organic solvents, with the exception of molten alkali metals and fluorine.

Despite PTFE having a thermoplastic character, it does not show conventional melting behavior, like flow during heating. PTFE, which can be used over a wide temperature range from –150°C to 280°C, is the material of choice as a high-performance substitute for the weaker and lower-melting-point polyethylene commonly used in low-cost applications. Also, this polymer is used at the industrial scale for the fabrication of gaskets, pump parts, and laboratory equipment.

1.1.3.14 Polyurethanes

Polyurethane polymers are most commonly formed by reacting a di- or polyisocyanate with a polyol. Both the isocyanates and the polyols used to make polyurethanes contain on average two or more functional groups per molecule. "Urethane" is the name given to the functional group formed from the reaction of an isocyanate group with a hydroxy group (Reaction 1.11) (Nicholson, 2006).

$$-N{=}C{=}O + HO- \longrightarrow -NHCOO- \quad (1.11)$$

Most polyurethanes, which have a backbone joined by carbamate (urethane) links, are thermosetting polymers, but thermoplastic polyurethanes are also available (Dodge, 2003).

The properties of polyurethane are greatly influenced by the types of isocyanates and polyols used. Long, flexible segments, contributed by the polyol, give a soft, elastic polymer. High amounts of crosslinking give tough or rigid polymers. Long chains and low crosslinking give a polymer that is very stretchy; short chains with lots of crosslinks produce a hard polymer, while long chains and intermediate crosslinking give a polymer useful for making foam. The crosslinking present in polyurethanes means that the polymer consists of a three-dimensional network and the molar mass is very high.

A number of different di-isocyanates are used commercially; typical examples are shown in Table 1.1.

Table 1.1 Some commercially important di-isocyanates

Name	Abbreviation	Structure
Hexamethylene di-isocyanate	HMDI	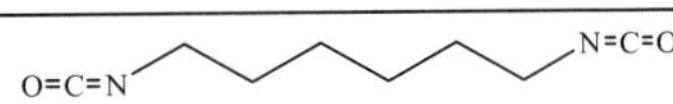
Diphenylmethane 4,4′-di-isocyanate	MDI	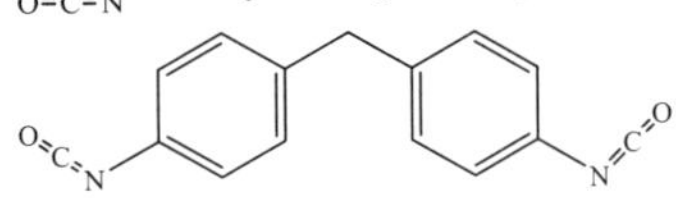
Isophorone di-isocyanate	IPDI	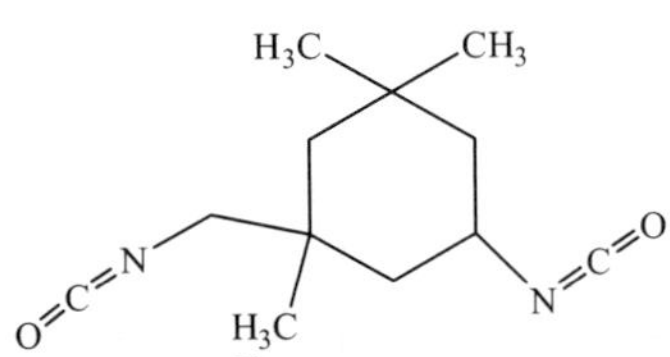
Tolylene di-isocyanate	TDI	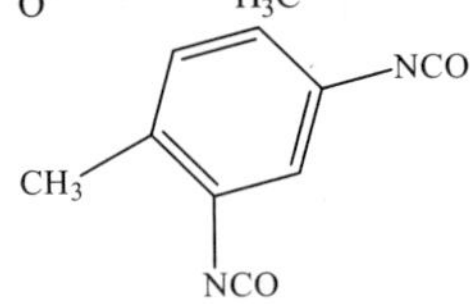

Polyurethanes are important commercial polymers. They are widely used as foams, fibers, adhesives, coatings, and medical materials because of their outstanding mechanical properties, abrasion resistance, and good biocompatibility. To date, commercial polyurethanes are commonly prepared from di-isocyanates. Di-isocyanates have high toxicity, and a major synthesis precursor of these compounds is the extremely toxic phosgene (Li et al., 2014). Recently, non-isocyanate-based polyurethanes have been developed as a new class of polyurethane polymers due to health and environmental concerns (Javni et al., 2013).

1.1.3.15 Poly(ether ether ketone)

Poly(ether ether ketone) (PEEK) possesses high mechanical properties, high-temperature tolerance, and good solvent resistance. PEEK, which is a semicrystalline thermoplastic, is produced by a step (condensation) process in which 4,4-difluorobenzophenone is reacted with the anion of hydroquinone. This reaction is carried out in a high-boiling-point polar solvent, such as *N*-cyclohexyl-2-pyrrolidone. The level of crystallinity achieved in a PEEK polymer depends on the processing regime and very rapid cooling can produce an amorphous polymer. This can subsequently be annealed to achieve any desired level of crystallinity. The optimum level of crystallinity for a PEEK resin is 25–40%. With respect to resistance to hostile environments, PEEK is generally considered to be outstanding in the field of polymeric resins.

PEEK is resistant to nonoxidizing acids (such as hydrochloric acid), alkalies, salts, and solvents. The only common material that will dissolve PEEK is concentrated sulfuric acid (Baker et al., 2004). PEEK has been used in a variety of industries, from aerospace and aviation to medical devices (Toth et al., 2006).

1.1.3.16 Silicones

Silicones or organosiloxanes have unique properties because of their flexible, inert backbone consisting of alternating silicon and oxygen atoms (Drobny, 2012). Typical structures found are shown in Fig. 1.8.

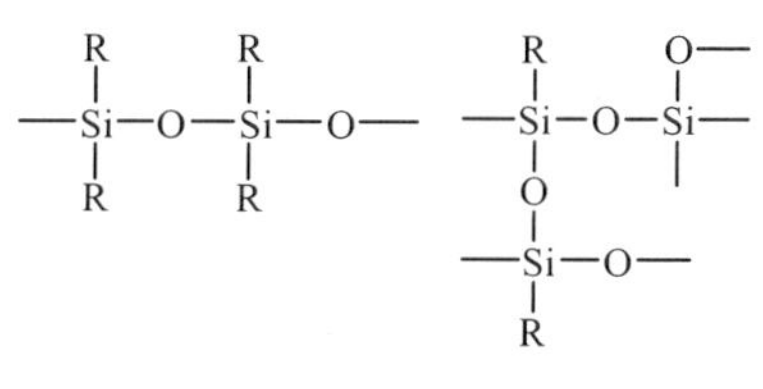

Figure 1.8 Two structures of silicones.

Silicones, which are typically heat resistant and rubber like, are inert, synthetic compounds with a variety of forms and uses. These polymers are used in sealants, adhesives, lubricants, medical applications, cooking utensils, and insulation.

1.1.4 Natural Polymers

Many natural polymers that have industrial applications are known. Cellulose and its derivatives, starch, and rubber are well-known examples of natural polymers. In addition, a number of important biological materials, most notably proteins, are natural polymers made up of macromolecules. These have been considered briefly next (Yu et al., 2006).

1.1.4.1 Cellulose

Cellulose is the structural basis of plant cells, and hence it is the most important natural substance produced by living organisms (Fengel and Wegener, 1989). Cellulose, which is built up from the D-glucose monomer, is a carbohydrate of the molecular formula $(C_6H_{10}O_5)_n$, where *n* is around thousands.

The polymer contains three reactive hydroxyl groups at C-2, C-3, and C-6 atoms, which are, in general, accessible to the typical conversions of primary and secondary alcoholic –OH groups (Heinze and Liebert, 2001). D-glucose itself is highly soluble in water, but cellulose is not. This is essentially a kinetic phenomenon; the hydroxy groups in cellulose would, in principle, readily form hydrogen bonds with water molecules, and hence the cellulose macromolecule would be carried off into aqueous solution. But these hydroxy groups interact with neighboring cellulose molecules, making it impossible for water molecules to penetrate, or solvate, the individual molecules. Cellulose can, however, somewhat dissolve in aqueous ammoniac cupric hydroxide, $Cu(NH_3)_4(OH)_2$. Cellulose is a linear polymer. Despite this, it is not thermoplastic, essentially because of its extensive intermolecular hydrogen bonding, which never allows the molecules to move sufficiently for the polymer to melt. Cellulose may be solubilized by treatment with sodium hydroxide and carbon disulfide. It can be regenerated by acidification of the solution. This is the basis of the production of regenerated cellulose fiber, so-called viscose rayon, which is a major textile fiber. The technique is also used for the production of continuous cellulose-derived film, so-called cellophane (Nicholson, 2006).

Cellulose is of enormous interest because of its applications in biosorption, biomedical packaging, biofiltration, and biocomposites (Thakur et al., 2013). Commercially cellulose ethers are also prepared, such as methylcellulose. This material is water soluble and gives a highly viscous solution at very low concentrations. Hence it is widely used as a thickener in latex paints and adhesives, in cosmetics, and for coating pharmaceutical tablets.

1.1.4.2 Starch

Starch is an extremely abundant edible polysaccharide present in a wide variety of tubers and cereal grains. For its commercial use, corn is the principal source, though wheat and potatoes are also used. In most of its manifestations, it is composed of two macromolecules bearing the same structural units, 1,4-D-glucopyranose, in linear (amylose) and highly branched architectures (amylopectin), present in different proportions according to the species that produces it (Belgacem and Gandini, 2008).

Amylopectin is the polymeric component of starch and consists mainly of glucose units joined at the 1, 4 positions. A variety of modified starches are used commercially, which are produced by derivatization to give materials such as ethanoates (acetates), phosphates, and hydroxyalkyl ethers. Starch, an omnipresent biomaterial, is one of the most abundant and inexpensive polysaccharide sources that have the unique characteristics of biodegradability and easy dissolution in water (Leja and Lewandowicz, 2010). Modified and unmodified starches are used in approximately equal tonnages, mainly in papermaking, paper coatings, paper adhesives, textile sizes, and food thickeners.

1.1.4.3 Natural rubber

All natural rubber currently used commercially arises from a single species, that is, the Brazilian rubber tree, *Hevea brasiliensis* (Tauer, 2004). The juice is latex consisting of a dispersion of a polymer phase at a concentration of about 35% by mass, together with traces of proteins, sterols, fats, and salts. The rubber is obtained either by coagulation of the latex with acid, either ethanoic or methanoic, or by evaporation in air or over a flame. The material that results from this process is a crumbly, cheese-like substance, sometimes called raw rubber or caoutchouc. To develop the mechanical properties that are considered characteristic of rubber, and called rubber-like elasticity, raw rubber needs further processing to provide, in particular, light crosslinking, and this process is known as *vulcanization*. The polymer in natural rubber consists almost entirely of *cis*-polyisoprene (Fig. 1.9). The molecules are linear, with a relative molar mass typically lying between 300,000 and 500,000. The macromolecular nature of rubber was established mainly by Staudinger in 1922, when he hydrogenated the material and obtained a product that retained its colloidal character, rather than yielding fragments of low relative molar mass (Zugenmaier, 2009).

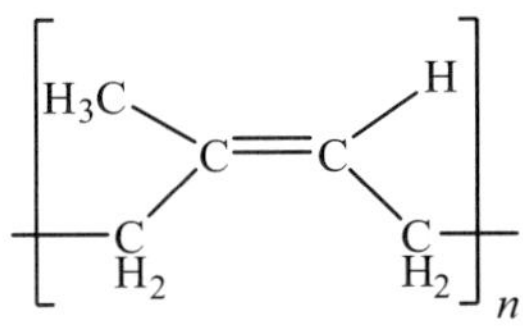

Figure 1.9 Structure of *cis*-polyisoprene.

The term "vulcanization" refers to the process of heating rubber, sulfur, and white lead. By terminology, the crosslinking process of rubber is often called vulcanization when it involves the utilization of sulfur or sulfur compounds (Akiba and Hashim, 1997). Sulfur reacts very slowly with rubber and so is compounded with rubber in the presence of accelerators and activators. Typical accelerators are thiazoles, and a typical activator is a mixture of zinc oxide and a fatty acid. The chemistry of the vulcanization reactions is complicated but generates a three-dimensional network in which rubber molecules are connected by short chains of sulfur atoms, with an average of about five atoms in each chain. A much more heavily crosslinked material can be obtained by increasing the amount of sulfur in the mixture so that it represents about a third of the mass of the product. Heating such a mixture of raw rubber and sulfur at 150°C until the reaction is complete gives a hard, thermoset material that is not at all elastic. This material is called ebonite and is used to make car battery cases (Coran, 1994; Nicholson, 2006).

1.1.4.4 Proteins

Proteins, forming an important group of biochemical macromolecules, appear in all cells, both animal and plant, and are involved in all cell functions. The word "protein" is derived from the Greek word *prôtos*, meaning "primary" or "first rank of importance" (Lau, 2005). Proteins are linear polyamides formed from α-amino acids (Vroman and Tighzert, 2009). An *α*-amino acid is one in which the carboxylic acid and amino groups reside on the same carbon atom (Fig. 1.10).

There are 20 naturally occurring species of amino acids, and to form proteins, they join end to end by peptide bonds. Hence, the alternative term used for them is "polypeptide" (Nielsen et al., 2006). They are arranged in a specific and characteristic sequence along the molecule. This sequence is generally referred to as the "primary structure" of the protein. Also part of the primary structure is the relative molar mass of the macromolecule.

The primary structure of a protein is the sequence of amino acids, and the secondary structure occurs when the sequence of amino acids is linked by hydrogen bonds. The tertiary structure results from long-range contacts within the chain. The quaternary structure is the organization of protein subunits, or two or more independent polypeptide chains (Prošková, 2014).

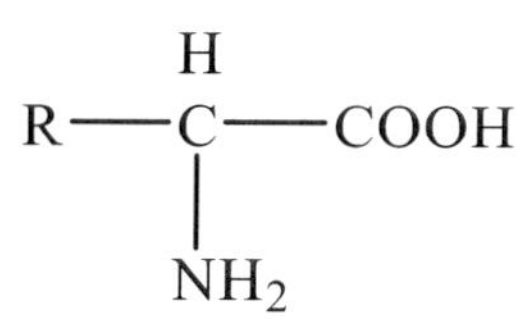

Figure 1.10 Structure of an α-amino acid.

The diversity in primary, secondary, tertiary, and quaternary structures of proteins means that few generalizations can be made concerning their chemical properties. Some fulfil structural roles, such as collagens (found in bone) and keratin (found in claws and beaks), and are insoluble in all solvents (Chitte et al., 1999; Korniłłowicz-Kowalska and Bohacz, 2011). Others, such as albumins or globulins of plasma, are very soluble in water. Still others, which form part of membranes of cells, are partly hydrophilic ("water loving," hence water soluble) and partly lipophilic ("lipid loving," hence fat soluble). Because proteins are not composed of identical repeating units, but of different amino acids, they do not fall within the previously made formal definition for polymers. They are, nevertheless, macromolecular, and techniques developed for the study of true polymers have been applied to them with success (Nicholson, 2006).

1.1.4.5 Poly-3-hydroxybutyrate

Poly-3-hydroxybutyrate (PHB) (Fig. 1.11) and its copolymers with up to 30% of 3-hydroxyvaleric acid are now commercially available under the trade name Biopol. The polymers can be degraded by soil bacteria but are relatively stable in ambient conditions (Ratner et al., 1996). The ratio of these two polymers in a given sample is determined by the ratio of glucose and propionic acid in the medium in which the bacteria live and carry out their metabolic processes.

The carbon atom that carries the methyl group is chiral, but biosynthesis is stereoselective and gives rise to a natural polymer with the R configuration. The polymer is a partially crystalline thermoplastic (Kulkarni et al., 2010).

PHB has attracted interest as an environmentally degradable thermoplastic that can be used in packaging, agriculture, and medicine. It undergoes enzymatic degradation quite readily, becoming less crystalline and rapidly decreasing in molar mass. The rate of degradation depends on a number of factors, such as moisture level, nutrient supply, temperature, and pH (Nicholson, 2006).

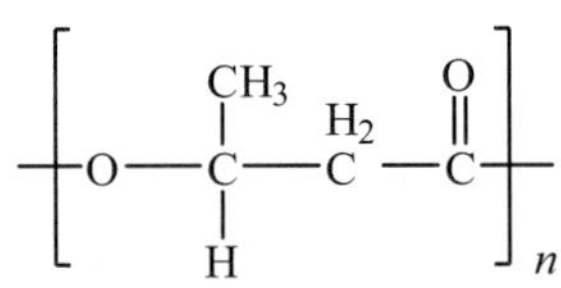

Figure 1.11 Structure of poly-3-hydroxybutyrate.

1.2 Polymerization Reactions

The mechanistic basis of classification made as chain and step-growth processes for polymerization reactions arises from major differences between these two processes.

There is no doubt about the nature of the reaction in almost all cases, as can be seen by considering the distinguishing features of the two mechanisms, which are summarized below.

- *Chain (chain-growth) polymerization*: Each polymer chain, after being started by a free-radical initiator, a cationic catalyst, or an anionic catalyst, grows rapidly, producing a high-molar-mass polymer (Stille, 1981). Long reaction times give higher yields but do not affect molar mass. The reaction mixture contains only a monomer, a high-molar-mass polymer, and a low concentration of growing chains.
- *Step-growth polymerization*: The molar mass of the polymer chain builds up slowly, and there is only one reaction mechanism for the formation of polymer. Long reaction times increase the molar mass and are essential to obtain very high molar masses. At all stages of the reaction, every possible molecular species from dimers to polymers of large degrees of polymerization are present in a calculable distribution. The distinct initiation, propagation, and termination steps of chain-growth polymerization are meaningless in step-growth polymerization (Stille, 1981).

1.2.1 Step-Growth Polymerization

The major classifications of step-growth polymers are given in Table 1.2 (Nicholson, 2006). Of these, the most important ones in terms of the size of the commercial market are aliphatic polyamides or nylons (nylon-6, nylon-6,6, and nylon-6,10) and polyester, for example, poly(ethylene terephthalate) (PET).

In step-growth polymerizations, any species (monomers, dimers, trimers, etc.) can react at any time in consecutive reactions, leading to an increase in macromolecules of high molar mass (Billiet et al., 2009).

Most step-growth polymerizations involve a classical condensation reaction such as esterification, ester interchange, or amidization. In step-growth polymerization, the stepwise reaction occurs between pairs of chemically reactive or functional groups on the reacting molecules. In most cases, step-growth polymerization is accompanied by the elimination of a small molecule, such as water, as a by-product (Ebewele, 2000).

Table 1.2 Classification of step-growth polymers and some examples

Classification	First precursor	Second precursor
Condensation		
Polyamide	Dicarboxylic acid	Diamine
Polycarbonate	Bisphenol	Phosgene
Polyester	Dicarboxylic acid	Diol or polyol
Polyimide	Tetracarboxylic acid	Di-imine
Polysulfone	Bisphenol	Dichlorophenylsulfone
Noncondensation		
Polyurethane	Di-isocyanate	Diol or polyol
Poly(phenylene oxide)	2,6-Disubstituted phenol	Oxygen

A typical step-growth polymerization of the condensation type is the formation of polyester through the reaction of a glycol and a dicarboxylic acid, as shown in Reaction 1.12:

$$n\mathrm{HO{-}R{-}OH} + n\mathrm{HOOC{-}R'{-}COOH} \longrightarrow n\mathrm{H}\left[\mathrm{O{-}R{-}O{-}\overset{\overset{\displaystyle O}{\|}}{C}{-}R'{-}\overset{\overset{\displaystyle O}{\|}}{C}}\right]_n\mathrm{OH} + n\mathrm{H_2O} \tag{1.12}$$

where R and R′ represent two different nonreactive parts of the molecules.

Step-growth polymerizations generally involve either one or more types of monomers (Kiparissides, 1996). In either case, each monomer has at least two reactive (functional) groups. In the case where only one type of monomer is involved, which is known as A–B step-growth polymerization, the functional groups on the monomer are different and capable of intramolecular reactions. The most prominent synthetic aliphatic polyesters, namely, poly(L-lactide) (PLA) and poly(ε-caprolactone) (PCL), are usually prepared by ring-opening polymerization (ROP) of the respective cyclic monomers, 1,1-dilactide (LA) and ε-caprolactone (CL) (Duda and Penczek, 2005). An example is the formation of aliphatic polyester by the self-condensation of ω-hydroxycaproic acid (Reaction 1.13).

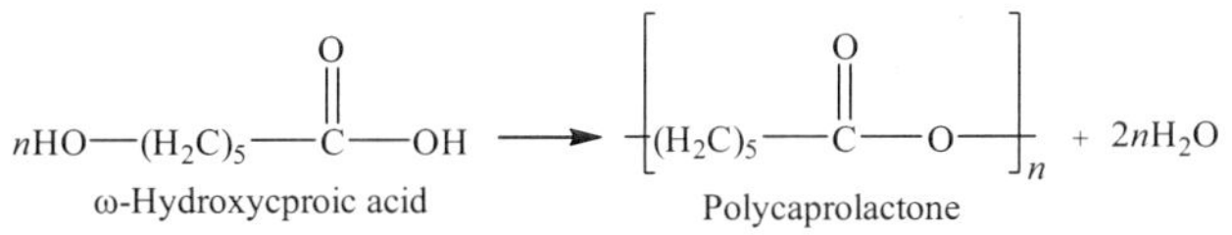

(1.13)

Each of these two molecules contains two different functional groups: a hydroxyl group (–OH) and a carboxylic acid group (–COOH). These react to form a series of ester linkages.

In the case where more than one type of molecule is involved, the functional groups on each type of monomer are the same but capable of intermolecular reaction with the other type of monomer. This is known as A–A/B–B step-growth polymerization and is exemplified by the preparation of PET and nylon-6,6 (Reactions 1.14 and 1.15) (Ebewele, 2000).

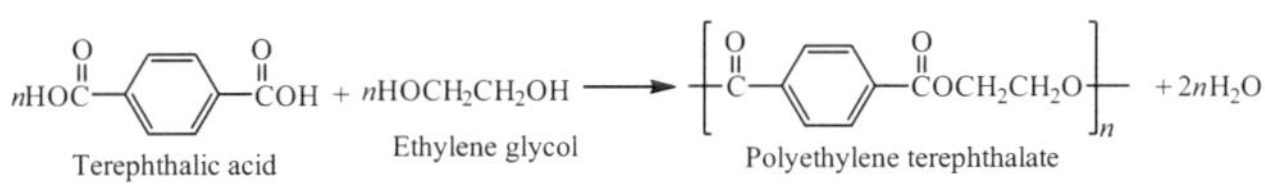

(1.14)

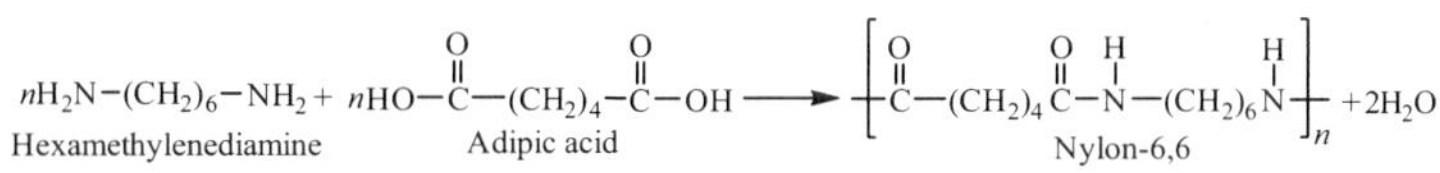

(1.15)

Step-growth polymerization can be divided into two main categories: polycondensation, in which a small molecule is eliminated at each step, and polyaddition (noncondensation), in which monomers react without the elimination of a small molecule. These two processes are schematized in Reactions 1.16 and 1.17, respectively, where R and R′ represent the two different nonreactive parts of the molecules.

$$A-R-A+B-R'-B \rightarrow A-R-R'-B+AB \quad (1.16)$$

Polycondensation

$$A-R-A+B-R'-B \rightarrow A-R-AB-R'-B \quad (1.17)$$

Polyaddition (noncondensation)

1.2.1.1 Typical step-growth polymers

1.2.1.1.1 *Polyesters*

A large class of commercially significant polymers includes polyesters. PET is a saturated polyester of terephthalic acid and ethylene glycol (Paszun and Spychaj, 1997). The traditional route for the production of commercial PET is through two successive ester interchange reactions, as shown in Reaction 1.18:

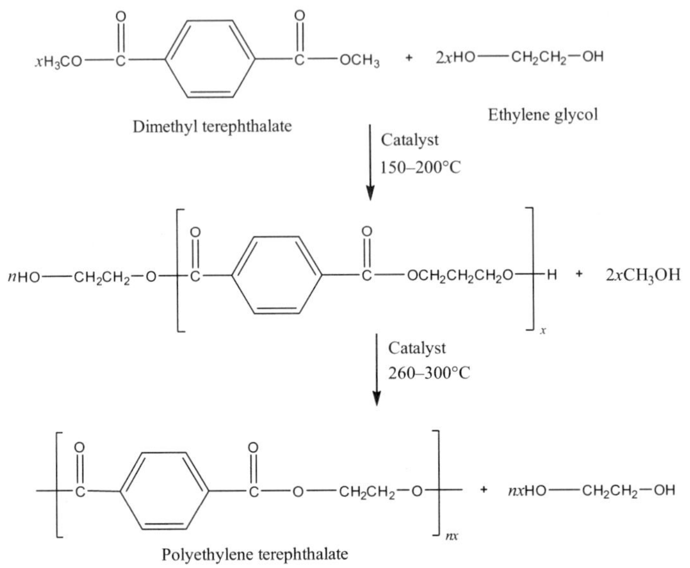

(1.18)

In the first step, a 1:2 molar ratio of dimethyl terephthalate to ethylene glycol is heated at temperatures near 200°C in the presence of a catalyst such as calcium acetate. During this stage, methanol is evolved and an oligomeric product (x = 1 to 4) is obtained. The second step involves a temperature increase to about 300°C. This results in the formation of a high polymer with the evolution of ethylene glycol (Ebewele, 2000; Paszun and Spychaj, 1997).

1.2.1.1.2 *Polycarbonates*

Polycarbonates are the second largest volume engineering thermoplastics after polyamides. Commercial polycarbonates have been produced mainly through a low-temperature solution polycondensation of 2,2-bis-(4-hydroxyphenyl) propane (bisphenol A) using phosgene (Reaction 1.19) or diphenyl carbonate as a carbonylation agent (Reaction 1.20) (Delpech et al., 2002; Ebewele, 2000; Nakamura and Nishimoto, 2002).

The reaction of bisphenol A with phosgene involves bubbling the phosgene into a solution of bisphenol A in pyridine at 20°C to 35°C and isolation of the resulting polymer by precipitation in water or methanol. In the reaction of bisphenol A with diphenyl carbonate, a prepolymer is formed initially by heating the mixture at 180°C–220°C in vacuum. Then the temperature is raised slowly to 280°C–300°C at reduced pressure to ensure the removal of the final traces of phenol.

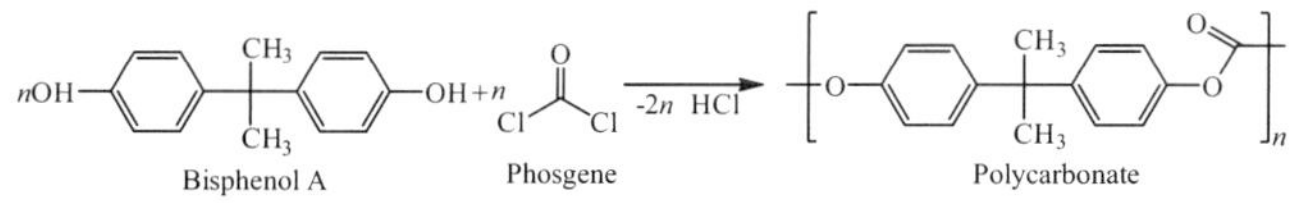

(1.19)

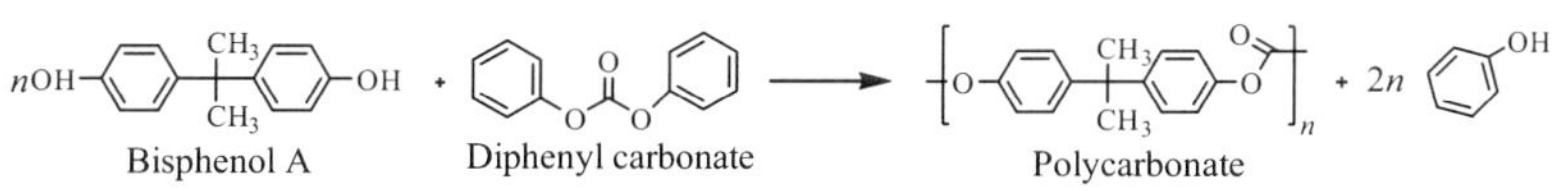

(1.20)

1.2.1.1.3 *Polyamides*

Heterochain polymers, such as polyamides or nylons, as they are commonly called, are widely represented in nature. A macromolecule of such polyamides is built up of individual units containing the characteristic amide linkages (–CONH–) and a hydrocarbon residue with various substituents (Patel, 1992).

Theoretically, a large number of polyamides can be synthesized on the basis of four main synthetic routes: (a) a condensation reaction between a dicarboxylic acid and a diamine, (b) the reaction between a diacid chloride and a diamine, (c) dehydration–condensation reactions of amino acids, and (d) ROP of lactams. Chemically, nylons may be divided into two categories, those based on synthetic routes (a) and (b) and those based on routes (c) and (d). The commercial use of nylons is centered around two products, nylon-6,6 from the first category and nylon-6 from the second category.

The classical route for the synthesis of nylon-6,6, as with other polyamides, is the direct reaction between a dicarboxylic acid (adipic acid) and a diamine (hexamethylenediamine) (Reaction 1.21).

In practice, however, to achieve an exact stoichiometric equivalence between the functional groups, a 1:1 salt of the two reactants is prepared initially and subsequently heated at high temperature to form the polyamide. For nylon-6,6, an intermediate hexamethylene diammonium adipate salt is formed. A slurry of 60% to 80% of the recrystallized salt is heated rapidly. The steam that is released is purged by air. The temperature is then raised to 220°C and finally to 270°C–280°C when the monomer conversion is about 80% to 90%, while maintaining the steam pressure generated during polymerization at 200–250 psi. The pressure is subsequently reduced to atmospheric pressure, and heating is continued until completion of polymerization. This polymerization process is known as melt polymerization because the polymerization reaction occurs above the melting points of both the reactants and the polymer (Ebewele, 2000).

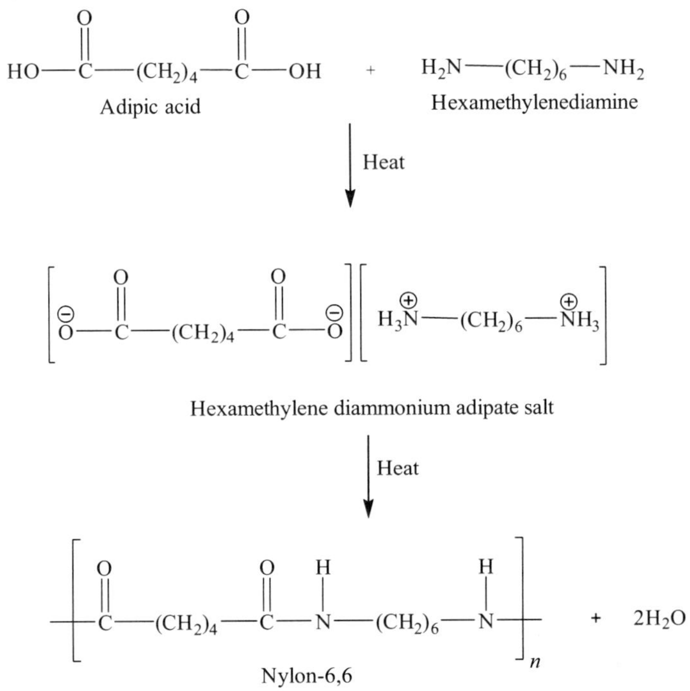

(1.21)

1.2.1.1.4 *Polyimides*

Polyimides involving aromatic and heterocyclic rings in the main chains, which are known as heat-resistant organic materials, are the condensation polymers obtained from the reaction of dianhydrides with diamines (Jin et al., 1993). Polyimides are synthesized generally from aromatic dianhydrides and aliphatic diamines or, in the case of aromatic polyimides, from the reaction of aromatic dianhydrides with aromatic diamines. Aromatic polyimides are formed by a general two-stage process (Reaction 1.22). The first step involves the condensation of aromatic dianhydrides and aromatic diamines in a suitable solvent, such as dimethylacetamide, to form a soluble precursor or poly(amic acid). This is followed by the dehydration of the intermediate poly(amic acid) at elevated temperature. Unlike the poly(amic acid), the cured or fully imidized polyimide exhibits insoluble and infusible behaviors, with high thermo-oxidative stability and good electrical insulation properties. Through an appropriate choice of the aromatic diamine, phenyl, or pendant alkyl groups, the main-chain aromatic polyether linkages can be introduced into the polymer.

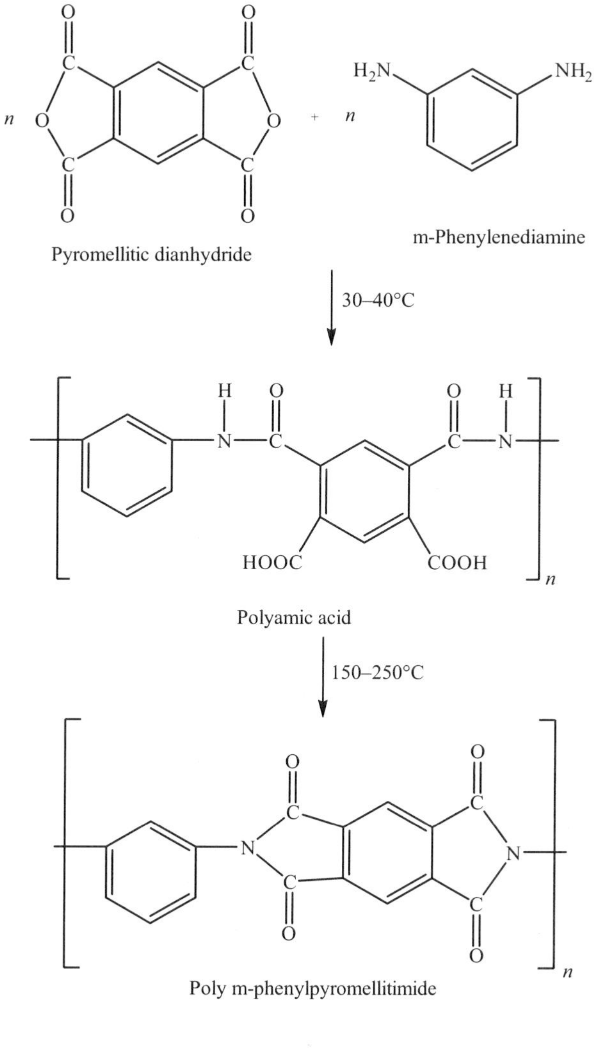

(1.22)

The resulting polyimides are soluble in partially nonpolar solvents (Ebewele, 2000). Common properties of most polyimides are high thermal stability, high mechanical strength, high glass transition temperature, and the typical two-step synthesis via the poly(amic acid) intermediate (Maier, 2001).

1.2.1.1.5 *Polybenzimidazoles and polybenzoxazoles*

The aromatic substituents of the chains of vinyl polymer, as a part of the main chain, have a profound influence on virtually all important properties of the resulting polymer. Aromatic polyamides are formed by the repetitive reaction of the aromatic amino group and the carboxyl group in the molar ratio of 1:1.

In aromatic polyamides as well as aromatic polyesters, the chain-stiffening aromatic rings are separated from each other by three consecutive single bonds. The two tetrahedral angles associated with these bonds permit some degree of chain flexibility, which limits the mechanical and thermal properties of the resulting polymers. One way of reducing flexibility and enhancing these properties is to reduce the number of consecutive single bonds between two aromatic units to two, one, or even zero. Also, in the case of polyether, polysulfide, and polysulfone, there are two consecutive single bonds as separated by only one tetrahedral angle (Fig. 1.12).

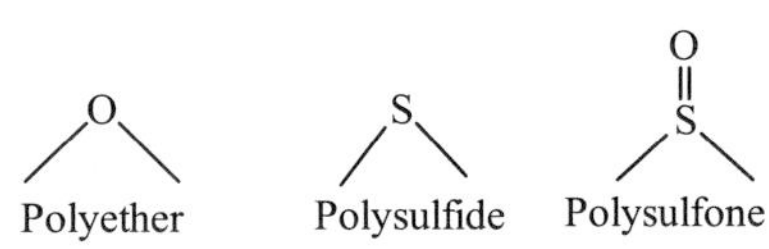

Figure 1.12 The single-bonded structures of polyether, polysulfide, and polysulfone, as separated by one tetrahedral angle.

Polyimides, polybenzimidazoles, and polybenzoxazoles are polymers where the number of these bonds has been reduced to one. In aromatic polyimides, two of the three consecutive single bonds between aromatic groups in polyamides are eliminated by the formation of a new ring. This is achieved by employing a 2:1 molar ratio of aromatic carboxyl and amino groups. When the molar ratio of carboxyl groups (e.g., terephthalic acid) to amino groups (e.g., 3,3′-diaminobenzidine) is 1:2, polybenzimidazoles are formed, whereas when the molar ratio of carboxyl, amino, and hydroxyl groups is 1:1:1, polybenzoxazoles are formed (Reaction 1.23). Silicon-containing polybenzimidazoles and silicon-containing polybenzoxazoles were prepared by the reaction of silylated diacid chlorides or diesters with aromatic bis(*o*-diamine)s or bis(*o*-aminophenol)s, respectively, in solution, at high temperature, by using high-boiling-point solvents such as dimethylaniline or diethylaniline (Bruma and Schulz, 2001; Ebewele, 2000).

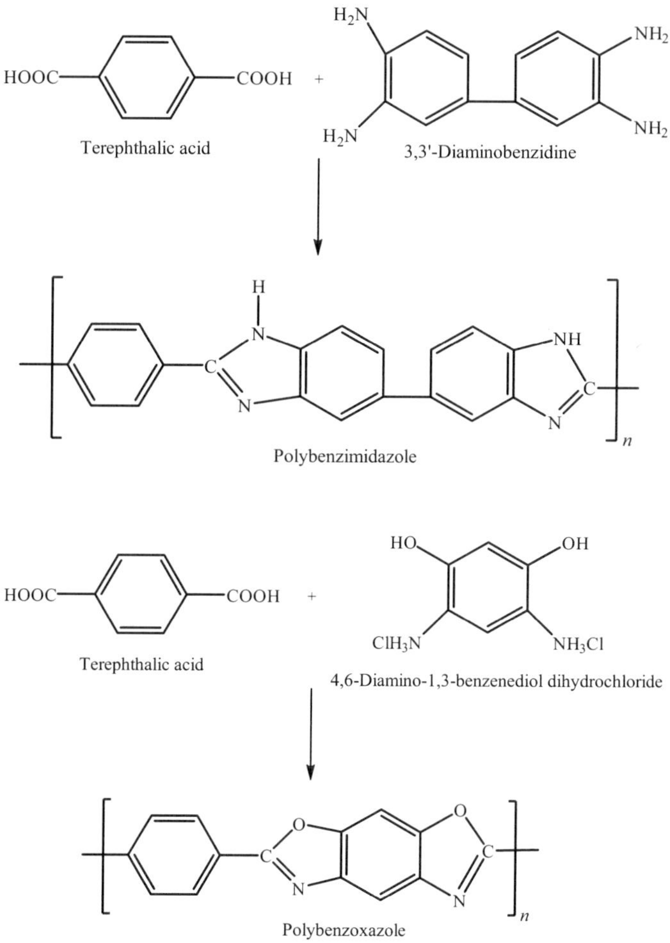

(1.23)

1.2.1.1.6 *Aromatic ladder polymers*

Ladder polymers typify the cases where there are not consecutive single bonds between the aromatic moieties in the main chain.

(Bruma and Schulz, 2001). To increase the rigidity of linear macromolecules, the elimination of single bonds in the main chain is necessary so that it is composed of only condensed cyclic units. The resulting polymer has the generalized structure shown in Fig. 1.13.

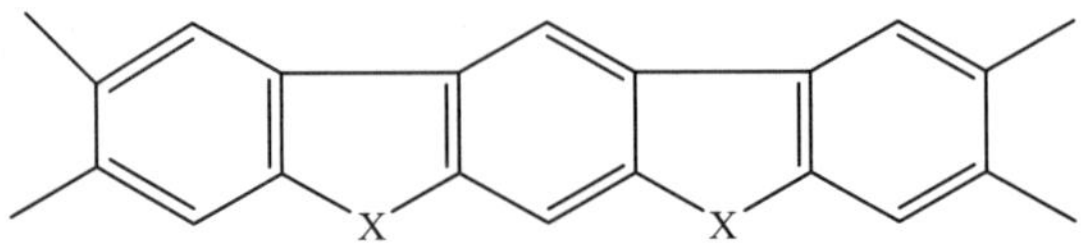

Figure 1.13 Structure of a ladder polymer.

The individual cyclic units in the polymers of this type may be either aromatic or cycloaliphatic and homocyclic or heterocyclic. Polymers of this type are known as ladder polymers. Polybenzimidazoles are typical aromatic ladder polymers, too. Longer segments of the ladder are present in polyimidazopyrrolones prepared by polymerization of aromatic dianhydrides or aromatic tetracarboxylic acids with o-*aromatic* tetramines, according to Reaction 1.24 (Bell and Pezdirtz, 1965).

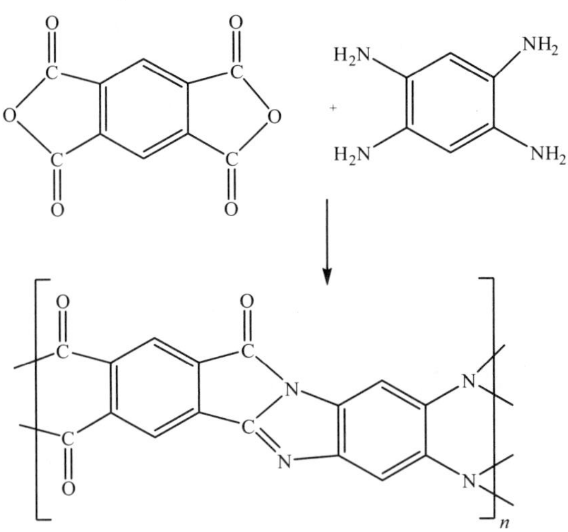

(1.24)

1.2.1.1.7 *Formaldehyde resins*

Formaldehyde is employed in the production of aminoplasts and phenoplasts, which are two different but related classes of thermoset polymers.

Aminoplasts are products of the condensation reaction of both urea (urea formaldehyde, or UF, resins) and melamine (melamine formaldehyde, or MF, resins) with formaldehyde. Phenoplasts or phenolic (phenol formaldehyde, or PF) resins are prepared from the condensation products of phenol or resorcinol and formaldehyde (Ebewele, 2000; Mindivan, 2013).

1.2.1.1.7.1 *Urea formaldehyde resins*

The synthesis of a UF resin takes place in two stages. In the first stage, urea is hydroxymethylolated by the addition of formaldehyde to the amino groups (Reaction 1.25).

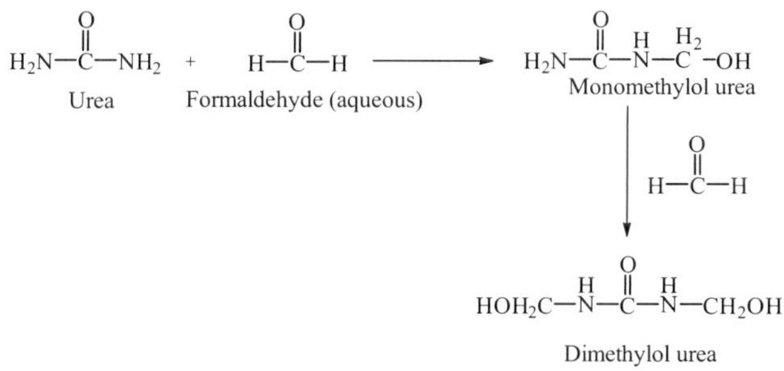

(1.25)

The second stage of UF resin synthesis consists of the condensation of methylol ureas to low-molar-mass polymers under acidic conditions, leading ultimately to the formation of a network structure (Reaction 1.26) (Conner, 1996):

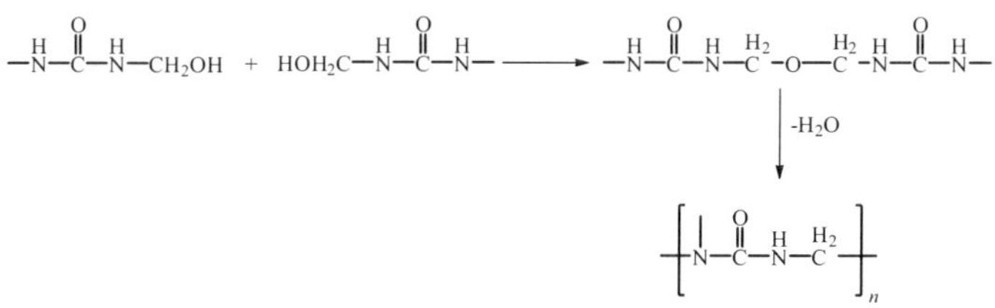

(1.26)

1.2.1.1.7.2 *Melamine formaldehyde resins*

The reactions occurring during the synthesis of an MF resin can be roughly divided into two different stages: methylolation and condensation. The addition of formaldehyde to melamine is the first step of reaction. The second step of reaction involves condensation, that is, bridge formation between the methylolated melamines, leading to the formation of many different oligomers with a rigid network structure (Reaction 1.27) (Jahromi, 1999; Meier et al., 1995).

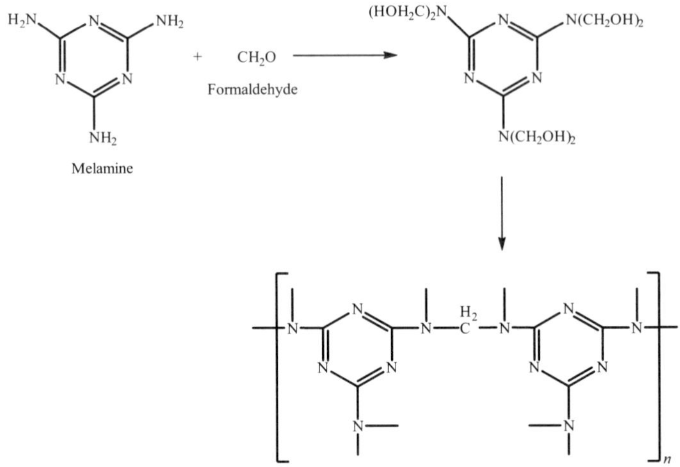

(1.27)

1.2.1.1.7.3 *Phenol formaldehyde resins*

Phenolic resins are synthesized from phenols or alkyl phenols and formaldehyde by two routes. A resole type of resin is produced by a base-catalyzed process (pH 9–12) and the molar ratio of phenol:formaldehyde is 1:1.2. The other variety, which is known as the novolac type, is made by an acid-catalyzed process (pH 3–5) and the molar ratio of phenol:formaldehyde is 1:0.85 (Setua et al., 2000). In the preparation of resoles, phenol and excess formaldehyde react to produce a mixture of methylol phenols. These condense on heating to yield soluble, low-molar-mass prepolymers or resoles (Reaction 1.28).

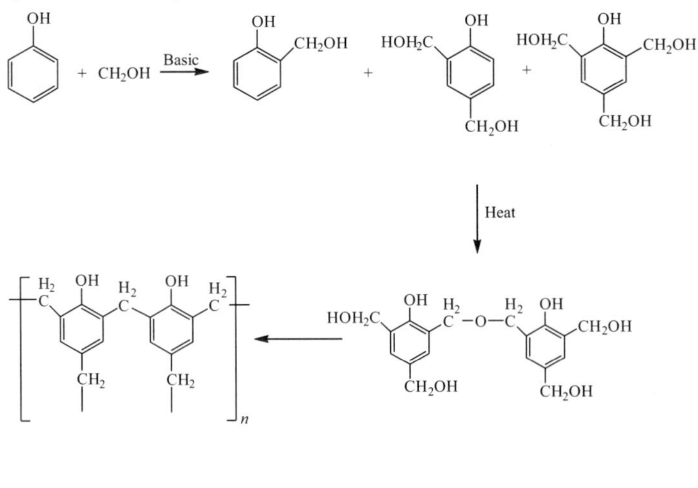

(1.28)

On heating resoles at elevated temperature under basic, neutral, or slightly acidic conditions, a high-molar-mass network structure or phenolic rings linked by methylene bridges (novolacs) (Reaction 1.29) are produced (Ebewele, 2000).

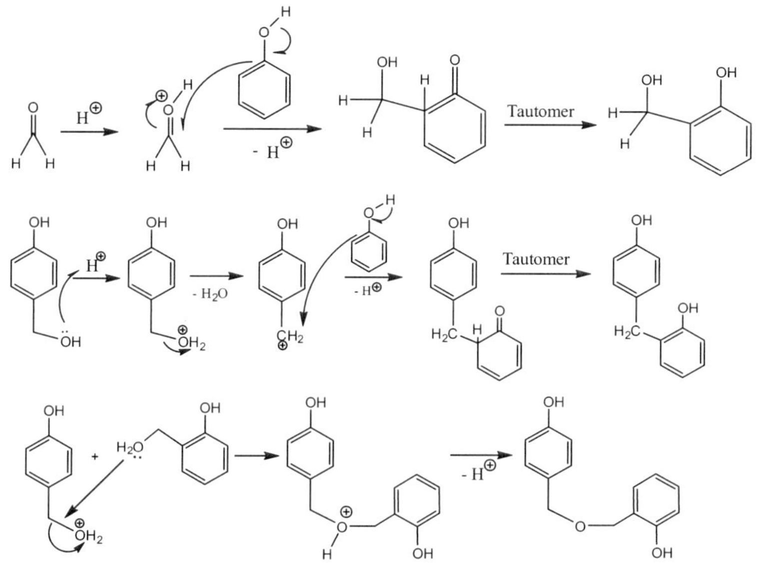

(1.29)

1.2.1.1.8 *Polyethers*

By the introduction of aromatic units into the main chain of polymers, polymers with better thermal stability than their aliphatic analogs result.

Poly(phenylene oxide) (PPO), which has many attractive properties, including high impact strength, resistance to attack by mineral and organic acids, and low water absorption, is one such polymer example (Reaction 1.30) (Zaheed and Jachuck, 2004). It is used, usually blended with high-impact polystyrene (HIPS), to ease processability in the manufacture of machined parts and business machine enclosures.

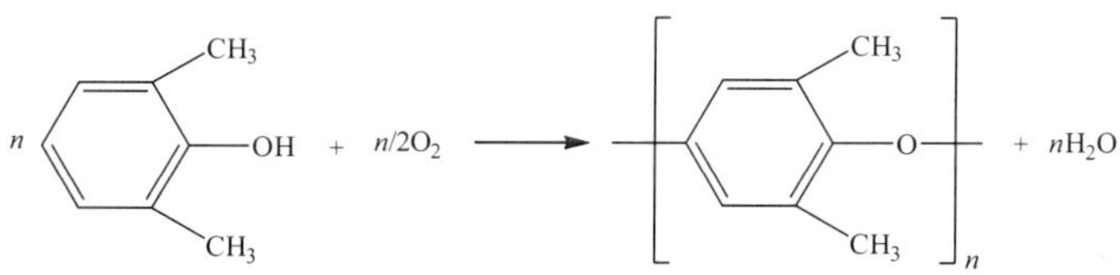

(1.30)

1.2.1.1.9 *Polysulfides*

Aromatic polythioethers or polysulfides are closely related to polyethers in structure and properties. A typical aromatic polysulfide is poly(phenylene sulfide) (PPS), which is used in electrical insulators and structural parts in the building of engines and vehicles (Tanthapanichakoon et al., 2006). PPS is prepared by the condensation reaction between *p*-dichlorobenzene and sodium sulfide (Reaction 1.31):

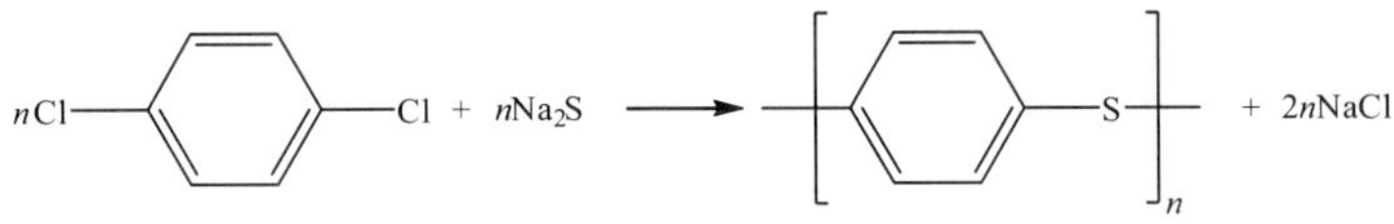

(1.31)

1.2.1.1.10 *Polysulfones*

Polysulfones are sulfur-containing, linear aromatic, high-temperature-resistant engineering thermoplastic polymers. Polysulfones may be synthesized by the nucleophilic substitution of alkali salts of biphenates with activated aromatic dihalides. A typical example is the preparation of bisphenol A polysulfone (Reaction 1.32) from the reaction of disodium salt of bisphenol A with dichlorodiphenyl sulfone (Arad-Yellin et al., 2003; Celebi et al., 2011; Li and McHugh, 2004):

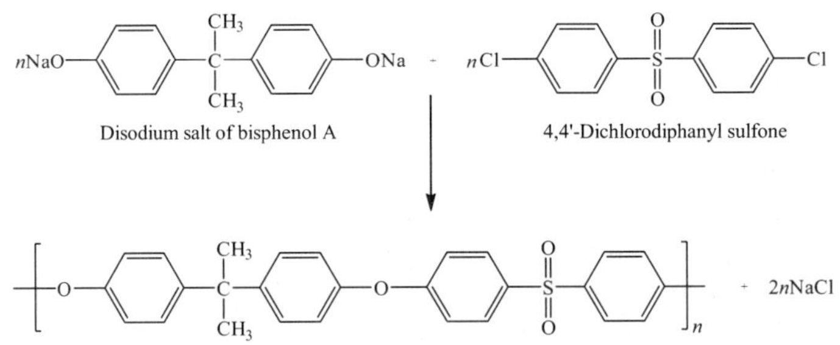

(1.32)

1.2.2 Chain Polymerization

Chain polymerization, or chain-growth polymerization, occurs by a chain reaction where unsaturated monomer molecules add onto the active site of a growing polymer chain one at a time and the active site is regenerated by the addition of each monomer unit. These active species are typically free radicals but may be anions or cations, and they react readily with other monomers. The fact that the reactivity of the active species and the rate constant of reaction for the chain polymerization are high explains the rapid formation of high-molar-mass polymers right away from the beginning of the reaction. A consequence of this is that almost no species intermediate between the monomer and the polymer are found in such reacting systems.

The chain reaction does not continue indefinitely, but the reactivity of free radical or ionic species will be lost as a result of their interaction with each other. For example, in radical polymerizations, two growing molecules may combine to extinguish both radical centers with the formation of a chemical bond. Alternatively, they may react in a disproportionation reaction to generate end groups in two molecules, one of which is unsaturated. Lastly, active centers may find other molecules to react with, such as solvent or impurity, and in this way the active center is destroyed and the polymer molecule ceases to grow.

Chain polymerization involves three fundamental steps: *initiation*, *propagation*, and *termination*. It is essentially a radical addition reaction. Free-radical polymerization is one of the most common ways of preparing organic polymers (Chandrasekhar, 2005).

1.2.2.1 Initiation

Initiation, which may occur spontaneously by the absorption of heat, light (ultraviolet), or high-energy irradiation, involves the acquisition of an active site by the monomer, but mostly initiation of free-radical polymerization is brought about by the addition, in small quantities, of compounds called initiators. The main role of an initiator is to determine the number of growing polymer chains (Matyjaszewski and Xia, 2001). Typical initiators include peroxides, azo compounds, Lewis acids, and organometallic reagents. However, while initiators trigger initiation of the chain and exert an accelerating influence on polymerization rate, they are not exactly catalysts, since they are changed chemically in the course of polymerization. An initiator is usually a weak organic compound that can be decomposed thermally or by irradiation to produce free radicals, whose molecules contain atoms with unpaired electrons. A variety of compounds decompose when heated to form free radicals.

Dialkyl peroxides (ROOR), diacylperoxides (RCO-O-O-CO-R), hydroperoxides (ROOH), and azo compounds (RN=NR) are typical organic compounds that can be decomposed thermally to produce free radicals. Benzoyl peroxide, azobisisobutyronitrile (AIBN), and di-*t*-butylperoxide are commonly used free-radical initiators, as illustrated in Reactions 1.33, 1.34, and 1.35.

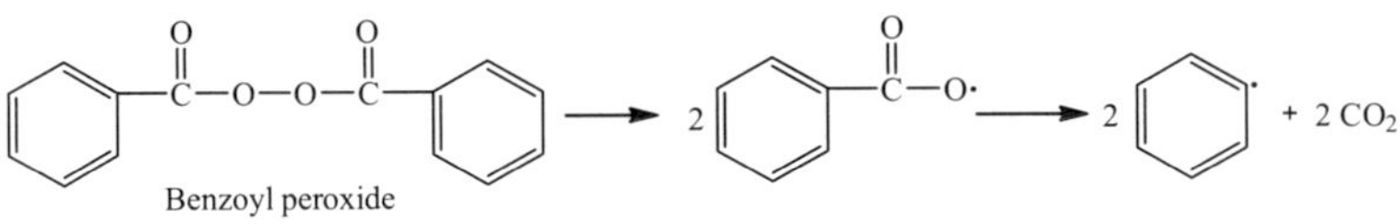

(1.33)

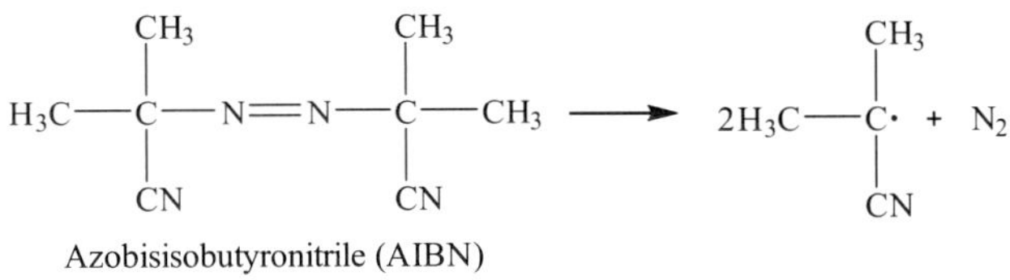

(1.34)

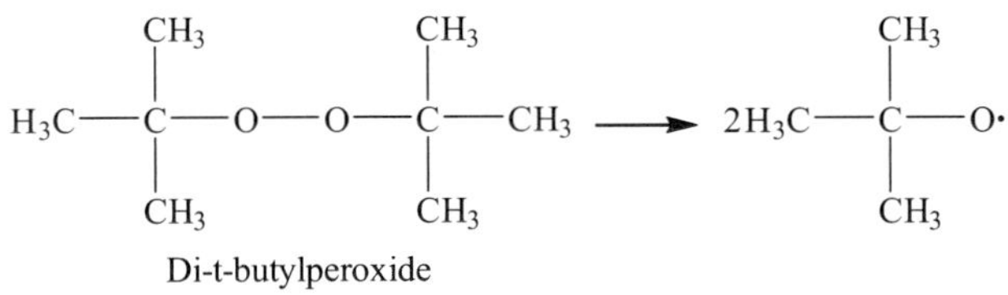

(1.35)

In addition to heat and light, generation of free radicals can be accomplished by using γ-rays, X-rays, or electrochemical means. Once produced, the free radical reacts rapidly with a molecule of monomer to yield a new species that is still a free radical, as shown in Reaction 1.36.

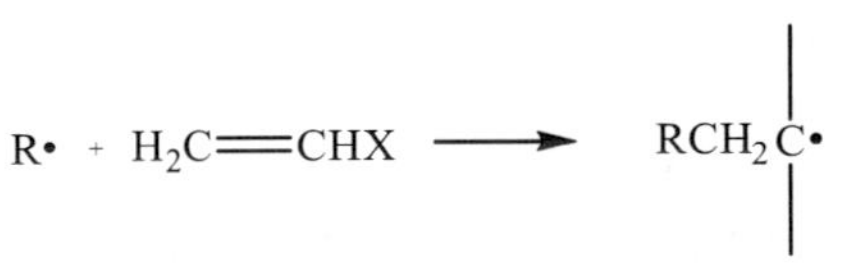

(1.36)

The efficiency of the initiator is a measure of the extent to which the number of radicals formed reflects the number of polymer chains formed. For example, typical initiator efficiencies for vinyl polymerizations lie between 0.6 and 1.0. The initiator is chosen so that the initiation occurs fast and is quantitative with the dormant polymer chain end that is stable during the polymerization. This means that the initiator should be carefully selected in accordance with the structure and reactivity of the monomers (Kamigaito et al., 2001).

1.2.2.2 Propagation

Propagation is a series of reactions in which the free-radical unit at the end of the growing polymer molecule reacts with a monomer to increase still further the length of the polymer chain. In this step, each addition creates a new radical that has the same identity as the previous one, except that it is larger by one monomer unit (Reaction 1.37) (Hasirci et al., 2011)

(1.37)

Propagation continues until the growing chain radical is deactivated by chain termination or transfer, as discussed below. The substituted carbon atom is regarded as the head and the unsubstituted carbon atom the tail of the vinyl monomer. There are, therefore, three possible ways for the propagation step to occur: head to tail, head to head (Reaction 1.38), and tail to tail (Reaction 1.39). A random distribution of these species along the molecular chain might be expected. It is found, however, that head-to-tail linkages in which the substituents occur on alternate carbon atoms predominate; only occasional interruptions of this arrangement by head-to-head and tail-to-tail linkages occur. In addition, exclusive head-to-head or tail-to-tail arrangements of monomers in the chain are now known.

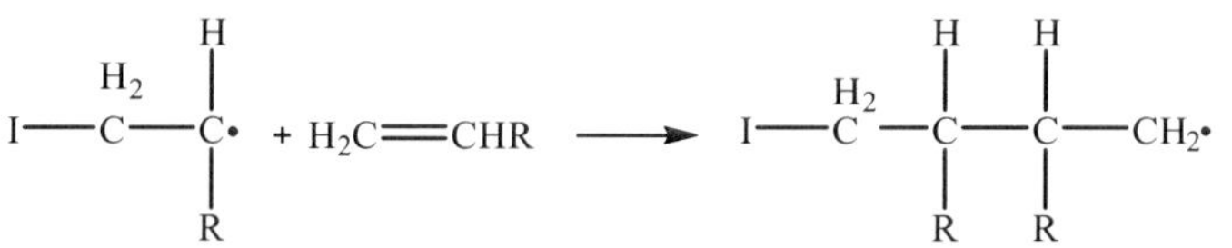
(1.38)

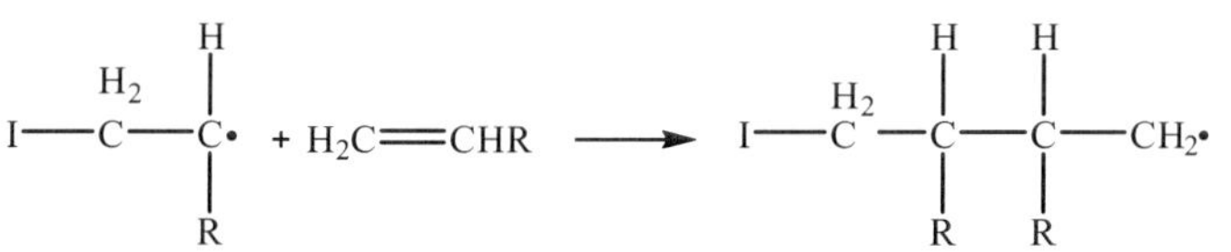
(1.39)

1.2.2.3 Termination

Termination occurs primarily by the bimolecular reaction of two growing polymer radicals. The two primary mechanisms observed are *radical–radical combination* and *disproportionation* (Gridnev and Ittel, 2001). The first of these occurs when two radical species react together to form a single bond and one reaction product, as in Reaction 1.40

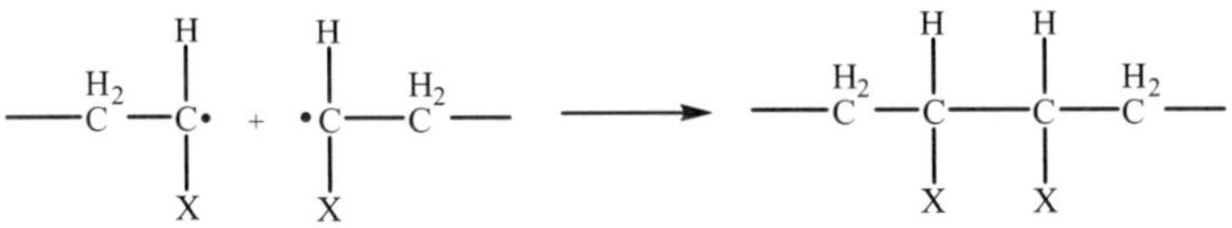 (1.40)

The second mechanism occurs by the abstraction of a hydrogen atom, owing to another radical chain by one from the two radical chains, resulting in one saturated chain end and one unsaturated chain end (Reaction 1.41).

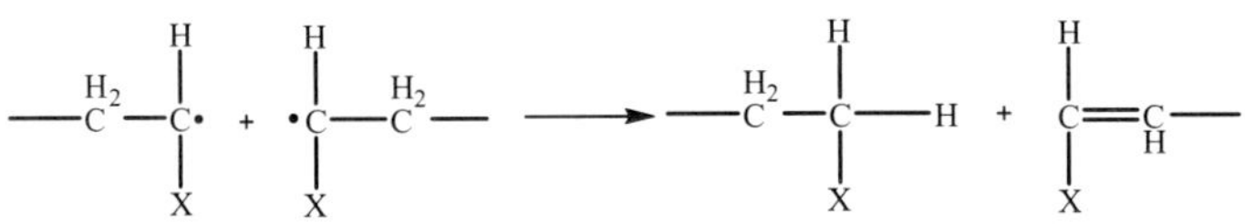 (1.41)

Both termination mechanisms have been shown to occur experimentally, the method being to examine the polymer molecules formed for fragments of the initiator. In such a way, polystyrene has been found to terminate mainly by combination and poly(methyl methacrylate) (PMMA) entirely by disproportionation at temperatures above 60°C.

1.2.2.4 Chain transfer

Free-radical polymerization essentially consists of three basic steps: initiation, propagation, and termination. However, a fourth step, called *chain transfer*, is taken into consideration, owing to the fact that the reactivity of a radical could be transferred to another species that would usually be capable of continuing the chain reaction (Billmeyer, 1971).

In chain transfer reactions, a growing polymer chain is deactivated or terminated by transferring its growth activity to a previously inactive species, as illustrated in Reaction 1.42.

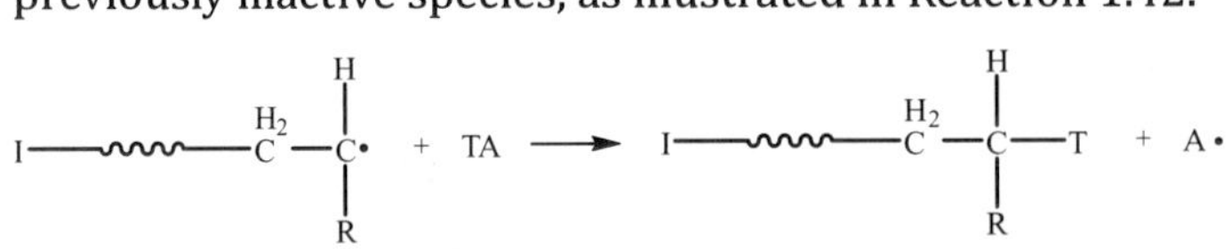 (1.42)

The species TA could be a monomer, polymer, solvent molecule, or other molecule deliberately or inadvertently introduced into the reaction mixture. Depending on its reactivity, the new radical A may or may not initiate the growth of another polymer chain. If the reactivity of A is comparable to that of the propagating chain radical, then a new chain may be initiated. If its reactivity toward a monomer is less than that of the propagating radical, then the overall reaction rate is retarded. If A is unreactive toward the monomer, the entire reaction could be inhibited. Transfer reactions do not result in the creation or destruction of radicals; at any instant, the overall number of growing radicals remains unchanged. However, the occurrence of transfer reactions results in the reduction in the average polymer chain length, and in the case of transfer to a polymer, it may result in branching.

1.2.3 Diene Polymerization

Butadiene (1), chloroprene (2), and isoprene (3) (Fig. 1.14) are the main and most distinctive members of conjugated dienes, the unsaturated organic compounds characterized by the presence of two conjugated double bonds (Ricci and Leone, 2014).

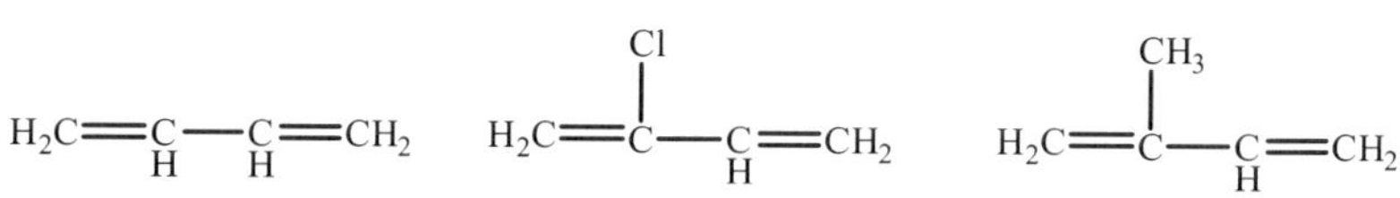

(1) (2) (3)

Figure 1.14 Conjugated dienes: (1) butadiene, (2) chloroprene, and (3) isoprene.

These structures contain double bonds in the 1,2 and 3,4 positions, each of which may participate independently in polymerization, giving rise to 1,2 and 3,4 units. A further possibility is that both bonds are involved in polymerization through conjugate reactions, resulting in 1,4 units. These structures are shown in Reaction 1.43.

Diene polymerization thus gives rise to polymers that contain various isomeric units. In the case of symmetrical dienes such as butadiene, the 1,2 and 3,4 units are identical. The 1,4 units may occur in the cis- or trans-configuration (Ricci and Leone, 2014). A

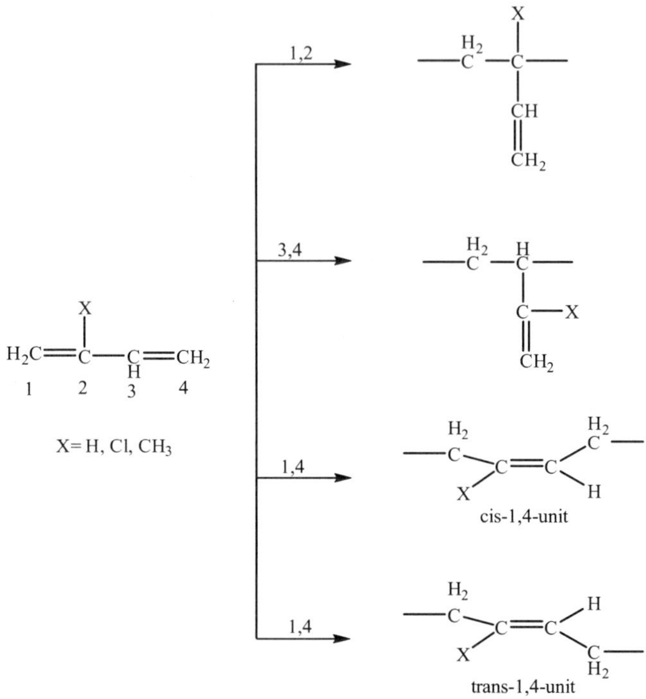

(1.43)

diene polymer contains more than one of these structural units. The relative abundance of each unit in the polymer molecule depends on the nature of the initiator, experimental conditions, and the structure of the diene. The proportion of each type of structure incorporated into the polymer chain influences both thermal and physical properties. For example, butadiene can be polymerized by free-radical addition at low temperature to produce a polymer that consists almost entirely of *trans*-1,4 units and only about 20% 1,2 units. As the temperature is increased, the relative proportion of *cis*-1,4 units increases, while the proportion of the 1,2 structure remains fairly constant. Anionic diene polymerization with lithium or organolithium initiators like *n*-butyllithium in nonpolar solvents such as pentane or hexane yields polymers with high *cis*-1,4 content. When higher-alkali-metal initiators or more polar solvents are used, the relative amount of *cis*-1,4 units decreases. Stereoregularity can also be controlled by the use of coordination catalysts like Ziegler–Natta catalysts (Ricci and Leone, 2014). Heterogeneous Alfin catalysts, which are combinations of alkenyl sodium compounds, alkali metal halides, and an alkoxide, give polymers with high molar mass and high content of *trans*-1,4 units. All chain reaction polymerizations involve essentially the same number of steps. However, the main distinguishing feature between chain reaction polymerizations is associated with the initiation mechanism, which may be free radical, ionic, or coordination (Ebewele, 2000).

1.2.4 Practical Methods of Chain Polymerization

In practice, four different methods, called bulk, solution, suspension, and emulsion, that employ chain reactions are used to prepare a variety of high-molar-mass polymers having commercial importance.

1.2.4.1 Bulk polymerization

At first sight, this method appears to be a very suitable method for producing high-molar-mass polymers by chain reactions, with the starting mix consisting of a pure polymer with only trace amounts of an initiator and a possible chain transfer agent. However, there are some problems associated with this process; for example, the increased viscosity during polymerization leads to difficulties in the processing of the product. Also, since chain reactions are generally exothermic, and the increased viscosity inhibits dissipation of heat, there can be localized overheating, leading to charring and possible degradation of the product.

Consequently, despite the apparent advantages, and the fact that it is often the method of choice for the laboratory-scale production of vinyl polymers, bulk polymerization is not widely used in industry (Saeki and Emura, 2002). At the industrial scale, only three polymers, namely, polyethylene, polystyrene, and PMMA, are produced by this method. Of these polymers, polyethylene is produced from a gaseous monomer under pressure, either high or low, and thus some of the disadvantages mentioned above for bulk polymerization hardly apply. They are certainly more severe for polymerization of liquid monomers, such as styrene and methyl methacrylate. In the case of polystyrene, bulk polymerization is nonetheless used for the commercial production of the polymer. However, the problems associated with the exothermic behavior of the process have been solved by considering the discrete stages in the manufacture of polymer.

The reaction is initiated in a tank, which is heated to a temperature of 80°C; styrene undergoes self-initiation on heating, so no extra initiator is required for this step, which is allowed to continue until about 35% conversion to polymer. At this conversion, the mixture still has a sufficiently low viscosity to enable fairly easy stirring and transport. From this stage of 35% polymerization, the mixture is passed down a tower in an atmosphere of nitrogen; there is a thermal gradient throughout the tower from 100°C at the top to 200°C at the bottom. This gradient is maintained by a complicated arrangement of heaters and coolers, which compensate for the exotherm that the polystyrene undergoes itself as increasing proportions of monomer are converted to polymer. At the bottom of the tower, the high-molar-mass polystyrene is extruded, granulated, and cooled prior to packaging (Ebewele, 2000).

1.2.4.2 Solution polymerization

Solution polymerization seems to a good alternative according to bulk polymerization to overcome the difficulties associated with the exothermic behavior during the polymerization. One way of overcoming some of the problems associated with bulk polymerization is to dissolve the monomer in an appropriate solvent. In particular, difficulties associated with the exotherm on polymerization may be overcome since temperature can be more readily controlled than in the bulk technique. If the right solvent is chosen, the product may form to give a solution suitable for casting or spinning. There are some disadvantages with solution polymerization, too.

First, the reaction temperature is limited by the boiling point of the solvent used, which, in turn, restricts the rate of reaction that may be achieved. Second, it is difficult to free the product of the last traces of the solvent. Finally, selection of a completely inert solvent cannot actually be done, which means that there is almost always chain transfer to the solvent and hence a restriction on the molar mass of the product. This last point is particularly important and is the one that is primarily responsible for the rarity of solution techniques in the manufacture of commercially important polymers. Solution polymerization is an exclusively used method for manufacturing copolymers containing vinyl acetate (Nicolson, 2006; Spyrou, 2012).

1.2.4.3 Suspension polymerization

For suspension polymerization, monomers are mixed with an initiator and dripped into a nonmiscible phase, most commonly water, with surfactants or detergents as stabilizers, to ensure formation of proper spheroid particles. The solution is heated to induce activation of the initiator and the start of the radical polymerization reaction. After the polymerization reaction has terminated, microspheres can be collected and washed free from the stabilizers.

The method is rather simple, and most monomer mixtures can be used. Also, monomers that are partly soluble in water can be used, although a more complex stabilizer solution is required in such cases. By this method, a wide range of microsphere sizes can be obtained, roughly ranging from 40–1000 μm (Saralidze et al., 2010).

There are some disadvantages associated with suspension polymerization. In particular, for polymers that are very soluble in their monomers, stirring has to be extremely vigorous; otherwise, the partially reacted droplets undergo agglomeration. Also, tacky polymers such as synthetic elastomers are very prone to agglomerate, so suspension polymerization cannot be used for these polymers. Despite these disadvantages, some typical polymers such as polystyrene, poly(vinyl chloride) (PVC), polyacrylates, poly(vinyl acetate) (PVAc), and their copolymers are produced by suspension polymerization (Qiu et al., 2001).

For the polymerization of PVC that is industrially produced by this method, the vinyl chloride monomer is suspended in demineralized water to which gelatin has been added as the suspending agent and caproyl peroxide as the initiator. Polymerization takes place under an atmosphere of nitrogen at 50°C, taking 12 h to reach 85–90% completion, at which point reaction is stopped. Pure PVC is obtained from the suspension by centrifugation (Nicolson, 2006).

1.2.4.4 Emulsion polymerization

Emulsion polymerization, which represents the next stage in development from the suspension method and is a versatile and widely used method of polymerization, is a complex process in which the radical addition polymerization proceeds in a heterogeneous system. This process involves emulsification of the relatively hydrophobic monomer in water by an oil-in-water emulsifier, followed by the initiation reaction with either a water-soluble (potassium persulfate) or an oil-soluble free-radical initiator (Fried, 2014; Yamak, 2013).

Emulsion polymerization is especially useful for the synthesis of small particles, nanospheres, and microspheres (range 0.1–10 μm). The obtained microspheres display low polydispersity, and high yields can be obtained (Saralidze et al., 2010). These microspheres or micelles contain a small quantity of monomer, the rest of the monomer being suspended in the water without the aid of any surfactant. This forms free radicals in solution, which may initiate some growing chains in solution. These radicals or growing chains pass to the micelles and diffuse into them, which causes the bulk of the polymerization to occur in these stabilized droplets.

As emulsion polymerization proceeds, like the suspension technique but unlike either the bulk or the solution techniques, there is almost no increase in viscosity. The resulting dispersed polymer is not a true emulsion any more but instead has become latex. The formation of latex particles may be considered an interfacial process in as much as it produces interfacial area. The particles of the latex do not interact with the water; hence viscosity is not found to change significantly up to about 60% solid content.

Emulsion polymerization is used in the commercial production of synthetic diene elastomers and also to produce commercial latexes of the type used in paints; these paints are known incorrectly as emulsion paints and are used extensively.

1.2.5 Ionic and Coordination Polymerizations

Depending on the initiator, the polymerization proceeds according to three different major reaction mechanisms, that is, cationic, anionic, and coordination–insertion mechanisms (Stridsberg et al., 2000). Polymer molecules in the ionic (cationic and anionic) and coordination polymerizations grow over the counterions and coordination complexes, respectively. Ionic polymerizations involve chain carriers or reactive centers that are organic ions or charged organic groups. In anionic polymerization, the growing chain end carries a negative charge or carbanions, while cationic polymerization involves a growing chain end with a positive charge or carbonium (carbenium) ion.

Coordination polymerization is thought to involve the formation of a coordination compound between the catalyst, monomer, and growing chain. The mechanisms of ionic and coordination polymerizations are more complex and are not as clearly understood as those of free-radical polymerization. Initiation of ionic polymerization usually involves the transfer of an ion or an electron to or from the monomer. Many monomers can polymerize by more than one mechanism, but the most appropriate polymerization mechanism for each monomer is related to the polarity of the monomers and the Lewis acid–base strength of the ion formed (Ebewele, 2000).

1.2.5.1 Cationic polymerization

Cationic polymerization is applied almost exclusively to monomers with olefinic double bonds. Monomers with electron-donating groups like isobutylene form stable positive charges and are readily converted to polymers by cationic catalysts. Any strong Lewis acid like boron trifluoride (BF_3) or Friedel–Crafts catalysts such as $AlCl_3$ can readily initiate cationic polymerization in the presence of a cocatalyst like water, which serves as a Lewis base or source of protons. During initiation, a proton adds to the monomer to form a carbonium ion, which forms an association with the counterion. This is illustrated for isobutylene and boron trifluoride in Reaction 1.44.

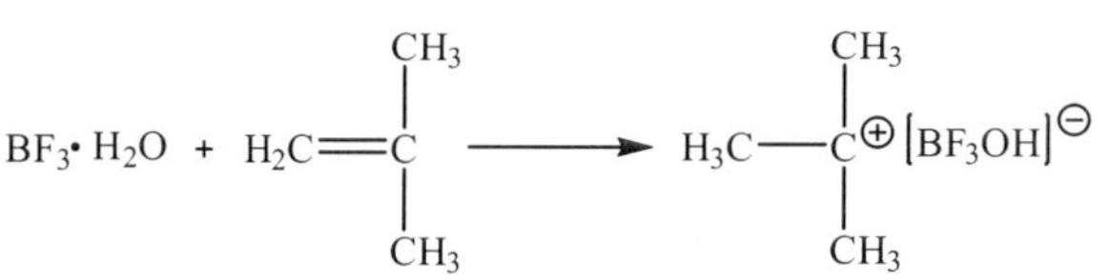

(1.44)

Propagation involves the consecutive additions of monomer molecules to the carbonium ion at the growing chain end. Termination in cationic polymerization usually involves rearrangement to produce a polymer with an unsaturated terminal unit and the original complex or chain transfer to a monomer and possibly to the polymer or solvent molecule. Unlike free-radical polymerization, termination by combination of two cationic polymer growing chains does not occur. Polymerization rates are high usually, although the polymerization is carried out frequently in solution and at temperatures as low as -80°C to -100°C. The cation and the counterion in cationic polymerization remain in close proximity. However, if the intimate association between the ion pair is too strong, monomer insertion during propagation will be prevented. Therefore, the choice of solvent in cationic polymerization has to be made carefully; a linear increase in polymer chain length and an exponential increase in the reaction rate usually occur as the dielectric strength of the solvent increases.

1.2.5.2 Anionic polymerization

Hydrocarbon monomers (dienes and styrene), polar vinyl monomers (vinyl pyridines, methacrylates, vinyl ketones, and acrylonitriles), and cyclic monomers containing oxirane, lactones, carbonates, and siloxanes are polymerized using anionic initiators (Baskaran and Müller 2010). The initiator in anionic polymerization may be any compound providing a strong nucleophile, including Grignard reagents and other organometallic compounds. Initiation involves the addition of the initiator to the double bond of the monomer, as illustrated for styrene and butyllithium in Reaction 1.45.

$$n\ \text{—}C_4H_9^-\ Li^+ \ +\ H_2C{=}CH_2 \longrightarrow n\ \text{—}C_4H_9\text{—}\overset{H_2}{C}\text{—}CH^-\ Li^+ \quad (1.45)$$

The reaction produces a carbanion at the head end to which is associated the positively charged lithium counterion. Propagation occurs by the successive insertion of monomer molecules by anionic attack of the carbanion. No chain transfer or branching occurs in anionic polymerization, particularly if reactions are carried out at low temperatures. Termination of the growth activity of the polymer chain takes place either by the deliberate or by the accidental introduction into the system of oxygen, carbon dioxide, methanol, water, or other molecules that are capable of reacting with the active chain ends.

In anionic polymerization as well as free-radical polymerization, the initiator or part of it becomes part of the resulting polymer molecule, as attached to the nongrowing chain end. This is different from cationic polymerization, where a catalyst is necessary for initiation and propagation but is regenerated at the termination step. In some systems, termination can be avoided if the starting reagents are pure and the polymerization reactor is purged of all oxygen and traces of water. This produces polymer molecules that can remain active even after all the monomer molecules are consumed. When a fresh monomer is added, polymerization resumes. Such polymeric molecules are referred to as "living polymers" because of the absence of termination. Since the chain ends grow at the same rate, the degree of polymerization (DP) of living polymers is determined simply by the ratio of monomer concentration to that of the initiator (Eq. 1.1).

$$\text{Degree of Polymerization (DP)} = \frac{[\text{monomer}]}{[\text{initator}]} \tag{1.1}$$

Polymers produced by living polymerization are characterized by a very narrow molar mass distribution (Poisson distribution). The polydispersity (D) *that can be used to determine the molar mass* is given in Eq. 1.2.

$$D = \frac{M_\text{w}}{M_\text{n}} = 1 + \frac{1}{\text{DP}} \tag{1.2}$$

where M_w and M_n are the mass-average molar masses and the number-average molar masses, respectively.

Living polymerization can also be employed to introduce a variety of desired functional groups at one or both ends of polymeric chains both in homo- and block polymers. In particular, living polymerization techniques are important in terms of the control of the architecture of a polymer (Ebewele, 2000).

1.2.5.3 Coordination polymerization

Monomers that have asymmetrically disposed side groups with respect to the double bond are capable of producing polymers in which the side groups have a specific stereochemical or spatial arrangement (isotactic or syndiotactic). In both cationic and anionic polymerizations, the association of the initiating ion and counterion permits a preferential placement of the asymmetric substituted monomers, depending on the polymerization conditions.

Some coordination polymers that may be unbranched and stereospecific polymers are also produced by using Ziegler–Natta catalysts, which are complex catalyst systems derived from a transition metal compound with an organometallic compound usually. A typical catalyst complex formed by trialkyl aluminum and titanium trichloride is shown in Fig. 1.15.

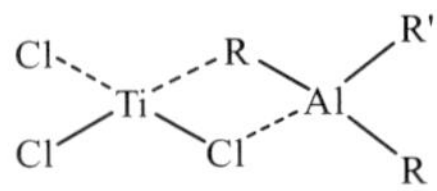

Figure 1.15 A typical catalyst complex formed by trialkyl aluminum and titanium trichloride.

Mono-olefins (propylene) and dienes (butadiene and isoprene) can be polymerized using Ziegler–Natta coordination catalysts, which function by forming transient complexes between the monomers and the transition metal species. The initiation produces a metal–alkyl complex, so propagation progresses with the consecutive insertion of monomer molecules into a polarized titanium–carbon bond.

Coordination polymerization may be terminated by introducing poisons such as water, hydrogen, aromatic alcohols, or metals like zinc into the reaction system.

1.2.6 Ring-Opening Polymerization

Ring-opening polymerization (ROP) is one of the three paths to polymerization, together with chain (radical and ionic) polymerization and condensation polymerization. Some ROP can be considered chain polymerization, in which the monomer binds to a growing chain end, but many reactions are more complicated with the activated monomer (Nuyken and Pask, 2013). While ROP shares certain features with condensation and addition polymerization mechanisms, it differs from each of them in at least a few important respect.

First, in contrast to condensation polymerization, no small molecule is split off in ROP. Second, unlike addition polymerization, the driving force of ROP is not the reduction of unsaturation. A significant number of polymers have been produced from the ring opening of cyclic organic compounds, including epoxides such as ethylene and propylene oxides and epichlorohydrin and other cyclic ethers like trioxane and tetrahydrofuran. Other important systems include cyclic esters (lactones), cyclic amides (lactams), cyclo-olefins, and siloxane. ROP involves essentially an initial ring opening of the cyclic monomer, followed by polyaddition. The resulting polymers are normally linear. Their structural units usually have the same composition as the monomer.

Major applications of polymers obtained from ROP are in coatings, fibers, elastomers, adhesives, and thermoplastic- and thermoset-based composite systems. ROPs may be represented broadly by Reaction 1.46:

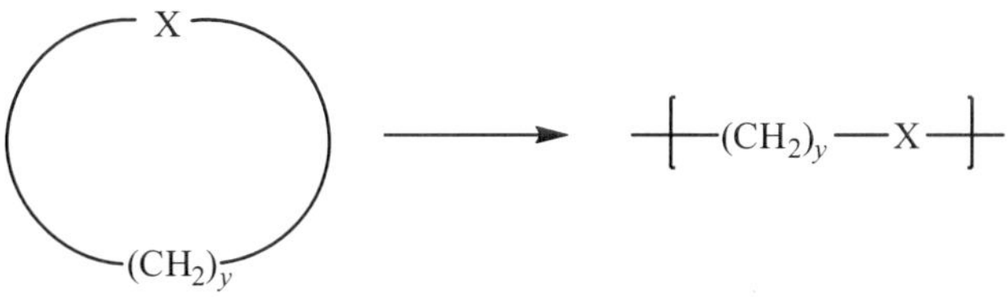

(1.46)

where X may be an atom (O and S) or group (NH, –O–CO–, and –NH–CO). All cyclic compounds are not suitable for ROP. Thus, Reaction 1.46 reveals that the cyclic structures are capable of ROP to produce linear polymer chains with *n* degrees. The nature of *X* is such that it provides a mechanism for a catalyst or initiator to form the initiating coordination intermediate with the cyclic ring.

1.2.6.1 Poly(propylene oxide)

The polymerization of propylene oxide that forms essentially linear polymer molecules represents an important example of industrial ROP. It involves an attack of the least sterically hindered carbon by the hydroxyl ion to produce the alkoxide (Reaction 1.47).

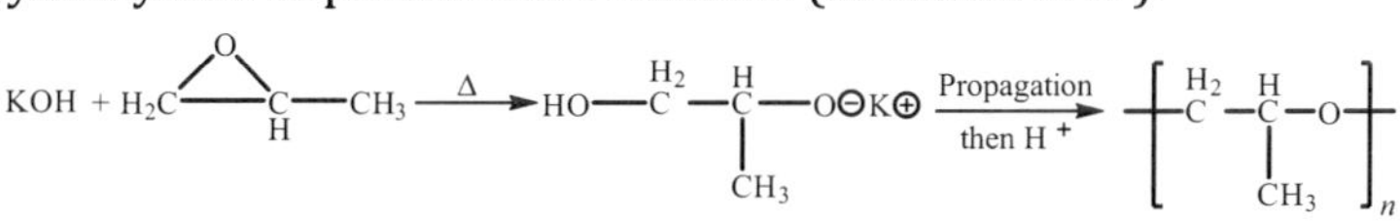

(1.47)

Poly(propylene oxide) has a great demand due to different applications. Some of the special applications are enclosed within their use as impact modifiers, surfactants, de-emulsifiers, dispersant agents, fuel additives, wetting agents, lubricants, rheological modifiers, biomedical applications, adhesives, and intermediaries in the manufacture of urethane elastomers or block copolymers

(Cendejas et al., 2008). It is frequently necessary to modify the growing species in propylene oxide polymerization with ethylene oxide in order to produce a polymer with acceptable reactivity with isocyanates and urethane product with desirable properties.

1.2.6.2 Epoxy resins

The first and still the most important class of commercial epoxy resins is the reaction product of bisphenol A and epichlorohydrin in the presence of sodium hydroxide (Reaction 1.48). It is called the diglycidyl ether of bisphenol A (DGEBA) (Unnikrishnan and Thachil, 2006). This produces resins ranging from liquid to semisolid to solid and varying molecular weights and softening points. The products are oligomers or prepolymers, which are hardly used as such; they have pendant hydroxyl groups and terminal epoxy groups. The epoxy prepolymer can be crosslinked or cured by reaction with a number of reagents, including primary and secondary amines.

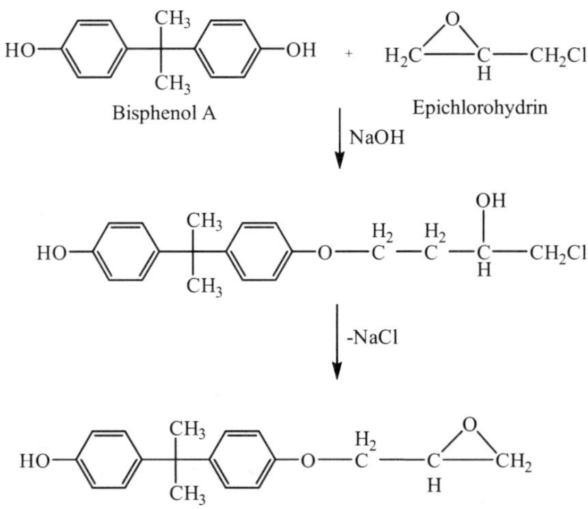

(1.48)

1.2.6.3 Polycaprolactam (nylon-6)

The two major commercial routes of the manufacturing of poly(caprolactam) (nylon-6) are water-initiated hydrolytic polymerization and a strong-base-initiated anionic polymerization (Fang et al., 2002). Anionic ROP of ε-caprolactam (CL) leads to the formation of polyamide 6 (PA6) (Ahmadi et al., 2010). Polymerization by cationic initiation is also known, but monomer conversion and attainable molar masses are inadequate for practical purposes, and as such there is no commercial practice of this process.

The most important commercial route for producing nylon-6 is the hydrolytic polymerization of CL. The process involves the polymerization of CL in the presence of water (Araromi and Adegbola, 2014). The following synthetic scheme (Reactions 1.49 and 1.50) outlines hydrolytic polymerization of caprolactam:

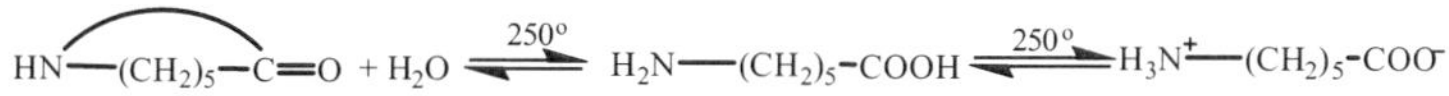

(1.49)

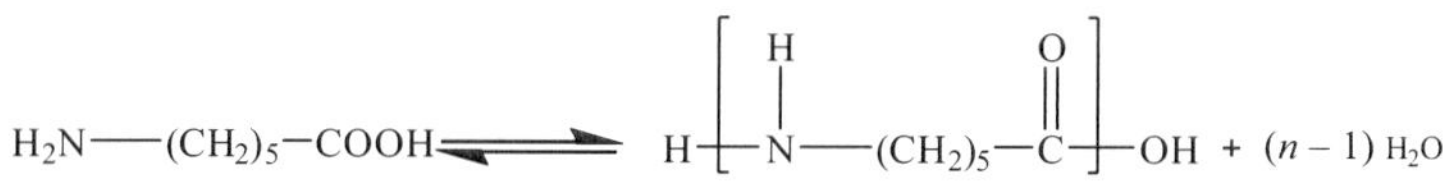

(1.50)

Water opens the caprolactam ring, producing aminocaproic acid, which is believed to exist as the zwitterion. The zwitterion interacts with and initiates the step polymerization of the monomer, with the ultimate generation of linear polymer molecules. In other words, the polymerization process involves an initial ring opening of the monomer, which is followed by step polymerization. The hydrolytic polymerization process may be carried out as a continuous operation, or it may be operated batchwise.

It involves heating caprolactam in an essentially oxygen-free atmosphere in the presence of water to temperatures in the range of 250°C to 270°C for periods ranging from 12 h to more than 24 h. Most of the water used to initiate the reaction is removed during the process after about 80% to 90% conversion. The overall polymerization involves various equilibria and does not result in complete conversion of the caprolactam. The quantity of the residual monomer depends on the reaction temperature, which under industrial conditions amounts to 8% to 10%. In addition, there are about 3% of predominantly cyclic low-molar-mass oligomers.

These impurities adversely affect the subsequent processing and end-use performance of the polymer and must, therefore, be removed from it. This is achieved either by hot-water extraction or by vacuum evaporation.

The second approach for the commercial synthesis of nylon-6, which accounts for up to 10% of the volume of the polymer, is the base-initiated anionic polymerization of CL. A small but important number of applications utilize this process, which is characterized by high conversion rates. It involves two techniques: high-temperature polymerization and low-temperature polymerization.

High-temperature polymerization is carried out at temperatures above the melting point of nylon-6 (i.e., 220°C), whereas low-temperature polymerization involves temperatures in the range of 140°C to 180°C, which are above the melting point of CL but below the melting point of the resulting polycaprolactam. The polymerization catalysts are strong bases such as sodium hydride or a Grignard reagent.

To obtain satisfactory low-temperature polymerization, a co-initiator such as *N*-acyl caprolactam or acyl-urea is employed in addition to the strong base. Sodium or magnesium caprolactam salt is produced by the reaction of sodium hydride or a Grignard reagent with caprolactam. A rapid polymerization occurs when the acylated lactam, the catalyst, and the monomer are mixed at temperature at or greater than 140°C.

This polymerization is usually carried out in a two-stream reactor, in which one stream contains the catalyst dissolved in the monomer and the second stream contains the initiator dissolved separately in the monomer (Reactions 1.51–1.53):

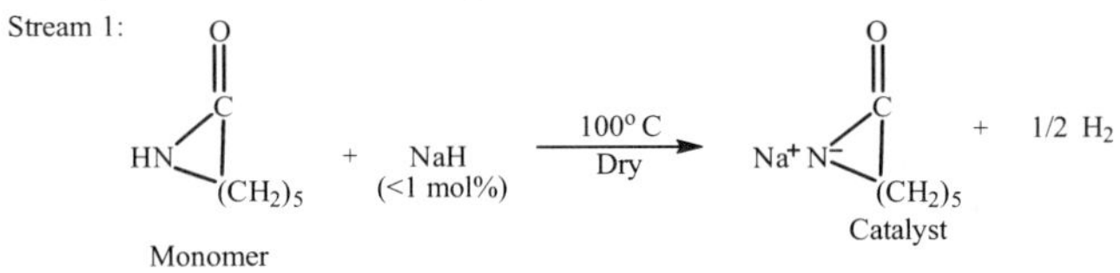

(1.51)

Stream 2:

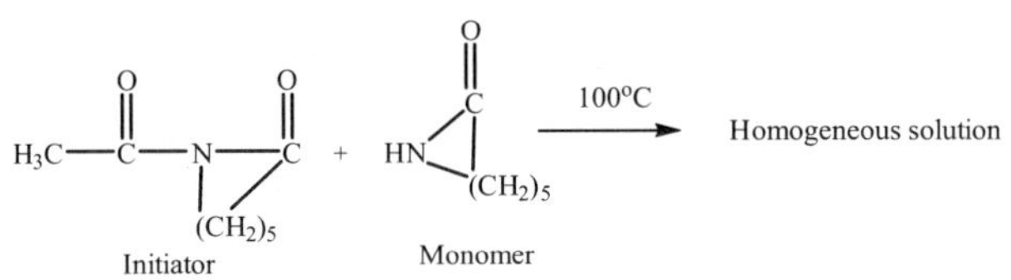

(1.52)

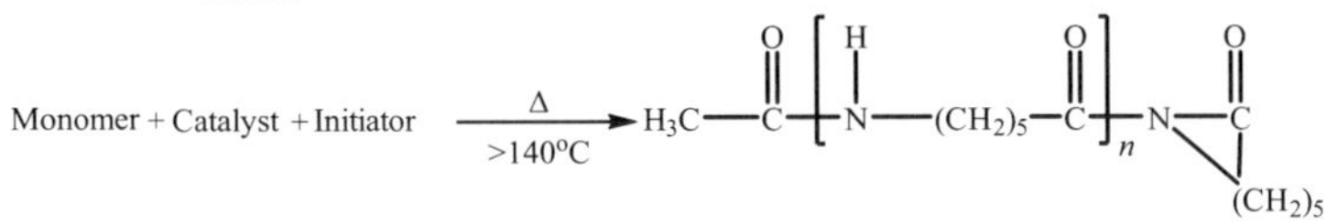

(1.53)

Unlike the high-temperature process where about 8% to 10% cyclics are generated, the equilibrium monomer content of nylon-6 resulting from polymerization at a temperature lower than 200°C (i.e., low-temperature polymerization) can be less than 2%. Therefore, the polymer does not usually require any additional purification. Also, the maximum rate of crystallization of nylon-6 falls within the range of temperature employed in low-temperature polymerization.

Consequently, the resulting polymers are characterized by a high degree of crystallinity. Such a polymer obtained by a single-stage polymerization is a highly crystalline solid (PA-6 or nylon-6) that is easily processable. Therefore, anionic polymerization is perfectly suitable for manufacturing parts directly from the monomer via casting, reaction injection molding (RIM), and rotational molding (Rijswijk, 2007; Tung, 1993).

1.3 Properties of Polymers

1.3.1 Mechanical Properties

Polymers are macromolecules that vary from liquids and soft rubbers to hard and rigid solids (Meijer and Govaert, 2005). The unique properties of polymers coupled with their light weight make them preferable alternatives to metallic and ceramic materials in many applications. In the selection of a polymer for a specific use, its mechanical properties must be taken into account.

The term "mechanical properties" is commonly used to denote stress–strain relationships for polymer systems. Unlike many more familiar materials where these relationships depend essentially only on temperature, in polymeric systems time dependence is also important (Aklonis, 1981). This is important, not only in those applications where the mechanical properties play a primary role, but also in other applications where other characteristics of the polymer, such as electrical, optical, or thermal properties, are crucial. In the latter cases, mechanical stability and durability of the polymer may be required for the part to perform its function satisfactorily.

The mechanical behavior of a polymer is a function of its microstructure or morphology. Polymer morphology itself depends on many structural and environmental factors. Compared with those of metals and ceramics, polymer properties show a much stronger dependence on temperature and time. This strong time and temperature sensitivity of polymer properties is a consequence of the viscoelastic nature of polymers. This implies that polymers exhibit a combined viscous and elastic behavior.

Depending on the temperature and stress levels, a polymer may show linear elastic behavior, yield phenomena, plastic deformation, or the cold-drawing behavior.

The most important feature of rheological behavior of polymeric systems is due to the combination of fluidity and high elasticity, that is, due to the ability to accumulate simultaneously unlimited irrecoverable and large recoverable deformations (Vinogradov, 1971).

An amorphous polymer with T_g (glass transition temperature) below ambient temperature may display nonlinear but recoverable deformation or even exhibit viscous flow.

Polymer components, like other materials, may fail to perform their intended functions in specific applications as a result of:

- excessive elastic deformation,
- yielding or excessive plastic deformation, and
- fracture.

Polymers show excessive elastic deformation, particularly in structural load-bearing applications, due to inadequate rigidity or stiffness. For such a failure, the elastic modulus is the controlling mechanical property. The elastic modulus of some polymers can be controlled to some measure through appropriate structural modification.

Failure of polymers in certain applications to carry design loads or occasional accidental overloads may be due to excessive plastic deformation resulting from the inadequate strength properties of the polymer. For the quantification of such failures, the primary mechanical property is the yield strength and the corresponding strain strength.

Cracks that constitute regions of the material discontinuity frequently precipitate failure through fracture. Fracture is a form of failure and is defined as the separation or fragmentation of a solid body into two or more parts under the action of stress. It may occur in a sudden, brittle manner or through fatigue (progressive fracture). Brittle fracture occurs in materials where the absence of local yielding results in a buildup of localized stresses, whereas fatigue failure occurs when parts are subjected to alternating or repeated loads. Fatigue fractures occur without visible signs of yielding since they occur at strengths well below the tensile strength of the material.

1.3.1.1 Mechanical tests

Mechanical tests measure the response or deformation of a material to periodic or varying forces. Generally, an applied force and its resulting deformation both vary sinusoidal with time. From such tests, it is possible to obtain simultaneously an elastic modulus and mechanical damping; the latter gives the amount of energy dissipated as heat during the deformation of the material (Gupta, 2010; Rosato and Rosato, 2003). Mechanical testing is frequently used for the study of structure–property correlations in polymers and modified polymer systems. The techniques, methods, and procedures used for homopolymers are often applied without any modification for the characterization of polymer blends.

1.3.1.1.1 *Tensile tests*

A tensile test is the most common technique for determining the various properties of materials, such as fracture and tear resistance, yield stress and strain (Kalpakjian and Schmid, 2009). This test, which is pretty simple, is used commonly for the mechanical characterization of all kinds of materials. Tensile characteristics of the prepared products from thermoplastic or thermoset polymers, as well as rubbers, are usually measured via the samples in the form of bars, films, and tubes. Although at a tensile test, typical stress–strain curves are obtained, Young's modulus, yield, and ultimate characteristics are usually considered tensile properties. Since the true stress and strain are difficult to measure in practice, the nominal or engineering stress and strain are usually obtained on the basis of the original dimensions of the test specimen. On the basis of typical stress–strain curves for plastics shown in Fig. 1.16, the tensile properties may be defined as follows:

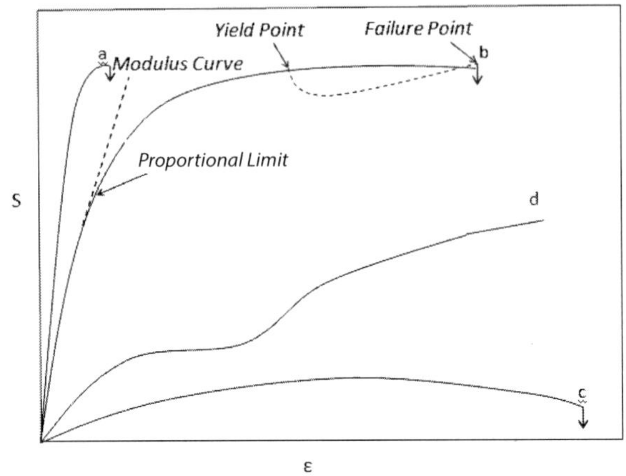

Figure 1.16 Typical stress–strain curves of plastics. (a) Brittle plastics, (b)plastics showing a yield point and cold drawing, (c) some elastomeric plastics, and (d) some rubbers.

- **Tensile stress** is the tensile load/unit area of minimum original cross section within the gauge boundaries carried by the test specimen at any given moment. It is expressed in force per unit area (Eq. 1.3) as follows:

$$\sigma = \frac{F}{A_0} \tag{1.3}$$

- **Tensile strength** is the maximum tensile stress sustained by a specimen during a tension test (Rosato and Rosato, 2003).
- **Tensile strength at break** is the tensile stress at the moment of rupture of the test specimen.
- **Elongation** is the deformation of a thermoplastic or thermoset material when a load is applied at the ends of the specimen test bar in the opposite axial direction (Tres, 2006). It is the increase in length produced in the gauge length of the test specimen by a tensile load, which is expressed in units of length, commonly as a percentage.
- **Elongation at break** is the elongation at the moment of rupture of the test specimen.
- **Strain** is the ratio of the elongation to the gauge length of the test specimen, that is, the change in length per unit of original length. It is expressed as a dimensionless ratio (Eq. 1.4) as follows:

$$\varepsilon = \frac{\Delta L}{L_0} \tag{1.4}$$

- **Yield point** is the first point on the stress–strain curve at which an increase in strain occurs without an increase in stress (Gere and Goodno, 2009).
- **Proportional limit** is the greatest stress or strain that a material is capable of sustaining without any deviation from proportionality of stress to strain.
- **Modulus of elasticity (Young's modulus)** is the ratio of stress to corresponding strain below the proportional limit of a material. It is expressed as force per unit area (Eq. 1.5) as follows:

$$E = \frac{\sigma}{\varepsilon} \tag{1.5}$$

- **Work to cause rupture** is the area under the stress–strain curve. It is a measure of the amount of energy or work that must be done to rupture secondary (yield point) and primary (rupture) bonds.

In some special cases, with crystalline polymers such as nylon, polypropylene, or linear polyethylene, the specimens may undergo what is called "necking," where highly localized strain takes place. The cold drawing that takes place in all such crystalline materials is usually accompanied by considerable increase in the strength of the specimen as orientation of molecules and realignment of crystallites take place. Necking, or cold drawing, invariably continues until all of the narrow center section has become oriented or until some serious flaw is encountered. If allowed to continue, the necking process encounters the greatly increased cross section of the ends of the specimen.

The strength of the highly oriented center section is usually not great enough to overcome the resistance of the enlarged sections at the ends of the specimen; at this point, the necking process stops and rupture occurs (Uçar, 2007).

1.3.1.1.2 *Flexural tests*

A flexural test measures the force required to bend a beam under three-point or four-point loading conditions. The data are often used to select materials for parts that will support loads without flexing. The flexural modulus is used as an indication of a material's stiffness when flexed (SoonKook, 2010). For the determination of flexural properties of polymers, a three-point or four-point loading system is used.

The three-point loading system utilizes center loading on a simply supported beam, and also the area of uniform stress concentrated under the center loading point is quite small (Grote and Antonsson, 2009). The bar rests on two supports and is loaded by means of a loading nose midway between the supports (Fig. 1.17).

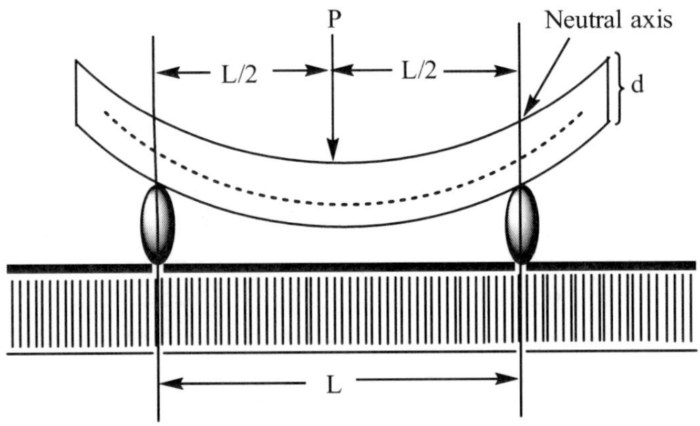

Figure 1.17 Schematic drawing of three-point bending test.

When the specimen is tested in flexure as a simple beam supported at two points and loaded at the midpoint, the maximum stress in the outer fibers occurs at midspan. This stress may be calculated for any point on the load-deflection curve by Eq. 1.6:

$$S = \frac{3PL}{2bd^2} \tag{1.6}$$

where S is stress in the outer fibers at midspan (MPa), P is load at a given point on the load-deflection curve (N), L is the support span (mm), b is the width of the beam tested (mm), and d is the depth of the beam tested (mm). The maximum strain in the outer fibers occurs at midspan as well and may be calculated by Eq. 1.7:

$$r = \frac{6Dd}{L^2} \tag{1.7}$$

where r is the maximum strain in the outer fibers (mm/mm), D is the maximum deflection of the center of the beam (mm), d is the depth of the beam tested (mm), and L is the support span (mm). The tangent modulus of elasticity, often called flexural modulus, is the ratio within the elastic limit of stress to corresponding strain and shall be expressed in megapascals. It is calculated by drawing a tangent to the steepest initial straight-line portion of the load-deflection curve and using Eq. 1.8:

$$E_B = \frac{L^3 m}{4bd^3} \tag{1.8}$$

where E_B is the modulus of elasticity (tangent modulus of elasticity) in bending (MPa), L is the support span (mm), b is the width of the beam tested (mm), d is the depth of the beam tested (mm), and m is the slope of the tangent to the initial straight-line portion of the load-deflection curve (N/mm)(Özkoç, 2007).

1.3.1.1.3 *Impact tests*

Another popular method associated with testing of the mechanical performance of polymers is the impact test. Impact tests measure the energy required for failure when a standard specimen receives a rapid stress loading. The impact strength of a polymer can be measured employing a number of techniques, including Izod and Charpy tests. For both Izod and Charpy tests, a weight is released, causing the specimens to be struck. The energy to break values is determined from the loss in the kinetic energy of the weight (Alyamaç, 2004).

In the Charpy method, the specimen is supported as a horizontal beam and is broken by a single oscillation of the pendulum, where the line of impact is central to the two supports. In the Izod method, the specimen is supported as an embedded beam and is broken by a single oscillation of the pendulum at a fixed distance from the specimen clamp and from the center line of the impact. In either case, the specimens may or may not have a stress concentrator notch.

The measuring method is based on determining the amount of energy, expressed in joules, needed to break the specimen under specified conditions, such as location of the specimen, shape of the notch, and speed of impact of the hammer on the specimen. As the maximum potential energy of the pendulum is known and is in relation to the weight of the hammer used and the drop height, the latter in turn being in relation to the starting angle (Nicholson, 2006).

1.3.2 Thermal Properties

Three factors affect the essential nature of a polymeric material and determine whether it is glassy, rubbery, or fiber forming under a given set of conditions. These are:

- flexibility of the macromolecule,
- magnitude of the forces between the molecules, and
- stereoregularity of the macromolecules.

These three factors influence the ability of the polymer to crystallize, the melting point of any resulting crystalline regions, and also the glass transition temperature.

1.3.2.1 Glass transition temperature

Most polymers are either completely amorphous or have an amorphous-like component even if they are crystalline. Such materials are hard, rigid glasses below a fairly sharply defined temperature known as the glass transition temperature, T_g. At temperatures above the glass transition temperature, at least at slow to moderate rates of deformation, an amorphous polymer is soft and flexible and is an elastomer or a very viscous liquid. Mechanical properties show profound changes in the region of the glass transition. For example, the elastic modulus may decrease by a factor of over 1000 times as the temperature is raised through the glass transition region. For this reason, T_g can be considered the most important matrixial characteristic of a polymer as far as mechanical properties are concerned.

Many other physical properties change rapidly with temperature in the glass transition region. These properties include coefficients of thermal expansion, heat capacity, refractive index, mechanical damping, nuclear magnetic and electron spin resonance behavior, electrical properties, and tensile strength and ultimate elongation in elastomers (Nielsen and Landel, 1994).

Glass transition that differs from the genuine phase changes, such as melting and boiling, known as first-order transitions, is a second-order transition. The first-order transitions are characterized by a distinct volume change, as well as changes in optical properties, and the existence of latent heat for the phase change. By contrast, no such changes occur at the glass transition, though the rate of change of volume with temperature alters at T_g, as illustrated in Fig. 1.18. As a material is heated to this point and beyond, molecular rotation around single bonds suddenly becomes significantly easier. A number of factors can affect the ease with which such molecular rotation takes place, and hence influence the actual value of the glass transition temperature. The inherent mobility of a single-polymer molecule is important, and molecular features that either increase or reduce this mobility will cause differences in the value of T_g.

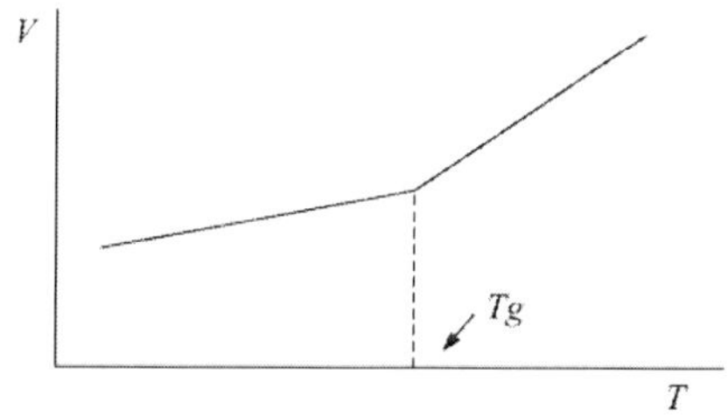

Figure 1.18 Rate of change of volume with temperature and T_g.

In addition, interactions between polymer molecules can lead to restrictions in molecular mobility, thus altering the T_g of the resulting material. Apart from the rate and method of measurement, there are other external factors by which the measured T_g is affected. These can be summarized as follows:

- *Structural and compositional factors.* The most fundamental of these are chain stiffness and interchain cohesive forces.
- *Thermal, thermo-oxidative, and/or photochemical degradation.* The onset of rapid degradation sometimes occurs in the temperature range of the glass transition, obscuring the distinction between glass transition and degradation.
- *Crosslinking.* It is another structural factor that can affect T_g. T_g increases with increasing crosslinking density (γ_c).
- *Pressure.* It also affects the value of T_g. Most measurements of T_g are performed under normal atmospheric pressure, so the effect of pressure is seldom considered in any detail in correlations for T_g.

- *Average molar mass of the polymer chains.* T_g increases asymptotically with increasing average molar mass to its limiting values for the "high polymer" at infinite average chain molar mass.
- *Presence of additives, fillers, unreacted residual monomers, and/or impurities.* These are either deliberately included in the formulation of a resin or left over as undesirable by-products of the synthesis. For example, plasticizers of low molar mass generally decrease T_g.
- *Progress of the thermal process.* Either the annealing or the physical ageing of test specimens at elevated temperatures below T_g usually results in an increase in T_g (Nakka, 2010).

The effects of these different factors can be seen by considering the T_g values of some typical polymers, shown in Table 1.3 (Arrese-Igor et al., 2010; Boiler et al., 1996; Coulon et al., 2013; Ganesan et al., 2012; Gaur and Nunderlich, 1980; Gracias et al., 1999; Lee et al., 1998; Noda et al., 2009).

The glass transition temperature is usually insensitive to the relative molar mass of the polymer above degrees of polymerization of about 200. Below this, T_g is related to relative-molar-mass, lower-molar-mass specimens having lower values of T_g. This is because the chain ends tend to have inherently greater degrees of freedom of motion than the other segments of the macromolecule, and hence the more chain ends per unit volume of material, the greater the overall freedom of the molecules, leading to reduced values of T_g.

Table 1.3 Glass transition temperatures of some typical polymers

Polymer	T_g (°C)	Contributing feature
Polyethylene	−20.0	Flexible backbone
Polypropylene	5.0	CH_3 group inhibiting freedom of rotation
PVC	80.0	Strong polar attraction between molecules
PTFE	115.0	Very stiff backbone

Random copolymers having an amorphous morphology generally exhibit a single T_g having a value between the T_g *values* of the individual homopolymers. The exact value depends on the relative proportions of the respective monomers in the copolymer.

In certain cases, polyblends of two very compatible polymers also show just one T_g. The value of such a T_g can be predicted from either of the following Eqs. 1.9 and 1.10 (Sandler, 1998):

$$(T_g)_{AB} = (T_g)_A f_A + (T_g)_B f_B \tag{1.9}$$

where f_A and f_B are the volume fractions of A and B, respectively, or

$$\left[\frac{1}{T_g}\right] = \left[\frac{M_A}{(T_g)_A}\right] + \left[\frac{M_B}{(T_g)_B}\right] \tag{1.10}$$

where M_A and M_B are the mass fractions of A and B.

The latter equation may also be used to predict the approximate T_g of a random copolymer using the mass fractions of the respective monomers from which the copolymer was prepared.

Both of these are approximate equations and are useful only for estimating the value of the T_g of the polyblend or copolymer. More accurate calculation of values of T_g requires additional information such as the coefficients of thermal expansion of both components in both their liquid and glassy states. The polyblends that are made from incompatible polymers having a two-phase structure show two T_g values, including a value for each phase. The temperatures of these transitions correspond closely to the T_g values of the respective homopolymers (Sandler, 1998)

1.4 Polymer Characterization

1.4.1 Nuclear Magnetic Resonance

Solution and solid-state two-dimensional nuclear magnetic resonance (NMR) are important tools for materials characterization and for understanding the structure and dynamics of polymers.

The power of NMR as a tool arises from its ability to visualize individual sites along the polymer main chain and side chain, making it possible to relate the structure and dynamics at the molecular level to the bulk macroscopic properties (Levine, 1975; Mirau, 1989). The time scale involved in NMR measurements makes it possible to study certain rate processes, including chemical reaction rates. Other applications are isomerism, internal relaxation, conformational analysis, and tautomerism.

1.4.2 Infrared Spectroscopy

Infrared (IR) radiation was discovered by Sir William Herschel in 1800 (Theophanides, 2012). Beginning in 1903, William W. Coblentz conducted comprehensive experiments, leading to the accurate measurement of IR spectra for hundreds of inorganic and organic compounds (Derrick et al., 1999). IR spectroscopy is probably the method most extensively used for the investigation of polymer structure and the analysis of functional groups.

IR spectrometers have been used to study samples in the gaseous, liquid, and solid states, depending on the types of accessories used (Fan et al., 2012). Fourier transform infrared (FTIR) spectroscopy can be used to characterize polymer blends, polymer complexes, dynamics, surfaces, and interfaces, as well as chromatographic effluents and degradation products. It provides information about the complexion interactions between the various constituents in the polymer electrolyte. It is capable of qualitative identification of the structure of unknown materials as well as the quantitative measurement of the components in a complex mixture (Jasmeet et al., 2012; Sandler, 1998).

1.4.3 Thermogravimetric Analysis

Thermogravimetric analysis (TGA) or thermogravimetry (TG) is a technique where the mass of a polymer or material associated with a transition or thermal degradation is measured as a function of temperature or time while the sample is subjected to a controlled temperature program in a controlled atmosphere (Prime et al., 2009).

TG curves are characteristic for a given polymer or compounds because of the unique sequence of the physicochemical reaction that occurs over specific temperature ranges and heating rates and are a function of the molecular structure. The changes in mass are a result of the rupture and formation of various chemical and physical bonds at elevated temperatures that lead to the evolution of volatile products or the formation of heavier reaction products.

From TGA curves, data concerning the thermodynamics and kinetics of the various chemical reactions, reaction mechanisms, and the intermediate and final reaction products are obtained.

Other processes that can be studied by TGA are adsorption and desorption phenomena, reactions with purge gases, ash content analysis, quantitative determination of additives, solid-state reaction composition of filled polymers, rates of evaporation, and sublimation. TGA has also been used to estimate the flame retardancy of polymers, as enhanced flame retardancy is often paralleled by increased amounts of residual char at high temperature (Sandler, 1998).

1.4.4 Differential Scanning Calorimetry

Thermal analysis (TA) has been defined as a group of techniques in which a physical property of a substance and/or its reaction products is measured as a function of temperature while the substance is subjected to a controlled temperature program. Both differential thermal analysis (DTA) and differential scanning calorimetry (DSC) are applicable to a wider range of processes according to TG (Warrington and Höhne, 2001). DTA/DSC curves reflect changes in the energy of the system that undergoes a physical or chemical change. DSC measures the heat required to maintain the same temperature in the sample versus an appropriate reference material in a furnace. Enthalpy changes associated with any physical or chemical conversion can be determined by this technique. DTA differs from DSC in which the temperature difference is determined, rather than enthalpy differences between the sample and the reference material.

The many important thermophysical properties of a polymer may be determined by DSC. These include the glass transition temperature (T_g), the crystallization temperature (T_c), the melt temperature (T_m), and the degradation or decomposition temperature (T_d). Chemical changes such as polymerization and degradation reactions, as well as other reactions affecting the sample, can also be investigated by this technique (Tant et al., 1995).

1.4.5 X-Ray Diffraction

X-ray diffraction techniques are a very useful characterization tool to study, nondestructively, the crystallographic structure, chemical composition, and physical properties of materials (Sharma et al.,

2012). The wavelengths of X-rays are comparable to the interatomic distance in crystals (0.5–2.5 Å). Two primary diffraction methods are used to study polymers: wide-angle X-ray scattering (WAXS) and small-angle X-ray scattering (SAXS).

Also called X-ray diffraction, WAXS is very frequently used in characterization of semicrystalline multiphase systems. The diffraction pattern contains information that is specific to each phase within the illuminated volume, including both geometric and structural parameters, many of which are inaccessible to other techniques (Denchev Viana, 2011). The value of angles used in WAXS is from 5° to 120°. As the crystallinity of polymers is often low, the width of the diffraction peaks in WAXS gives information on the size of the crystals. By the measurement of relative intensities of diffraction peaks in the crystalline and amorphous parts, the crystalline content of the polymer may be deduced. Also, WAXS is used to provide information about the number of repeat units per the turn at the helical structures, as well as the length of the repeat unit along the fiber axis, and the degree of orientation.

The value of angles used in SAXS is at the range of 1° to 5°. SAXS can provide information on greater interatomic distances from 50 to 700 Å. Consequently, SAXS is available for detecting larger periodic repeats in a structure. For example, many polymeric materials crystallize with individual chains folding back and forth within a given crystalline region or crystallite. Other examples of the utility of SAXS are in the study of lamellae crystallites or in the distribution of particles or voids in the material (Simon, 2003).

1.4.6 Optical Microscopy

Microscopy involves obtaining magnified images to examine the morphology, structure, and shape of various samples, including grains, phases, embedded phases, and particles (Mukhopadhyay, 2003).

In polymer science, the term "morphology" generally refers to an organization related to surface in a size scale above the atomic arrangement but smaller than the size and shape of the whole sample.

It is known that the structures present in a polymer reflect the processing variables that greatly influence the physical and mechanical properties. Therefore, the properties of polymeric materials are influenced by the chemical composition, processing regime, and morphology.

A morphological study usually requires two preparatory steps: selection of instrumentation and specimen preparation technique. Structural observations must be correlated with the properties of the material in order to develop the various applications with the material.

Polymers are considered either amorphous or crystalline, although they may not be completely one or the other. It is possible to grow single crystals of polymers. The units of organization in polymer crystals are lamellae or crystals and spherulites. Lamellae are thin, flat platelets often 100 Å thick and a few micrometers in lateral dimensions. The size, shape, and regularity of the crystals depend on their growth conditions, including solvent type, temperature, stress, contaminants, and growth rate. The types of solvent, temperature, stress, contaminants, and growth rate are important factors.

The thickness of the lamellae depends on crystallization temperature and any subsequent annealing treatment. Electron diffraction measurements indicate that polymer chains are generally oriented normal or almost normal to the plane of the lamellae. The length of the molecules in a polymer is at least 1000 Å, and also the lamellae are folded. The most prominent organization in the polymers on a scale larger than the lamellae is the spherulite, a spherical aggregate having a diameter ranging from the submicroscopic size to millimeters.

1.4.7 Dynamic Mechanical Analysis

Polymers vary from liquids and soft rubbers to very hard and rigid solids (Landel and Nielsen, 1994). The mechanical behavior of polymeric materials depends on many structural factors.

Dynamic mechanical analysis (DMA) or dynamic mechanical thermal analysis (DMTA) is known as a very efficient technique for the examination of time-, frequency-, and temperature-dependent mechanical properties of polymeric materials (Kazakevičiūtė-Makovska et al., 2014).

The dynamic mechanical properties, called dynamic modulus, dynamic loss modulus, and mechanical damping, generally refer to the response of the material subjected to a periodic force (Kaliappan, 2007). The term "viscoelasticity" describes the time-dependent mechanical properties of polymers that can exhibit either elastic solid or viscous liquid behavior. The knowledge of the molecular structures of polymers and their viscoelastic behavior is essential of understanding both processing and usage properties. DMA can be applied to a wide range of materials using different sample fixture configurations and deformation modes.

This procedure can be used to evaluate, by comparing to known materials, (a) the degree of phase separation in multicomponent systems, (b) the amount and dispersion of the filler, (c) the degree of polymer crystallinity, (d) effects of certain pretreatment, and (e) the stiffness of polymer composites. Dynamic mechanical experiments yield both the elastic modulus of the material and its mechanical damping, or energy dissipation, characteristics. These properties can be determined as a function of frequency (time) and temperature.

References

Abdel-Kader, N. S., Mohamed, R. R. (2013). Synthesis, characterization and thermal investigation of some transition metal complexes of benzopyran-4-one Schiff base as thermal stabilizers for rigid poly(vinyl chloride) (PVC). *J. Therm. Anal. Calorim.*, **114**(2), pp. 603–611.

Ahmadi, S., Morshedian, J., Hashemi, S. A., Carreau, P. J., Leelapornpisit, W. (2010). Novel anionic polymerization of ε-caprolactam towards polyamide 6 containing nanofibrils. *Iranian Polym. J.*, **19**(3), pp. 229–240.

Akiba, M., Hashim, A. S. (1997). Vulcanization and crosslinking in elastomers. *Prog. Polym. Sci.*, **22**, pp. 475–521.

Aklonis, J. J. (1981). Mechanical properties of polymers. *J. Chem. Educ.*, **58**(11), pp. 892–897.

Alwaan, I. M. (2014). Kinetics of thermal degradation of recycled polyvinyl chloride resin. *Int. J. Chem. Eng.*, **2014**, Article ID 701092, 7 pages.

Alyamaç, E. (2004). *Impact Modified Poly(Ethylene Terephthalate)-Organoclay Nanocomposites*. Master thesis. Middle East Technical University, Graduate School of Natural and Applied Science, Ankara, Turkey.

Arad-Yellin, R., Firer, M., Kahana, N., Green, B. S. (2003). Functionalized polysulfone as a novel and useful carrier for immunization and antibody detection. *React. Funct. Polym.*, **54**, pp. 5–16.

Araromi, D. O., Adegbola, B. (2014). System identification of nylon-6 caprolactam polymerization process. *ARPN J. Sci. Technol.*, **4**(2), pp. 99–107.

Arrese-Igor, S., Alegría, A., Colmenero, J. (2010). Comparison of calorimetric and dielectric single component glass transitions in PtBS–PI blends. *Macromolecules*, **43**(15), pp. 6406–6413.

Baker, A. A., Hodgkin, J., Hou, M. (2004). Polymeric matrix materials. In Baker, A., Dutton, S., Kelly, D. (Eds.), *Composite Materials for Aircraft Structures*, 2nd Ed. (p. 109). Reston, Virginia: American Institute of Aeronautics and Astronautics. ISBN: 1-56347-540-5.

Baskaran, D., Müller, A. H. E. (2010). Anionic vinyl polymerization. In Müller, A. H. E., Matyjaszewski, K. (Eds.), *Controlled and Living Polymerizations: From Mechanisms to Applications* (p. 5). Weinheim, Germany: Wiley-VCH Verlag. ISBN: 978-3-527-32492-7.

Belgacem, M. N., Gandini, A. (2008). The state of the art. In Belgacem, M. N., Gandini, A. (Eds.), *Monomers, Polymers and Composites from Renewable Resources* (p. 10). Amsterdam, Netherlands: Elsevier. ISBN: 978-0-08-045316-3.

Bell, V. L., Pezdirtz, G. F. (1965). Polyimidazopyrrolones: A new route to ladder polymer. *J. Polym. Sci., Part B: Polym. Lett.*, **3**(12), pp. 977–984.

Benhui, S. (1994). Study of the mechanism of nylon 6,6 dissolving process using $CaCl_2$/MeOH as the solvent. *Chin. J. Polym. Sci.*, **12**(1), pp. 57–65.

Bernal, P. J. (2006). From "greasy chemistry" to "macromolecule": Thoughts on the historical development of the concept of a macromolecule. *J. Chem. Educ.*, **83**(6), pp. 870–879.

Bhattacharya, A., Ray, P. (2009). Basic features and techniques. In Bhattacharya, A., Rawlins, J. W., Ray, P. (Eds.), *Polymer Grafting and Crosslinking* (p. 27). Hoboken, New Jersey: John Wiley & Sons. ISBN: 978-0-470-40465-2.

Billiet, L., Fournier, D., Prez, F. D. (2009). Step-growth polymerization and "click" chemistry: The oldest polymers rejuvenated. *Polymer*, **50**, pp. 3877–3886.

Billmeyer, F. (1971). *Textbook of Polymer Science*, 2nd Ed. New York: John Wiley & Sons.

Boiler, A., Okazaki, I., Wunderlich, B. (1996). Modulated differential scanning calorimetry in the glass transition region. Part III. Evaluation of polystyrene and poly(ethylene tetrephthalate). *Thermochim. Acta,* **284**, pp. 1–19.

Bruma, M., Schulz, B. (2001). Silicon-containing aromatic polymers. *J. Macromol. Sci. Polym. Rev.*, *C***41**(1–2), pp. 1–40.

Brydson, J. A. (1999). *Plastic Materials,* 7th Ed. (p. 434). Oxford, UK: Butterworth-Heinemann. ISBN: 0-7506-4132-0.

Carpentier, C. E. (2013). *Three-Dimensional Visualization of Contact Networks in Granular Material.* PhD thesis. University of Amsterdam, Van der Waals-Zeeman Institute, Leiden, Netherlands.

Celebi, O., Lee, C. H., Lin, Y., McGrath, J. E., Riffle, J. S. (2011). Synthesis and characterization of polyoxazoline–polysulfone triblock copolymers. *Polymer,* **52**, pp. 4718–4726.

Cendejas, G., Flores-Sandoval, C. A., Huitrón, N., Herrera, R., Zamudio-Rivera, L. S., Beltrán, H. I., Vázquez, F. (2008). Theoretical and experimental studies of the initiator influence on the anionic ring opening polymerization of propylene oxide. *J. Mol. Struct.*, **879**, pp. 40–52.

Chandrasekhar, V. (2005). *Inorganic and Organometallic Polymers* (p. 30). Berlin, Germany: Springer. ISBN: 3-540-22574-9.

Charles, E., Carraher, Jr. (2003). *Polymer Chemistry,* 6th Ed. (p. 543). USA: Marcel Dekker. ISBN: 0-8247-0806-7.

Chitte, R. R., Nalawade, V. K., Dey, S. (1999). Keratinolytic activity from the broth of a feather degrading thermophilic *Streptomyces thermoviolaceus* strain SD8. *Lett. Appl. Microbiol.*, **28**(2), pp. 131–136.

Conner, A. H. (1996). Urea-formaldehyde adhesive resins. In Salamone, J. C. (Ed.), *Polymeric Materials Encyclopedia.11 T-Z* (p. 1597). USA: CRC Press.

Coran, A. Y. (1994). Vulcanization. In Mark, J. E., Erman, B., Eirich, F. R. (Eds.), *Science and Technology of Rubber* (pp. 339–385). London, UK: Academic Press. ISBN: 0-12-472525-2.

Coulon, J. F., Dorget, M., Poncin-Epaillard, F. (2013). Thermal annealing as a new simple method for PTFE texturing Patrice Glaris. *Polymer,* **54**, pp. 5858–5864.

Datye, K. V. (1991). Recycling processes and products in nylon 6 fibre industry. *Ind. J. Fibre Text. Res.*, **16**, pp. 46–51.

Delpech, M. C., Coutinho, F. M. B., Habibe, M. E. S. (2002). Bisphenol A-based polycarbonates: Characterization of commercial samples. *Polym. Test.*, **21**, pp. 155–161.

Denchev, Z. Z., Viana, J. C. (2011). X-ray scattering studies on multiphasic polymer systems. In Boudenne, A., Ibos, L., Candau, Y., Thomas, S. (Eds.), *Handbook of Multiphase Polymer Systems* (pp. 667–701). Hoboken, New Jersey: John Wiley & Sons. ISBN: 0-470-71420-4.

Derrick, M., Stulik, R. D., Landry, J. M. (1999). *Infrared Spectroscopy in Conservation Science* (p. 1). USA: J. Paul Getty Trust. ISBN: 0-89236-469-6.

Dodge, J. (2003). Polyurethanes and polyureas. In Rogers, M. E., Long, T. E. (Eds.), *Synthetic Methods in Step-Growth Polymers* (p. 197). Hoboken, New Jersey: John Wiley & Sons. ISBN: 0-471-38769-X.

Doğar, Ç., Gürses, A., Köktepe, S., Mindivan, F., Güneş, K., Aktürk, S. (2014). Synthesis and characterization of phenol/urea/formaldehyde (PUF) and clay composites. *Acta Polonica A*, **125**(2), pp. 374–378.

Drobny, J. G. (2012). *Polymers for Electricity and Electronics: Materials, Properties, and Applications*. Hoboken, New Jersey: John Wiley & Sons. ISBN: 978-047-04-5553-1.

Duda, A., Penczek, S. (2005). *Mechanisms of Aliphatic Polyester Formation. Biopolymers Online* (p. 373). Weinheim, Germany: Wiley-VCH. ISBN: 978-352-76-0003-8.

Ebewele, R. O. (2000). *Polymer Science and Technology*. New York, USA: CRC Press.

Ebnesajjad, S. (2000). *Fluoroplastics: Non-Melt Processible Fluoroplastics: The Definitive User's Guide and Data Book* (p. 48). New York: William Andrew. ISBN: 1-884201-84-7.

Fan, M., Dai, D., Huang, B., (2012). Fourier transform infrared spectroscopy for natural fibres. In Salih, M. S. (Ed.), *Materials Analysis*. Crotia: InTech Press. ISBN: 978-953-51-0594-7.

Fang, X., Simone, C. D., Vaccaro, E., Huang, S. J., Scola, D. A. (2002). Ring-opening polymerization of ε-caprolactam and ε-caprolactone via microwave irradiation. *J. Polym. Sci., Part A: Polym. Chem.*, **40**, pp. 2264–2275.

Fengel, D., Wegener, G. (Eds.) (1989). *Wood: Chemistry, Ultrastructure, Reactions* (p. 66). Berlin, Germany: Walter de Gruyter. ISBN: 3-11-012059-3.

Fenichell, S. (1996). *Plastic: The Making of a Synthetic Century*. New York: Harper Collins.

Fink, J. K. (2005). *Reactive Polymers Fundamentals and Applications: A Concise Guide to Industrial Polymers* (p. 139). Norwich: William Andrew. ISBN: 0-8155-1515-4.

Flory, P. J. (1953). *Principles of Polymer Chemistry* (p. 136). Ithaca, New York: Cornell University. ISBN: 0801401348.

Fratini, C. M. (2006). *Study of the Morphology and Optical Properties of Propylene/Ethylene*. PhD thesis. Virginia State University, Polytechnic Institute, Blacksburg, Virginia, USA.

Fried, J. R. (2014). *Polymer Science and Technology,* 3rd Ed. Upper Saddle River, New Jersey: Prentice Hall.

Ganesan, A., Affi, J., Yamada, M., Fukumoto, M. (2012). Bonding behavior studies of cold sprayed copper coating on the PVC polymer substrate. *Surf. Coat. Technol.*, **207**, pp. 262–269.

Gardziella, A., Pilato, L. A., Knop, A. (2000). *Phenolic Resins: Chemistry, Applications, Standardization, Safety and Ecology,* 2nd Ed. Berlin, Germany: Springer. ISBN: 3-540-65517-4.

Gaur, U., Nunderlich, B. (1980). The glass transition temperature of polyethylene. *Macromolecules*, **13**, pp. 445–446.

Gere, J. M., Goodno, B. J. (Eds.) (2009). *Mechanics of Materials*, 7th Ed. (p. 20). USA: Cengage Learning Press. ISBN: 0-534-55397-4.

Gracias, D. H., Zhang, D., Lianos, L., Ibach, W., Shen, Y. R., Somorjai, G. A. (1999). A study of the glass transition of polypropylene surfaces by sum-frequency vibrational spectroscopy and scanning forcemicroscopy. *Chem. Phys.*, **245**, pp. 277–284.

Gregorova, A. (2013). Application of differential scanning calorimetry to the characterization of biopolymers, applications of calorimetry in a wide context: Differential scanning calorimetry. In Elkordy, A. A. (Ed.), *Isothermal Titration Calorimetry and Microcalorimetry* (pp. 3–20). Croatia: InTech Press. ISBN: 78-953-51-0947-1.

Gridnev, A. A., Ittel, S. D. (2001). Catalytic chain transfer in free-radical polymerizations. *Chem. Rev.*, **101**, pp. 3611–3659.

Grote, K.-H., Antonsson, E. K. (Eds.) (2009). *Springer Handbook of Mechanical Engineering*, 10th Ed. Würzburg, Germany: Stürtz Gmbh. ISBN: 978-3-54049131-6.

Gupta, A. L. (2010). *Polymer Chemistry*. Meerut, India: Pragati.

Hallensleben, M. (2000). Polyvinyl compounds, others. In *Ullmann's Encyclopedia of Industrial Chemistry*, Vol. 29. (pp. 605–621). Weinheim, Germany: Wiley-VCH.

Hasirci, V., Yilgor, P., Endogan, T., Eke, G., Hasirci, N. (2011). Polymer fundamentals: Polymer synthesis. In Ducheyne, P., Healy, K. E., Hutmacher, D. W., Grainger, D. W., Kirkpatrick, C. J. (Eds.), *Comprehensive Biomaterials*, Vol. 1 (pp. 349–371). Amsterdam, Netherlands: Elsevier.

Heeger, A. J. (2001). Nobel lecture: Semiconducting and metallic polymers; the fourth generation of polymeric materials. *Rev. Mod. Phys.*, **73**(3), pp. 681–700.

Heinze, T., Liebert, T. (2001). Unconventional methods in cellulose functionalization. *Prog. Polym. Sci.*, **26**, pp. 1689–1762.

Hussain, R., Mahmood, F. (1994). Effect of rare earth (Ho, Tb and Dy) laurates on the thermal degradation of poly(vinyl chloride). *J. Chem. Soc. Pak.*, **16**(4), pp. 225–228.

Jahromi, S. (1999). The storage stability of melamine formaldehyde resin solutions: III. Storage at elevated temperatures. *Polymer*, **40**, pp. 5103–5109.

Jasmeet, K., Harikumar, S. L., Amanpreet, K. (2012). Interpolyelectrolyte complexes as prospective carriers for controlled drug delivery. *Int. Res. J. Pharm.*, **3**(4), pp. 58–62.

Javni, I., Hong, D. P., Petrovic, Z. S. (2013). Polyurethanes from soybean oil, aromatic, and cycloaliphatic diamines by nonisocyanate route. *J. Appl. Polym. Sci.*, **128**(1), pp. 566–571.

Jin, Q., Yamashita, T., Horie, K., Yokota, R., Mita, I. (1993). Polyimides with alicyclic diamines. I. Syntheses and thermal properties. *J. Polym. Sci., Part A: Polym. Chem.*, **31**, pp. 2345–2351.

Kaliappan, S. K. (2007). *Characterization of Physical Properties of Polymers Using AFM Force-Distance Curves*. PhD thesis. University of Siegen, Siegen, Germany.

Kalpakjian, S., Schmid, S. (2009). *Manufacturing Engineering and Technology*, 6th Ed. (p. 57). New York, USA: Prentice Hall. ISBN-10: 0136081681.

Kamigaito, M., Ando, T., Sawamoto, M. (2001). Metal-catalyzed living radical polymerization. *Chem. Rev.*, **101**, pp. 3689–3745.

Kauffman, G. B. (1988). Wallace Hume Carothers and nylon, the first completely synthetic fiber. *J. Chem. Educ.*, **65**(9), pp. 803–808.

Kavitha, B., Dasharatham, D., Srinivasu, D., Srinivas, C. H., Narsimlu, N. (2011). Synthesis and characterization of TiO_2 doped polyvinyl alcohol polymer composites. *J. Chem. Pharm. Sci.*, **4**(4), pp. 155–157.

Kazakevičiūtė-Makovska, R., Özarmut, A. Ö., Steeb, H. (2014). Characterization of shape memory polymer estane by means of dynamic mechanical thermal analysis technique. *Smart Mater. Res.*, **2014**, Article ID 250258, p. 9.

Khandal, R. K. (2006). *Organic Chemistry, Synthetic Polymers and Dyes* (pp. 1–72). Delhi, India: ShriRam Institute for Industrial Research.

Kiparissides, C. (1996). Polymerization reactor modeling: A review of recent developments and future directions. *Chem. Eng. Sci.*, **51**(10), pp. 1637–1659.

Korniłłowicz-Kowalska, T., Bohacz, J. (2011). Biodegradation of keratin waste: Theory and practical aspects. *Waste Manage.*, **31**, pp. 1689–1701.

Koutsos, V. (2009). Polymeric materials: An introduction. In Forde, M. (Ed.), *ICE Manual of Construction Materials* (pp. 571–577). London, UK: Thomas Telford.

Kulkarni, S. O., Kanekar, P. P., Nilegaonkar, S. S., Sarnaika, S. S., Jog, J. P. (2010). Production and characterization of a biodegradable poly(hydroxybutyrate-co-hydroxyvalerate) (PHB-co-PHV) copolymer by moderately haloalkalitolerant Halomonas campisalis MCM B-1027 isolated from Lonar Lake, India. *Bioresour. Technol.*, **101**, pp. 9765–9771.

Kulshreshtha, A. K., Talapatra, S. (2000). Competitive new technologies in polyolefin synthesis and materials. In Vasıle, C. (Ed.), *Handbook of Polyolefins*, 2nd Ed. (pp. 1–70). New York: Marcel Dekker. ISBN: 0-8247-8603-3.

Landel, R. F., Nielsen, L. E., (1994). *Mechanical Properties of Polymers and Composites*, 2nd Ed. (p. 16). New York: Marcel Dekker. ISBN: 0-8247-8964-4.

Lau, J. Y. (2005). *Protein Structure Database for Structural Genomics Group.* Master thesis. The State University of New Jersey, New Brunswick, New Jersey.

Lee, S. G., Lee, J. H., Choi, K. Y., Rhee, J. M. (1998). Glass transition behavior of polypropylene/polystyrene/styrene-ethylene-propylene block copolymer blends. *Polym. Bull.*, **40**(6), pp. 765–771.

Leja, K., Lewandowicz, G. (2010). Polymer biodegradation and biodegradable polymers: A review. *Polish J. Environ. Stud.*, **19**(2), pp. 255–266.

Levine, I. N. (1975). *Molecular Spectroscopy*. New York: Wiley.

Li, C., Li, S., Zhao, J., Zhangi, Z., Zhang, J., Yang, W., Res, J. P. (2014). Synthesis and characterization of aliphatic poly(amide urethane)s having different nylon 6 segments through non-isocyanate route. *J. Polym. Res.*, **21**, pp. 498–507.

Li, D., McHugh, M. A. (2004). Limited polysulfone solubility in supercritical dimethyl ether with THF and DMF cosolvents. *J. Supercrit. Fluids*, **28**, pp. 79–83.

Maier, G. (2001). Low dielectric constant polymers for microelectronics. *Prog. Polym. Sci.*, **26**, pp. 3–65.

Matar, S., Hatch, L. F. (2000). *Chemistry of PetroChem Processes*, 2nd Ed. (p. 362). Houston, Texas: Gulf.

Matyjaszewski, K., Xia, J. (2001). Atom transfer radical polymerization. *Chem. Rev.*, **101**, pp. 2921–2990.

May, C. A. (1987). *Epoxy Resins: Chemistry and Technology*, 2nd Ed. (p. 794). New York: Marcel Dekker. ISBN: 0-8247-7690-9.

Meier, R. J., Tiller, A., Vanhommerig, S. A. M. (1995). Molecular modeling of melamine-formaldehyde resins. 2. Vibrational spectra of methylolmelamines and bridged methylolmelamines. *J. Phys. Chem.*, **99**(15), pp. 5457–5464.

Meijer, H. E. H., Govaert, L. E. (2005). Mechanical performance of polymer systems: The relation between structure and properties, *Prog. Polym. Sci.*, **30**, pp. 915–938.

Mindivan, F. (2013). *The Synthesis of the Composites of Phenolic Resin and the Modified Clay/Resin, and the Investigation of Their Structural, Mechanical and Thermal Properties*. PhD thesis. Atatürk University, Institute of Pure and Applied Sciences, Turkey.

Mirau, P. A. (1989). 2D NMR studies of synthetic polymers. *Bull. Magn. Reson.*, **13**(3/4), pp. 109–137.

Mukhopadhyay, S. M. (2003). Sample preparation for microscopic and spectroscopic characterization of solid surfaces and films. In Mitra, S. (Ed.), *Sample Preparation Techniques in Analytical Chemistry*, Vol. 162 (pp. 377–411). Hoboken, New Jersey: John Wiley & Sons.

Nakamura, S., Nishimoto, Y. (2002). Synthesis and properties of fluorine-containing aromatic condensation polymers obtained from bisphenol AF and its derivatives. In Hougham, G., Cassidy, P. E., Johns, K., Davidson, T. (Eds.), *Fluoropolymers-1 Synthesis* (p. 128). New York: Kluwer Academic. ISBN: 0-306-46060-2.

Nakka, J. S. (2010). *Tailoring of Epoxy Material Properties*. PhD thesis. Delft University of Technology, Netherlands.

Nicholson, J. W. (2006). *The Chemistry of Polymers*, 3rd Ed. Cambridge, UK: RSC.

Nielsen, B. G., Røgen, P., Bohr, H. G. (2006). Gauss-integral based representation of protein structure for predicting the fold class from the sequence. *Math. Comp. Modell.*, **43**, pp. 401–412.

Nielsen, L. E., Landel, R. F. (1994). Mechanical tests and polymer transition. In *Mechanical Properties of Polymers and Composites*, 2nd Ed. (p. 1). New York: CRC Press. ISBN: 978-0824789640.

Nkwachukwu, O. I., Chima, C. H., Ikenna, A. O., Albert, L. (2013). Focus on potential environmental issues on plastic world towards a sustainable plastic recycling in developing countries. *Int. J. Ind. Chem.*, **4**(34), pp. 1–13.

Noda, I., Dowrey, A. E., Marcott, C. (2009). Glass transition of atactic polystyrene probed at the submolecular level by dynamic infrared linear dichroism (DIRLD) spectroscopy. *Vib. Spectrosc.*, **51**, pp. 22–27.

Nuyken, O., Pask, S. D. (2013). Ring-opening polymerization: An introductory review. *Polymers*, **5**, pp. 361–403.

Okieimen, F. E. (2000). Thermal stabilisation of polyvinylchloride: Effect of rubber seed oil derivatives on the thermal degradation of polyvinylchloride. *J. Sci. Ind. Res.*, **59**(07), pp. 563–568.

Özkoç, G. (2007). *Abs/Polyamide-6 Blends, Their Short Glass Fiber Composites and Organoclay Based Nanocomposites: Processing and Characterization*. PhD thesis. Middle East Technical University, Graduate School of Natural and Applied Science, Ankara, Turkey.

Parameswaran, P. S. (2009). *Modification of Phenol Formaldehyde Resin for Improved Mechanical Properties*. PhD thesis. Cochin University of Science and Technology, India.

Paszun, D., Spychaj, T. (1997). Chemical recycling of poly(ethylene terephthalate). *Ind. Eng. Chem. Res.*, **36**, pp. 1373–1383.

Patel, S. N. (1992). *Thermal and Oxidative Degradation of an Aromatic Polyamide*. PhD thesis. University of Leeds, UK.

Petre, A. L., Budrugeac, P., Segal, E. (1999). Thermal degradation of polyvinyl chloride. *J. Therm. Anal. Calorim.*, **56**, pp. 1065–1070.

Pilato, L. (Ed.) (2010). *Phenolic Resins: A Century of Progress*. Berlin, Germany: Springer-Verlag. ISBN: 978-3-642-04713-8.

Prime, R. B., Bair, H. E., Vyazovkın, S., Gallagher, P. K., Riga, A. (2009). Thermogravimetric analysis (TGA). In Menczel, J. D., Prime, R. B. (Eds.), *Thermal Analysis of Polymers: Fundamentals and Applications* (pp. 241–318). Hoboken, New Jersey: John Wiley & Sons. ISBN: 978-0-471-76917-0.

Prošková, J. (2014). Description of protein secondary structure using dual quaternions. *J. Mol. Struct.*, **1076**, pp. 89–93.

Qiu, J., Charleux, B., Matyjaszewski, K. (2001). Controlled/living radical polymerization in aqueous media: Homogeneous and heterogeneous systems. *Prog. Polym. Sci.*, **23**, pp. 2083–2134.

Ratner, B. D., Hoffman, A. S., Schoen, F. J., Lemons, J. E. (Eds.) (1996). *Biomaterials Science: An Introduction to Materials in Medicine* (p. 66). USA: Academic Press. ISBN: 0-12-582460-2.

Ricci, G., Leone, G. (2014). Recent advances in the polymerization of butadiene over the last decade. *Polyolefins J.*, **1**(1), pp. 43–60.

Rijswijk, K. V. (2007). *Thermoplastic Composite Wind Turbine Blades Vacuum Infusion Technology for Anionic Polyamide-6 Composites.* PhD thesis. Delft University of Technology, Netherlands.

Rocks, J. (2004). *Characterization of Novel Co-Anhydride Cured Epoxy Resin.* PhD thesis. Queensland University of Techonology, Brisbane, Australia.

Rosato, D., Rosato, D. (2003). *Plastics Engineered Product Design* (p. 89). Oxford, UK: Elsevier. ISBN: 1856174166.

Saeki, Y., Emura, T. (2002). Technical progresses for PVC production. *Prog. Polym. Sci.*, **27**, pp. 2055–2131.

Sandler, S. R., Karo, W., Bonesteel, J., Pearce, E. M. (1998). *Polymer Synthesis and Characterization: A Laboratory Manual.* San Diego, California: Academic Press.

Saralidze, K., Koole, L. H., Knetsch, M. L. W. (2010). Polymeric microspheres for medical applications. *Materials*, **3**, pp. 3537–3564.

Setua, D. K., Shukla, M. K., Nigam, V., Singh, H., Mathur, G. N. (2000). Lignin reinforced rubber composites. *Polym. Compos.*, **21**(6), pp. 988–995.

Sharma, R., Bisen, D. P., Shukla, U., Sharma, B. G. (2012). X-ray diffraction: A powerful method of characterizing nanomaterials. *Recent Res. Sci. Technol.*, **4**(8), pp. 77–79.

Simon, G. P. (2003). *Polymer Characterization Techniques and Their Application to Blends.* New York: Oxford University Press.

Singh, V. K., Annu, A., Singh, U., Singh, P., Pandey, S. P., Bhattacharya, B., Singh, P. K. (2013). Dye sensitized solar cell based on poly(vinyl alcohol) doped with ammonium iodide solid polymer electrolyte. *J. Optoelectron. Adv. Mater.*, **15**(9–10), pp. 927–931.

Sinha, S., Kumar, V. (2010). *Polymeric Systems and Applications.* New Delhi, India: Studium Press.

SoonKook, H. (2010). *Assessment of Durability of Carbon/Epoxy Composite Materials after Exposure to Elevated Temperatures and Immersion in Seawater for Navy Vessel Applications.* PhD thesis. University of California, USA.

Spyrou, E. (Ed.) (2012). *Powder Coatings Chemistry and Technology,* 3rd Ed. (p. 24). Hanover, Germany: Vincentz Network. ISBN: 978-3-86630-824-4.

Stille, J. K. (1981). Step-growth polymerization. *J. Chem. Educ.*, **58** (11), pp. 862–866.

Stridsberg, K. M., Ryner, M., Albertsson, A. C. (2000). *Controlled Ring-opening Polymerization: Polymers with Designed Macromolecular Architecture* (p. 5). Stockholm, Sweden: Kista Snabbtryck AB. ISBN: 91-7170-522-8.

Stuart. B. H. (2002). *Polymer Analysis*. Chichester, UK: John Wiley & Sons.

Tabhane, P., Chimankar, O. P., Tabhane, V. A. (2012). Phase separation studies in polyvinyl chloride-polyvinyl acetate blend by ultrasonic technique. *J. Chem. Pharm. Res.*, **4**(6), pp. 3051–3056.

Taghizadeh, M. T., Dasdar ligvan, H. (2007). The study of the thermal stability of poly(vinyl- chloride) with different molecular weight. *J. Sci. (Univ. Tehran)*, **33**(2), pp. 25–32.

Tant, M. R., McManus, H. L. N., Rogers, M. E. (1995). High-temperature properties and applications of polymeric materials. In Tant, M. R., Connell, J. W., McManus, H. L. N. (Eds.), *High-Temperature Properties and Applications of Polymeric Materials. ACS Symposium Series*, **603** (pp. 1–20), Washington, DC: American Chemical Society. ISBN-13: 9-780-841-233-133.

Tanthapanichakoon, W., Hata, M., Nitta, K., Furuuchi, M., Otani, Y. (2006). Mechanical degradation of filter polymer materials: Polyphenylene sulfide. *Polym. Degrad. Stab.*, **91**, pp. 2614–2621.

Tauer, K. (2004). Latex particles. In Caruso, F. (Ed.), *Colloids and Collloid Assemblies Synthesis Modification Organization and Utilization of Colloid Particles* (pp. 1–45). Weinheim, Germany: Wiley-VCH. ISBN: 3-527-30660-9.

Thakur, V. K., Thakur, M. K., Gupta, R. K. (2013). Graft copolymers from cellulose: Synthesis, characterization and evaluation. *Carbohydr. Polym.*, **97**, pp. 18–25.

Theophanides, T. (Ed.) (2012). Introduction to infrared spectroscopy. In *Infrared Spectroscopy: Materials Science, Engineering and Technology* (pp. 1–10). Croatia: InTech Press. ISBN: 978-953-51-0537-4.

Toth, J. M., Wang, M., Estes, B. T., Scifert, J. L., Seim, H. B., Turner, A. S. (2006). Polyetheretherketone as a biomaterial for spinal applications. *Biomaterials*, **27**, pp. 324–334.

Tres, P. A. (2006). Strength of material for plastics. In *Designing Plastics Parts for Assembly* (p. 49). Cincinnati, Ohio: Hanser. ISBN: 9781569904015.

Tung, J. F. (1993). *Synthesis and Characterisation of Polyamide 6 Blends Made by Reactive Extrusion*. PhD thesis. Brunel University, Department of Materials Technology, London, UK.

Uçar, E. (2007). *Ternary Nanocomposites of High Density, Linear Low Density and Low Density Polyethylenes*. Master thesis. Middle East Technical University, Graduate School of Natural and Applied Science, Ankara, Turkey.

Unnikrishnan, K. P., Thachil, E. T. (2006). Toughening of epoxy resins. *Des. Monomers Polym.*, **9**(2), pp. 129–152.

Vijayaram, T. R. (2009). A technical review on rubber. *Int. J. Des. Manuf. Technol.*, **3**(1), pp. 25–37.

Vinogradov, G. V. (1971). Flow and rubber elasticity of polymeric systems. *Pure Appl. Chem.*, **26**(3), pp. 423–450.

Vroman, I., Tighzert, L. (2009). Biodegradable polymers. *Materials*, **2**, pp. 307–344.

Warrington, S. B., Höhne, G. W. H. (2001). Thermal analysis and calorimetry. In Gunzler, H., Williams, A. (Eds.), *Handbook of Analytical Techniques* (pp. 827–850). Weinheim, Germany: Wiley-VCH. ISBN: 3-527-30165-8.

Whiteley, K. S. (2012). Polyethylene. In *Ullmann's Encyclopedia of Industrial Chemistry* (pp. 1–38). Weinheim, Germany: John Wiley & Sons. ISBN: 978-352-73-0673-2.

Wyatt, L. M., Coveney, V., Riddiford, C., Sharpe, R. (1994). Materials, properties and selection. In Smith, E. H. (Ed.), *Mechanical Engineer's Reference Book*, 12th Ed. (pp. 7/1–7/180). Oxford, UK: Butterworth-Heinemann. ISBN: 0-7506-4218-1.

Yamak, H. B. (2013). Emulsion polymerization: Effects of polymerization variables on the properties of vinyl acetate based emulsion polymers. In Yilmaz, F. (Ed.), *Polymer Science* (pp. 35–67). Crotia: InTech Press. ISBN: 978-953-51-0941-9.

Yu, L., Dean, K., Li, L. (2006). Polymer blends and composites from renewable resources. *Prog. Polym. Sci.*, **31**, pp. 576–602.

Zaheed, L., Jachuck, R. J. J. (2004). Review of polymer compact heat exchangers, with special emphasis on a polymer film unit. *Appl. Therm. Eng.*, **24**, pp. 2323–2358.

Zugenmaier, P. (2009). Contribution to the historical development of macromolecular chemistry: Exemplified on cellulose. *Cellul. Chem. Technol.*, **43**(9-10), pp. 351–378.

Chapter 2

Clay Minerals, Surface Chemistry of Clays and Organoclays, and the Mechanisms of Organoclay Synthesis

2.1 Introduction

Since ancient times, clay minerals have been investigated because of their importance in agriculture, ceramics, building, and other uses such as intercalation, quasi-two-dimensional magnetism, structural phase transition, fractal characteristics, and mixed-crystal behavior. In the past few decades, natural clay minerals, such as montmorillonite (MMT), kaolinite, and palygorskite (attapulgite), have been widely used in catalysis, as adsorbents, in nanocomposites, in sensors, in electrodes, as antibacterial materials, in nuclear waste storage, as pesticide carriers, etc. (Cavani et al., 1991; Darder et al., 2005; Galan, 1996; Hanley et al., 1997; Holeman, 1965; Huh et al., 2000; Jaber et al., 2005; Li et al., 2010; Murray, 2000; Sànchez-Martìn et al., 2008). Nowadays, surface modification of clay minerals has become increasingly important for improving the practical applications of clays and clay minerals. Surface modifications of clay minerals have received intensive attention because they allow the creation of new materials and new applications. Organically modified layered silicates can be produced by replacing the cations originally present in the galleries with organic cations. Ion exchange with

Introduction Polymer–Clay Nanocomposites
Ahmet Gürses

ISBN 978-981-4613-02-6 (Hardcover), 978-981-4613-03-3 (eBook)
www.panstanford.com

alkylammonium ions is well known and the preferential method to prepare organoclays. As a result, the study of clays and organoclays is a large field and shows immense potential to be explored. The physical and chemical properties of clay minerals determine their practical use in process industries. Especially the development of inorganic–organic hybrid materials was widespread in the past few decades. The main focus of surface modification of clays is materials science because organoclays are essential for developing polymer nanocomposites. Nanocomposites constitute one of the most developed areas of nanotechnology. Modified clays are also used in other applications such as adsorbent of organic pollutants in soil, water, and air; rheological control agents; paints; cosmetics; refractory varnish; and thixotropic fluids (DePaiva et al., 2008; Gao, 2004; Gürses et al., 2011; Lan and Pinnavaia, 1994; Moraru, 2001; Paiva et al., 2008; Ruiz-Hitzky and Meerbeek, 2006; Zampori et al., 2010).

2.2 Clay Minerals

Clay minerals occur from the hydrothermal exchange of basic volcanic ash and rocks of the Cretaceous (limestone, chalk) period (85–125 million years ago). Several ideas have been proposed about the mechanisms of the ash-to-bushel transformation. The most important of these ideas, the conversion, began in marine water in reactions including enough amounts of Mg^{2+} and Na^{+} (Annabi-Bergaya, 2008; Gürses et al., 2011; Lagaly and Ziesmer, 2003; Utracki, 2004).

Clay is a sediment, which is geological material, transported somewhere by water and subsequently deposited. Sediments can be subdivided into fractions based on the grain size, which can vary from blocks with diameters >630 mm to clay with grains <2 μm. Boulders, stones, gravel and sand are between these two extremes (Bormans, 2003). The grain size is an aspect of the so-called texture, a collective of characteristic features among which are the shape of the individual grain, the grain size, the packing of the grains, and the density of the mass.

Clay minerals, hydrated oxides, and hydroxides, mainly of Si, Al, Fe, and Mn, and some geo-organic polymers are responsible for most surface and colloid reactions in the earth, including adsorption of organic matter by its components (Tombàcz et al., 1998).

2.2.1 Structure of Clay Minerals

The term "clay minerals" is used for a group of hydrous layered aluminosilicates or magnesium-aluminosilicates. In many of these minerals, various metallic cations, such as lithium, magnesium, and aluminum, act as proxies, wholly or in part for the magnesium, aluminum, or silicon, respectively, with alkali metal and alkaline earth metal cations present as exchangeable cations. Iron (di- or trivalent) is also a common substituent of aluminum and magnesium (Malucelli et al., 2007; Manjanna, 2008). Each magnesium- or aluminophyllosilicate is essentially composed of two types of sheets—octahedral and tetrahedral—designated as O and T, respectively. Each sheet is composed of planes of atoms, arranged one above the other, a plane of hydroxyls and oxygens above a plane of aluminums and magnesiums or silicons, the latter above another plane of hydroxyls and oxygens, and so on. Variations among clay minerals and the differences in their physical and chemical properties arise from the various combinations of octahedral (O) and tetrahedral (T) sheets and the electrostatic effects that chemical substitutions have on the units (Yariv and Cross, 2002).

In the **tetrahedral sheets**, each silicon atom is surrounded by four oxygen atoms, and the tetrahedrally coordinated silicon cations are linked to one another via covalent bonding through shared oxygens. These shared oxygens form a basal plane, and the remaining apical oxygens are shared with another layer of cations. The tetrahedral units arrange as a hexagonal network (Fig. 2.1) (Auerbach et al., 2004) along this basal plane. A side view of the tetrahedral sheet shows that it is composed of three parallel atomic planes, which are composed of oxygens, silicons, and oxygens, respectively. The tetrahedra are arranged so that all of their apices point in the same direction, with their bases in the same plane. The oxygens form an open hexagonal network in this plane, often referred to as the hexagonal or perforated oxygen plane (or O-plane) (Uskarcı, 2006). Aluminum atoms can replace silicon atoms in the tetrahedral sheet, thereby contributing a negative charge to the sheet. In many minerals, this substitution is very small, but in micas about 25% of the silicons are substituted by aluminums. The substitution of Al for Si changes the surface properties of the perforated oxygen plane, since Si–O–Al groups are better donors of electron pairs than are Si–O–Si groups (Yariv and Cross, 2002).

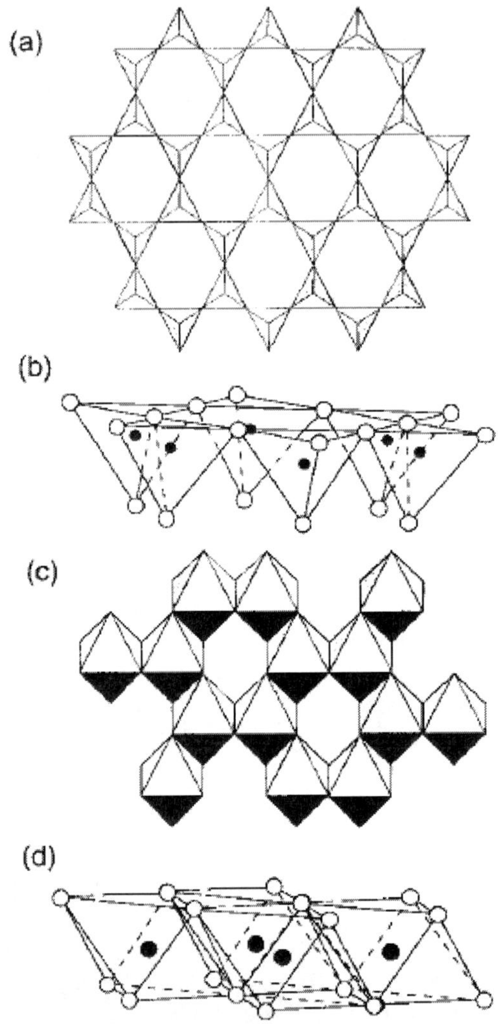

Figure 2.1 A clay $(Si_4O_{10})_4$ tetrahedral (T) sheet in (a) top view and (b) side view and a clay octahedral (O) sheet in (c) top view and (d) side view. (The $(Al_4O_{12})_{12}$ dioctahedral top view is shown in (c); a $(Mg_6O_{12})_{12}$ trioctahedral top view would show a continuous sheet of octahedral units).

The **octahedral sheets** have cations (usually aluminum or magnesium) that are coordinated with six oxygens or hydroxyls, and these units are covalently linked into a sheet structure as well.

In clays, T and O sheets are also covalently linked through the apical tetrahedral oxygens. Natural minerals made up of only O sheets include gibbsite $Al(OH)_3$ and brucite $Mg(OH)_2$. In the former, just two-thirds of the octahedra are filled with Al atoms. Clay derivatives of gibbsite are, therefore, referred to as dioctahedral. Brucite and its clay derivatives, on the other hand, are referred to as trioctahedral because all of the octahedra are filled with Mg atoms (Auerbach et al., 2004).

2.2.2 Classification of Clay Minerals

Many different classifications of clay minerals have been suggested. A committee on the terminology of the Clay Minerals Group of the Mineralogy Society of Great Britain suggested that the crystalline clay minerals be divided into chain and layer structures and that the layer structures be divided into 2:1 (TOT) and 1:1 (TO) families. The classification shown in Table 2.1 is based on distinctions in the shape of clay minerals and the expandable or nonexpendable character of the 2:1 and 1:1 layer silicates (Grim, 1968; Gürses et al., 2011).

2.2.2.1 The 1:1 layer type

The 1:1 layer type usually has no layer charge or a very small layer charge because the tetrahedral cation sites usually are all occupied by Si^{4+} and the octahedral sites by all Al^{3+} or all Mg^{2+}. If there is substitution in one sheet (see Figs. 2.1 and 2.3) of a 1:1 layer silicate, there almost always is a compensating substitution in the other sheet so that neutrality is maintained. There are dioctahedral and trioctahedral varieties of 1:1 layer silicates, and within these subgroups there are individual species.

Serpentine minerals, $Mg_6Si_4O_{10}(OH)_8$. Serpentine minerals are usually larger in particle size than clay minerals, but they do cross the boundary into the clay-size range. They are found as silky asbestos fibers and as smaller fibers in more massive or splintery material (chrysolite), as platy material (lizardite and amestite), or as the iron-rich variety berthierine. Some materials are only apparently platy. Upon investigation it proves to be a series of alternating waves with a wavelength of 30–100 Å (antigorite). These structures are called modulated.

Table 2.1 Classification of clay minerals

I. Amorphous
   Allophane group

II. Crystalline

   A. Two-layer type (sheet structures composed of units of one layer of silica tetrahedrons and one layer of alumina octahedrons
      1. Equidimensional
         kaolinite group (kaolinite, nacrite, etc.)
      2. Elongate
         halloysite group

   B. Three-layer types (sheet structures composed of two layers of silica tetrahedrons and one central dioctahedral or trioctahedral layer)
      1. Expanding lattice
         a. Equidimensional
            montmorillonite group (montmorillonite, sauconite, etc.), vermiculite
         b. Elongate
            montmorillonite group (nontronite, saponite, hectorite)
      2. Nonexpanding lattice,
         illite group

   C. Regular mixed-layer types (ordered stacking of alternate layers of different types): Chlorite group

   D. Chain-structure types (hornblende-like chains of silica tetrahedrons linked together by octahedral groups of oxygens and hydroxyls containing Al and Mg atoms):
      Attapulgite, sepiolite, palygorskite

Berthierine, $(Fe^{2+}{}_3Mg_{0.75}Al_2)(Si_{2.5}Al_{1.5})O_{10}(OH)_8$ (a representative formula). Berthierine is a mineral that has been recognized for a long time but has often been overlooked. Its identity also was confused by the misunderstanding of the nature of chamosite, a name, at times, used for a 7 Å Fe-bearing mineral.

Odinite, $(Fe^{3+}{}_{1.6}Mg_{1.6}Al_{1.1}Fe^{2+}{}_{0.6})(Si_{3.6}Al_{0.4})O_{10}(OH)_8$ (a representative formula). Odinite is a second 1:1 mineral similar in composition to berthierine but with more ferric than ferrous iron and currently forming on continental shelves and in reef lagoons.

Odinite is part of the verdine facies, a group of green marine clay minerals extensively studied (Odin, 1988).

Kaolin ($Al_4Si_4O_{10}(OH)_8$) and serpentine ($[Mg_6Si_4O_{10}](OH)_8$) minerals. In general, there is no recognized substitution in any of the species of kaolin minerals: kaolinite, dickite, nacrite, and halloysite-endellite. The kaolinite lattice consists of one sheet of tetrahedrally coordinated Si (with O) and one sheet of octahedrally coordinated Al (with O and OH) (see Fig. 2.2) (Ke and Stroeve, 2005), hence a 1:1, or a two-layer, structure. The basal oxygen atoms of the tetrahedral sheet form a hexagonal pattern, and the apical or top oxygen atoms of all tetrahedra are perpendicular to the sheet. The tetrahedral and octahedral sheets are connected with shared oxygen atoms. Kaolinite and the two halloysites are single-layer structures, whereas dickite and nacrite are double-layer polytypes; the repeat distance along the direction perpendicular to *d* (001) for dickite and nacrite is two 7 Å layers, or 14 Å (Moore and Reynolds, 1997).

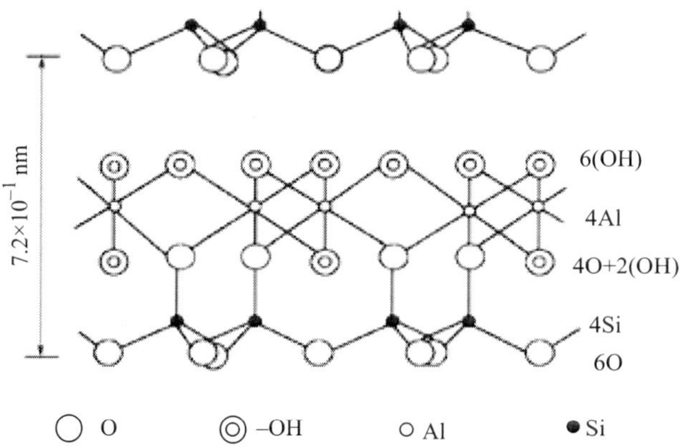

Figure 2.2 Schematic representation of the crystal structure of kaolinite.

There is confusion over whether there are two halloysites or only one and what the relation of halloysite is to kaolinite. The nature of halloysite and its relation to kaolinite are particularly intriguing problems. Kaolinite, in contrast to halloysite, has not been observed to swell in water, but treatment with some organic chemicals, for example, formamide or dimethyl-sulfoxide, will open up kaolinite, or cause it to swell.

Four names for clay materials are associated with kaolinite: ball clay, fireclay, flint clay, and underclay. These names come from the people that mine them and use them commercially. There is considerable overlap and flexibility in the properties and uses associated with these names. Ball clays yield a white product when fired. They have the plasticity and the green and dry strength for making sanitary ware, ceramic tiles, etc. Fireclays are similar to ballclays but are usually found immediately beneath coals. When underclays have been referred to as fireclays, they are kaolinite-rich underclays. Flint clay also can be used to make refractory items. This term is more specific and applies to clay that is hard and smooth, breaks with conchoidal fracture, and develops no plasticity when ground up and wetted as other clays do, that is, it does not slake in water. Kaolinite is prominent in various underclays, but underclays do not necessarily have kaolinite in them. Underclay is the unit immediately below a coal bed that was the soil upon which the vegetation that formed the coal grew (Moore and Reynolds, 1997).

The serpentine subgroup is interesting in that these 1:1 layers can be of three types: planar, continuously curved to produce rolled or tubular structures, and planar structures comprising layers curved about the mean plane. The various situations arise as a result of misfits between T and O layers due to isomorphous substitutions and the subsequent strains from size differences. Large amounts of Al (III) and Fe (III) lead to minerals of planar structure, while correspondingly low levels lead to rolls from curved layers, as occurs for chrysotile (Auerbach et al., 2004).

2.2.2.2 The 2:1 layer type

2.2.2.2.1 *Pyrophyllite-talc group*

The 2:1 layer type is one in which one octahedral sheet is sandwiched between two tetrahedral sheets (Figs. 2.3 and 2.4) (Meunier, 2005). The commonly used layered silicates belong to the same general family of 2:1 layers or phyllosilicates. Their crystal structure consists of layers made up of two tetrahedrally coordinated silicon atoms fused to an edge-shared octahedral sheet of either aluminum or magnesium hydroxide. The layer thickness is around 1 nm, and the lateral dimensions of these layers may vary from 30 nm to several microns or larger, depending on the particular

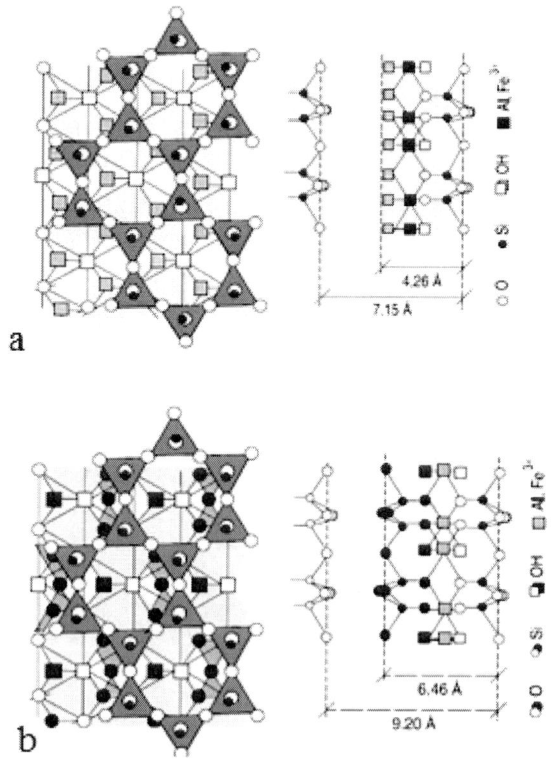

Figure 2.3 Crystal structure of dioctahedral phyllosilicates: (a) 1:1 layer and (b) 2:1 layer.

layered silicate (Roelofs and Berben 2006). Stacking of the layers leads to a regular van der Waals gap between the layers, called the interlayer or gallery. Isomorphic substitution within the layers (e.g., Al^{3+} replaced by Mg^{2+} or by Fe^{2+} or Mg^{2+} replaced by Li^{+}) generates negative charges that are counterbalanced by alkali or alkaline earth cations situated in the interlayer (Wing and Zhong-Zhen, 2006).

2.2.2.2.2 *Smectites*

Smectite is the name for a group of minerals, both dioctahedral and trioctahedral, all of which display the startling property of being able to expand and contract their structures, while maintaining two-dimensional crystallographic integrity (Moore and Reynolds, 1997). Smectite minerals also contain 2:1

TOT layers, but they differ from pyrophyllite-talc in that a more significant amount of isomorphous substitutions take place, enough to raise the charge per unit cell to 0.5–1.2 (Lee and Tiwari, 2012). The net negative charge that arises from this is compensated for by interlayer, hydrated cations (primarily K, Na, Ca, and Mg) (Auerbach et al., 2004; Newman and Brown, 1987). These ions are hydrated due to the fact that, in nature, smectites are formed in aqueous environments. Because they are hydrated, these cations are only loosely held by the negatively charged clay layers. Smectites saturated with other cations dissociate in aqueous suspensions into exchangeable cations and tactoids, which are composed of several parallel TOT layers, held together by electrostatic forces by some of the exchangeable cations that remain in the interlayer space. Because they dissociate, the original cations are exchangeable by other inorganic and organic cations. In smectite tactoids, electrostatic forces keep the layers together. Parallel TOT layers are packed one above the other, and the exchangeable hydrated cations are located between the layers (Yariv and Cross, 2002).

2.2.2.2.3 *Montmorillonite*

MMT has been known by several names: smectite, sodium MMT, sodium bentonite or swelling bentonite, and sodium-activated bentonite, while the nonswelling bentonite is calcium MMT or bentonite, sub-bentonite (Unnikrishnan et al., 2011). MMT is a 2:1 smectite, that is, it has three atomic lattice layers in each of the nanolayers with an aluminum–oxygen–hydroxyl octahedral sheet sandwiched between two silicon–oxygen tetrahedral sheets (Işık, 2007).

The crystal structure of MMT is shown in Fig. 2.4 (Ke and Stroeve, 2005). Individual nanolayers are 1 nm thick, and their lateral dimensions are about 100 nm. The formula for pyrophyllite is $Al_4[Si_4O_{10}](OH)_2$. The difference between the crystal structure of MMT and pyrophyllite is that the latter is neutral, while the former has layer charges due to isomorphic substitution. Isomorphic substitution occurs when some atoms in the crystal structure are replaced with other atoms with different valence without any change in the sketch of the crystal structure (Liu et al., 2008). For instance, when there is an isomorphic substitution of Mg^{2+} for Al^{3+}

in the octahedral lattice, a negative charge of minus one (−1) is generated, which is normally counterbalanced by cations adsorbed from the ambient solution (Ke and Stroeve, 2005).

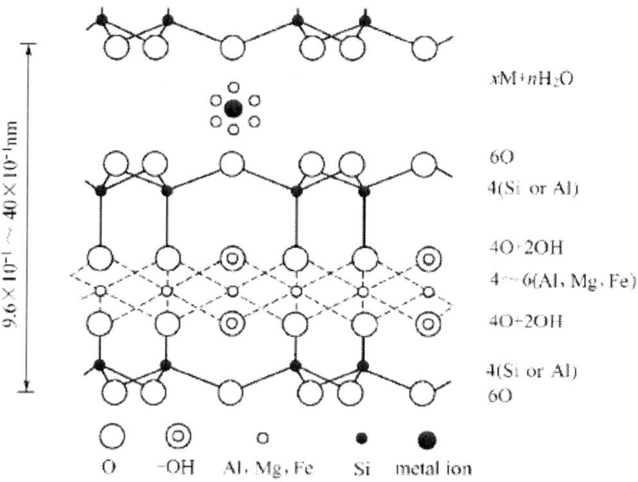

Figure 2.4 Schematic representation of the crystal structure of MMT.

The commonly used layered silicates for the preparation of polymer-layered silicate (PLS) nanocomposites belong to the same general family of 2:1 layers or phyllosilicates (i.e., smectites, especially MMT) (Alexandre and Dubois, 2000; Meng et al., 2008; Ray and Okamoto, 2003; Rooj et al., 2012).

2.2.2.2.4 *Illite group (micas)*

"Mica" is a generic term applied to a group of complex alumino-silicates having a sheet- or plate-like structure with a wide range of chemical compositions and physical properties. The theoretical formula of illite is (K, Na, Ca)m(Al, Fe, Mg)$_4$(Si, Al)$_8$O$_{20}$(OH)$_4$·nH$_2$O where $m < 1$. Illite's host minerals are muscovite and biotite. During the process that gradually turns mica into illite, the mica particle gets finer by degree, and as a result, its specific surface increases. Potassium ions exposed on the surface are more easily hydrated and exchanged with other cations than those inside the layer structures. A portion of K^+ between the layers is replaced by Ca^{2+}, Mg^{2+}, and $(H_3O)^+$. Chemical analysis shows that illite contains less potassium and more water molecules than its host mineral—thus the name "hydromic." The lattice of illite is difficult to expand because water molecules have great difficulty in approaching the interlayer region. The reason for this is that the negative charge is situated mainly on the tetrahedral sheet close to the surface, resulting in strong electrostatic attraction between K^+ and the negative charge in the layers. In addition, K^+ ions are embedded in the hole formed by the surface oxygen atoms in adjacent layers and are 12-number-coordinated with the surrounding oxygen atoms, causing the K^+ ions to link the adjacent layers. As a result, the linkage is usually very firm.

Generally, the properties of clay minerals are closely associated with the crystal structure, and it is essential to know the surface properties. A comparison of the structural characteristics of three common layered silicates of clay minerals is given in Fig. 2.5 (Ke and Stroeve, 2005).

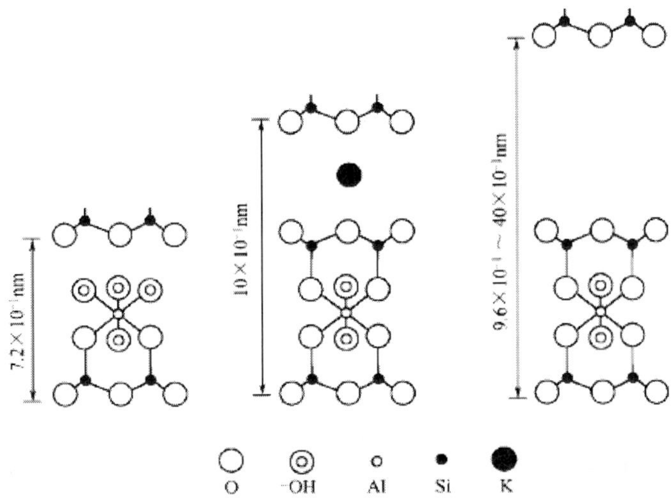

Figure 2.5 Structural characteristics of kaolinite, illite, and MMT.

2.2.2.2.5 *Vermiculite group*

These minerals are very similar to 2:1 TOT smectites, except their layer charge densities are higher (1.2–1.8) due to greater levels of isomorphous substitution, which arises primarily from tetrahedral replacements (Abollino et al., 2008). While water and polar organic molecules will intercalate, swelling is more limited than it is for smectites. In natural minerals, the exchangeable cation is Mg, with small amounts of Ca and Na. Vermiculites are trioctahedral and occur frequently as large crystals with platy morphology, similar to micas, although they are softer and contain interlayer water. Relative to the parent micas from which they are derived, the negative layer charge of vermiculites is lower and the iron is oxidized. They have a positive octahedral charge (0.3–1.2) and a large negative tetrahedral charge (>2), which distinguishes them from high-charge saponites (Auerbach et al., 2004; Mottona et al., 2002).

2.2.2.2.6 *Sepiolite, palygorskite, and attapulgite*

Sepiolite, attapulgite, and palygorskite contain a continuous two-dimensional tetrahedral sheet and thus differ from the other layer silicates by the absence of the octahedral sheet (Utracki, 2004). Sepiolite is a type of Al–Mg silicate that contains water and has a fibrous structure. Its formula cannot be accurately determined, but it is usually expressed as $4MgO{\cdot}6SiO_2{\cdot}2H_2O$. Sepiolite has a unique fibrous crystal structure. In a suspension of sepiolite, the fibers intersect each other to form a network frame, which is an important factor to keep the suspension stable. As a result, the rheological behavior of the suspension of the sepiolite group depends on the fibrous structure rather than the electrostatic attraction between particles. Figure 2.6 gives the sketch map of the palygorskite crystal structure (Ke and Stroeve, 2005).

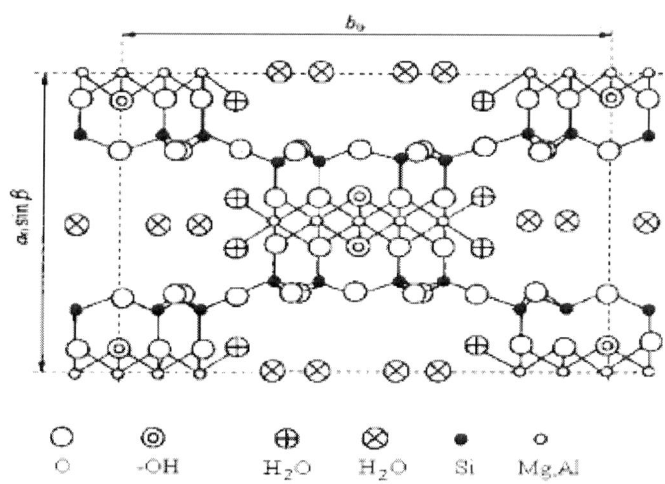

Figure 2.6 Schematic representation of the crystal structure of polygorskite.

2.2.2.2.7 *Chlorites*

The layer structure of chlorite consists of alternately stacked pyrophyllite-like sheets and brucite sheets (Fig. 2.7). The negative charge is created by the substitution of Al^{3+} for Si^{4+} in the tetrahedral sheet, but the net charge is very small. Al^{3+} substitutes for Mg^{2+} in the brucite sheet and produces some positive charge, which is balanced by the negative charge above it. The formula is $2[(Si, Al)_4(Mg, Fe)_3O_{10}(OH)] + (Mg, Al)_6(OH)_{12}$ (Ke and Stroeve, 2005).

2.2.2.2.8 *Mica and brittle mica groups*

Micas are 2:1 TOT minerals (Giannelis, 1996). A variety of isomorphous substitutions yield dozens of specific micas and brittle micas divided into di- and trioctahedral subgroups (Nishimura et al., 1994). True micas normally have K or Na exchangeable cations, while the main interlayer cation is Ca in brittle micas. The high charge densities of micas and brittle micas lead to electrostatic forces that are so strong that interlayer polar molecules are not present, and they are all consequently nonswelling (Auerbach et al., 2004).

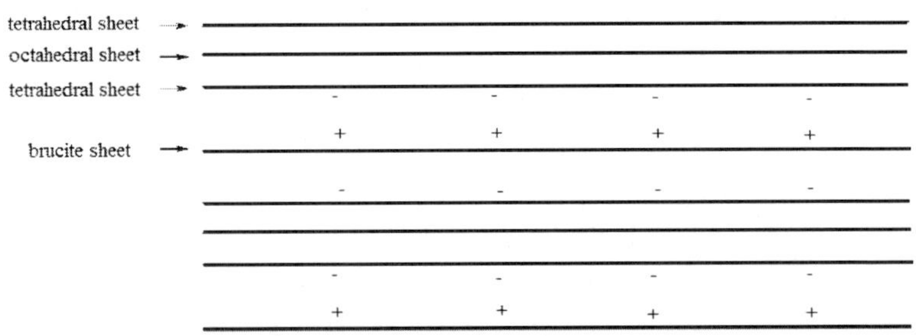

Figure 2.7 Schematic representation of the layer structure of chlorite.

2.2.2.2.9 *Mixed-layer clay minerals*

In some clays, different types of clay layers are stacked up in the crystal structure of the same clay mineral. These clays are called mixed-layer clay minerals. In the mixed-layer structure, different types of layers mutually overlap. The layer sequence is generally irregular. Occasionally, however, regular stacking can be found. Illite/MMT and chlorite/vermiculite layers are the most familiar mixed-layer structures. Mixed-layer clay minerals usually disperse and expand in water more easily than other clay minerals, especially when one component of the mixed-layer clay is expandable (Ke and Stroeve, 2005).

2.2.3 Characterization Techniques of Clay Minerals

2.2.3.1 X-ray diffraction

The advent of the uniaxial powder diffractometer along with using oriented clay samples has allowed for routine and reliable analysis of clay minerals. Prior to these developments, the small coherence domains in clay crystals (compounded by their sheet structure) and their small particle size (most are up to 10–20 times smaller than 2 μm) made them difficult to study. Diffraction effects are enhanced for the small crystallites by orienting them or by having them lie on the same plane by exploiting their sheet structure. As a result, the basal spacing (001) and higher orders of reflection (00*l*) are often enhanced relative to other diffraction planes (*hkl*, *hk*0), especially those without an *l* component. There are two standard polar molecules that are intercalated to determine the swelling properties of clays: water yields a 5 Å increase in basal spacing, and ethylene glycol expands layers by 7 Å. To enhance the (*hk*) planes, nonoriented clay specimens are employed, where the least physical pressure possible is used to ensure that the powders are not packed. The two most useful x-ray diffraction (XRD) regimes for clay characterization are the basal spacing region (2°–10° 2θ, Cu radiation) and the (060) reflection (near 60° 2θ, Cu radiation), which is diagnostic for several structural and compositional types.

Figure 2.8 illustrates the principle known as Bragg's law (Kirk-Othmer, 1993). The two paths, A and B, for the incident and diffracted beams differ in length by $2d \sin \theta$, where d is the perpendicular distance between two adjacent parallel planes. The parameter θ is equal to the angle between the incident beam and each plane and is equal to the angle between the diffracted beam and each plane. For constructive interference to take place, this difference must be equal to a whole number of wavelengths. Bragg showed that diffraction takes place only if all the lattice points of the crystal are on the parallel planes.

$$n\lambda = 2d \sin \theta \tag{2.1}$$

where n is the degree of diffraction, λ is the wavelength, θ is the measured diffraction angle, and d refers to interlayer spacing.

2.2.3.2 Nuclear magnetic resonance spectroscopy

Nuclear magnetic resonance (NMR) spectroscopy has been used successfully in the study of clay minerals and clay organic interactions, providing information on both structural and dynamic aspects. NMR spectroscopy is sensitive to the local environment of atoms, and information concerning crystallographic sites, coordination number, tetrahedral sheet distortions, and cation distribution has been obtained from ^{29}Si and ^{27}Al spectra. In pillared clays, where diffractometric methods give limited information, NMR has been used to study pillar formation between clay layers. Finally, ^{1}H, ^{13}C, and ^{31}P NMR spectra (and of other atoms) provide information on the orientation, interaction, and mobility of species sorbed in the interlamellar space of clay minerals (Yariv and Cross, 2002).

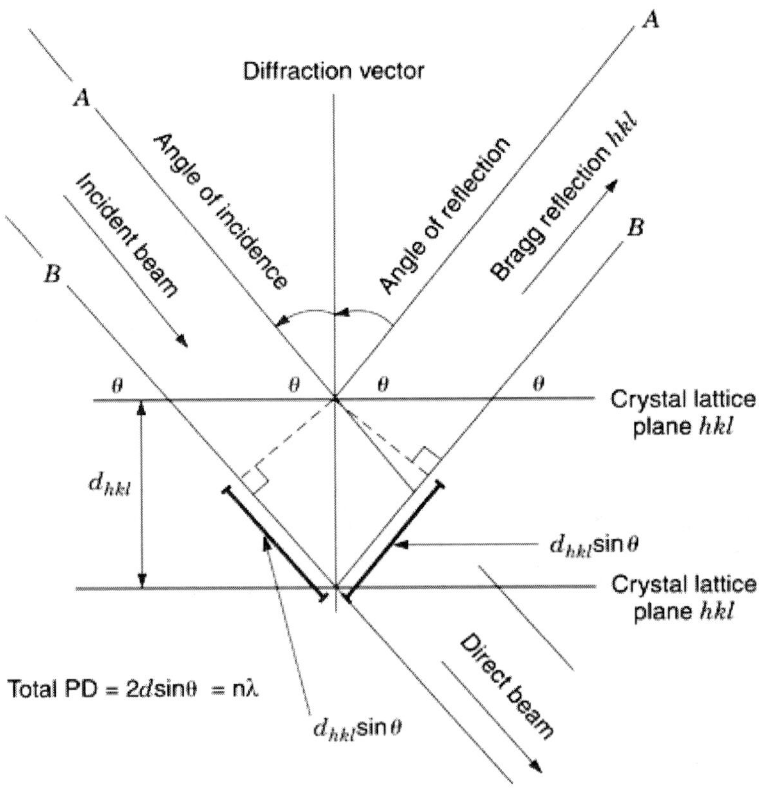

Figure 2.8 Geometrical representation of diffraction and its relationship to Bragg's law.

2.2.3.3 Thermal analysis

Several different types of water and hydroxyl groups are associated with a clay matrix that is liberated with increasing temperature. Thermogravimetric analysis (TGA) is often used to qualitatively determine the binding energies of these different types of water, which can then be used to help characterize a mineral. In TGA, the mass of a sample is monitored versus temperature while it is heated according to a program in a specific atmosphere (oxidizing, reducing, inert). At lower temperatures, adsorbed water is lost from clay surfaces in defect sites or at broken bond sites; typical values for this are 1 wt% at 80°C–90°C. Zeolitic water is observed only for sepiolite–palygorskites, at 100°C–150°C. Interlayer water, which is associated with interlayer cations in smectites as monolayers (2.5 Å) or bilayers (5 Å), is liberated between 100°C and 200°C. Dehydroxylation of the inner 2:1 or 1:1 lattice hydroxyl groups, also lost as water, is observed at 500°C–700°C. Derivative thermogravimetry (DTG), where the first derivative of the TG curve is plotted against temperature, is often more useful.

In differential scanning calorimetry (DSC), the heat flow rate (power) to a sample is monitored against temperature during programmed heating. Differential thermal analysis (DTA) allows for the observation of endothermic and exothermic events. In DTA, the difference in temperature between a sample and a reference material is monitored against temperature, while the temperature of the sample in a specific atmosphere is programmed. Water losses are normally endothermic, and recrystallization (phase changes, etc.) is exothermic. Organic transformations from clay–organic complexes, under certain atmospheres (oxidizing or reducing), will also yield diagnostic endo- or exothermic events. Combining evolved gas analysis (EGA), mass spectrometry (MS), or infrared (IR) with a thermal method can be a very powerful method of characterization.

2.2.3.4 Fourier transform infrared spectroscopy

Much of the power of IR techniques does not apply to clay minerals unless they are complexed with organic molecules, because IR is less sensitive to differences in clay structures than methods such as XRD and NMR. Only broad absorbances are observed for the Si–O and Si–O–Si tetrahedral stretching vibrations near 1000–1100 cm^{-1},

Al–O stretching near 600 cm^{-1}, and octahedral Fe, Al, or Mg bending frequencies near 800–900 cm^{-1} (Cole, 2008; Tzavalas and Gregoriou, 2009). One of the most useful IR regions for pure clay minerals contains the OH vibrational modes from 3500 cm^{-1} to 4000 cm^{-1}. Here the bonding energies of structural hydroxyl groups, and therefore octahedral site occupancy, are easily determined. Samples must first be heated to >400°C and kept dry to remove any associated water (surface, zeolitic, interlayer), which would interfere with the signal.

The Fourier transform infrared (FTIR) technique is used in the analysis of fingerprinting contaminants, characterizing chemical property gradients, or detecting physical property anisotropy (Velde, 1992). The FTIR technique corresponds to the vibrational energies of atoms or specific groups of atoms within a molecule, as well as rotational energies. It identifies components by comparing the spectrum of a sample to reference spectra. An IR spectrum is commonly obtained by passing IR radiation through a sample and determining what fraction of the incident radiation is absorbed at a particular energy (Suga and Rusling, 1993). The energy, at which any peak in an absorption spectrum appears, corresponds to the frequency of a vibration of a part of a sample molecule. For a molecule to show IR absorptions, it must possess a specific feature: an electric dipole moment of the molecule must change during the movement. The interactions of IR radiation with matter may be understood in terms of changes in molecular dipoles associated with vibrations and rotations. The atoms in molecules can move relative to one another, that is, bond lengths can vary or one atom can move out of its present plane. This is a description of stretching and bending movements, that is, vibrations. Vibrations can involve a change in either bond length (stretching) or bond angle (bending). Some bonds can stretch in phase (symmetrical stretching) or out of phase (asymmetric stretching). If a molecule has different terminal atoms, the amount of coupling will vary. Varying proportions of the stretching motion of each group are observed (Kirk-Othmer, 1993).

2.2.3.5 Electron and scanning probe microscopy

In scanning electron microscopy (SEM), the flux of secondary and backscattered electrons from a clay sample bombarded by an electron beam is used to form an intensity image. The resulting three-dimensional maps are very useful for identifying textures and morphologies, with resolution on the order of 0.01 micrometer. SEM shows clay crystal morphologies very clearly, as evidenced by the micrographs in Fig. 2.9. Transmission electron microscopy (TEM) is used to image crystal shapes and sizes in essentially a two-dimensional plane (Wang et al., 2008). Clays are normally deposited on carbon-coated copper grids. In TEM the electron beam is absorbed to a greater extent by the clay than the carbon, creating a shadow image of the clay particle. Resolution is greater than for SEM. Further, the composition of individual crystallites can be estimated using energy-dispersive techniques.

Figure 2.9 shows the SEM images of some clay samples (Zbik and Smart, 2002; Fang et al., 2012; Kevadya et al., 2014; Zhang et al., 2012; Li et al., 2014). Many clay structures have hexagonal symmetry and yield forms that are either hexagons or those derived from hexagonal prisms. High-resolution TEM (HRTEM) has a resolution of <2 Å, allowing the imaging of atomic planes in clays.

Figure 2.9 Scanning electron micrographs of (a) kaolinite (Reprinted from Zbik and Smart, Copyright 2002, with permission from Elsevier), (b) Ca-MMT (Reprinted from Fang et al., Copyright 2012, with permission from Elsevier), (c) Na-MMT (Reprinted from Kevadya et al., Copyright 2014, with permission from Elsevier), (d) halloyste (Reprinted from Zhang et al., Copyright 2012, with permission from Elsevier), (e) illite (Reprinted from Li et al., Copyright 2014, with permission from Elsevier).

Atomic force microscopy (AFM) allows the observation of individual atoms on mineral surfaces and their interactions (Lindgreen, 2000). Fluid cells allow one to follow the progress of reactions between minerals and solutions in real time (Occelli and Gould, 2001; Occelli et al., 1994; Occelli et al., 1995; Occelli et al., 1993). Minerals with layered structures usually display good cleavage planes, which make them easy to study by AFM (Henderson et al., 1994). Many materials have in fact been imaged on mica because of the excellent qualities of their freshly cleaved surfaces (Liu et al., 2000). This is not a necessary condition, however, and studies on ultrafine clay particles also have been carried out (Lindgreen et al., 1991). AFM studies of various clay mineral surfaces themselves are available, including illite and MMT (Hartman et al., 1990); lizardite, muscovite, and clinochlore (Wicks et al., 1993); and chlorite (Vrdoljak and Henderson, 1994). Finally, the molecular-scale structure of cations, surfactants, and polymers on various clay surfaces has also been examined with these scanning force techniques (Nishimura et al., 1994; Patrick et al., 1999; Porter et al., 1996; Schulz and Warr, 2000; Vrdoljak et al., 1994; Xiao et al., 1996). Lindgreen et al. (1991) have recently published a study of electrical conduction in micas by AFM combined with scanning tunneling microscopy.

2.3 Surface Chemistry of Clay and Organoclays

Particle size, shape, and distribution are important physical properties, which are intimately related to the applications of clay minerals. Other important properties are surface chemistry, surface area, and surface charge. These, along with color and brightness, affect many use properties such as low and high shear viscosity; absorption; plasticity; green, dry, and fired strength; casting rate; permeability; and bond strength. In almost every application, clays and clay minerals are functional and are not just inert components in the system (Kirk-Othmer, 1993).

Most substances acquire a surface electric charge when brought into contact with a polar (e.g., aqueous) medium, possible charging mechanisms being ionization, ion adsorption, and ion dissolution or isomorphic substitution. This surface charge influences the distribution of nearby ions in the polar medium. Ions of opposite charge (**counterions**) are attracted toward the surface, and ions of like charge (**co-ions**) are repelled away from the surface. This, together with the mixing tendency of thermal motion, leads to the formation of an **electric double layer** made up of the charged surface and a neutralizing excess of counterions over co-ions distributed in a diffuse manner in the polar medium (Shaw, 1992).

Some substances acquire their charge mainly through the ionization of carboxyl and amino groups to give COO^- and NH^+ ions. The net molecular charge owing to the ionization of these groups depends strongly on the pH of the solution. At low pH, such a molecule will be positively charged and negatively charged at high pH (Karube, 1998). The pH at which the net charge (and electrophoretic mobility) is zero is called the **isoelectric point**.

A net surface charge can be acquired by the unequal adsorption of oppositely charged ions. Ion adsorption may involve positive or negative surface excess concentrations. Surfaces in contact with aqueous media are more often negatively charged than positively charged. This is a consequence of the fact that cations are usually more hydrated than anions and so have a greater tendency to reside in the bulk aqueous medium, whereas the smaller, less hydrated, and more polarizing anions have a greater tendency to be specifically adsorbed.

Surfaces that are already charged (e.g., by ionization or isomorphic substitution) usually show a preferential tendency to adsorb counterions, especially those with a high charge number. It is possible for counterion adsorption to cause a reversal of charge. If surfactant ions are present, their adsorption will usually determine the surface charge. Hydrated surfaces adsorb ions less readily than hydrophobic surfaces.

Ionic substances, minerals, and oxides can acquire a surface charge by virtue of unequal dissolution of the oppositely charged ions that ionic substances are composed of and isomorphic substitution of the like-charged ions at the mineral or oxide lattices. In a similar way, hydrogen and hydroxyl ions for hydrous metal oxide sols are **potential-determining ions**, since their concentrations determine the electric potential at the particle surface (Reactions 2.1 and 2.2).

$$\text{M–OH} + \text{H}^+ = \text{M–OH}_2^+ \tag{2.1}$$

$$\text{M–OH} + \text{OH}^- = \text{M–O}^- + \text{H}_2\text{O} \tag{2.2}$$

Adsorption of dipolar molecules will not contribute to a net surface charge, but the presence of a layer of orientated dipolar molecules at the surface may make a significant contribution to the nature of the electrical double layer. The electric double layer can be regarded as consisting of two regions, an inner region, which may include adsorbed ions, and a diffuse region, in which ions are distributed according to the influence of electrical forces and random thermal motion. The requirement of overall electroneutrality dictates that for any dividing surface, if the charge per unit area is $+\sigma$ on one side of the surface, it must be $-\sigma$ on the other side. It follows, therefore, that the magnitude of σ will depend on the location of the surface. Surface location is not a straightforward matter owing to the geometric and chemical heterogeneity that generally exists. It follows, furthermore, that electric double-layer parameters (potentials, surface charge densities, and distances) are not amenable to unequivocal definition. The surface potential ψ_o, therefore, depends on both the surface charge density σ_o and on the ionic composition of the medium. If the double layer is compressed, then either σ_o must increase or ψ_o must decrease or both. Specifically adsorbed ions are those that are attached to the surface by electrostatic and/or van der Waals forces strongly enough to overcome thermal agitation. They may be dehydrated, at least in the direction of the surface. The centers of any specifically adsorbed ions are located in the **Stern layer**, that is, between the surface and the Stern plane. Ions located beyond the Stern plane form the diffuse part of the double layer with ψ_o replaced by ψ_d. The potential changes from ψ_o (the surface or wall potential) to ψ_d (the Stern potential) in the Stern layer and decays from ψ_d to zero in the diffuse double layer. In the absence of specific ion adsorption, the charge densities at the surface and at the Stern plane are equal. When specific adsorption takes place, counterion adsorption usually predominates over co-ion adsorption. It is possible, especially with polyvalent or surface-active counterions, for reversal of charge to take place within the Stern layer, that is, for ψ_o and ψ_d to have opposite signs. Adsorption of surface-active co-ions could create a situation in which ψ_d has the same sign as ψ_o and is greater in magnitude. Overall electrical neutrality throughout the whole of the double layer can be expressed as follows (Eq. 2.1):

$$\sigma_o + \sigma_1 + \sigma_2 = 0 \tag{2.1}$$

where σ_o, σ_1, and σ_2 are the surface charge density of surface, the surface charge density of the Stern layer, and the surface charge density of the diffuse part of the double layer, respectively.

ψ_d can be estimated from electrokinetic measurements. Electrokinetic behavior depends on the potential at the surface of shear between the charged surface and the electrolyte solution. This potential is called the **electrokinetic** or **ζ (zeta) potential**. Figure 2.10 shows the electrical double layer as schematical (Möckel et al., 1998; Shaw, 1992; Taroco et al., 2011).

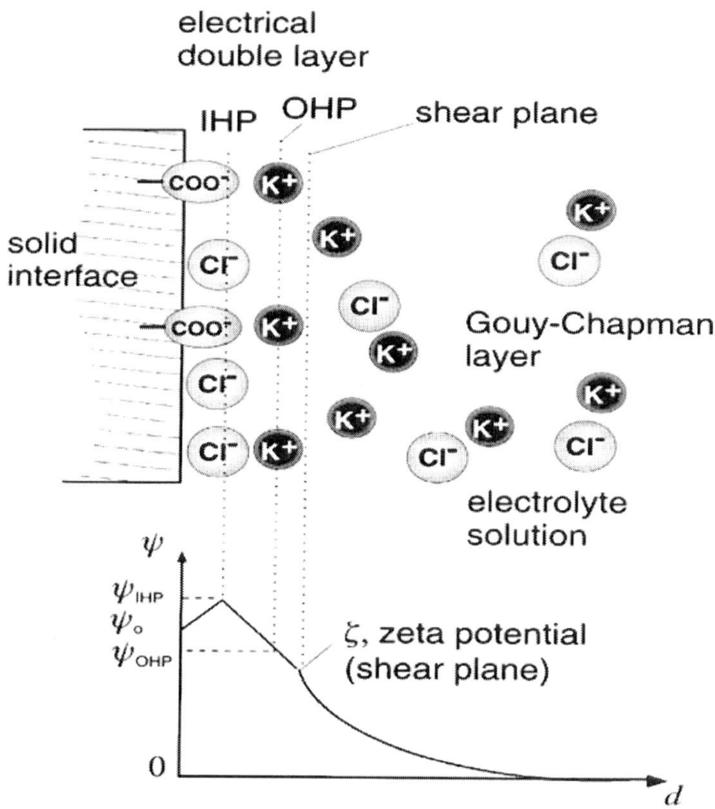

Figure 2.10 Schematic representation of electrical double layer, according to the GCSG (Gouy-Chapman-Stern-Grahame) model (IHP: inner Helmholtz plane and OHP: outer Helmholtz plane). Reprinted from Möckel et al., Copyright 1998, with permission from Elsevier.

The exact location of the shear plane (which, in reality, is a region of rapidly changing viscosity) is another unknown feature of the electric double layer. In addition to ions in the Stern layer, a certain amount of solvent will probably be bound to the charged surface and form a part of the electrokinetic unit. It is, therefore, reasonable to suppose that the shear plane is usually located at a small distance further out from the surface than the Stern plane and that ζ is, in general, marginally smaller in magnitude than ψ_d (Shaw, 1992).

The layered silicates commonly used in nanocomposites belong to the structural family known as 2:1 phyllosilicates (Zen and Senthil Kumar, 2004). Their crystal lattice consists of two-dimensional layers where a central octahedral sheet of alumina or magnesia is fused to two external silica tetrahedron by the tip so that the oxygen ions of the octahedral sheet do also belong to the tetrahedral sheets. The layer thickness is around 1 nm, and the lateral dimensions of these layers may vary from 300 Å to several microns and even larger depending on the particular silicate. These layers organize themselves to form stacks with a regular van der Walls gap in between them called the interlayer or the gallery. Isomorphic substitution within the layers (e.g., Al^{3+} replaced by Mg^{2+} or by Fe^{2+} or Mg^{2+} replaced by Li^{1+}) generates negative charges that are counterbalanced by alkali or alkaline earth cations situated in the interlayer (Baldassari et al., 2006). As the forces that hold the stacks together are relatively weak, the intercalation of small molecules between the layers is easy.

The charge of the layer is not locally constant, as it varies from layer to layer, and must rather be considered an average value over the whole crystal. Proportionally, even if a small part of the charge-balancing cations is located on the external crystallite surface, the majority of these exchangeable cations is located inside the galleries (Pavlidou and Papaspyrides, 2008).

Kaolin is usually white or near white in color. There is very little substitution in the structural lattice, and thus it has a minimal layer charge and a low exchange capacity. The kaolinite crystals, which make up most of the kaolin deposits, are pseudohexagonal along with plates, some larger books, and vermicular stacks. Other properties that are important are the relatively low surface area in comparison to smectite and palygorskite–sepiolite and the low absorption capacity that relates to the minimal layer charge and the low surface area. A relatively low viscosity at high solid concentrations is an important characteristic of some kaolins (Murray, 2000). Viscosity of kaolin products is dependent on several factors, including mineral content; particle size, shape, and distribution; presence of soluble salts; and dispersability (Murray, 2003). Kaolins are hydrophilic and are easily dispersed in water by using small amounts of a chemical dispersing agent such as sodium hexametaphosphate or sodium polyacrylate. Ionic and/or polar nonionic surfactants can be applied to the surface of the kaolin to make it hydrophobic or organophilic. The two most important industrially are sodium MMT and calcium MMT. Smectites have very different properties than kaolins. Both octahedral and tetrahedral sheets can have substitutions, which creates a charge imbalance in the 2:1 layer. Alumina substitutes for silica in the tetrahedral sheet, and iron and magnesium substitute for aluminum in the octahedral sheet. Many analyses have shown that the charge imbalance in smectite is about 0.66 per unit cell. This net positive charge deficiency is balanced by exchangeable cations absorbed between the unit layers and on the edges. If the dominant exchangeable cation is sodium, then the mineral in the smectite group is sodium MMT, and if it is predominantly calcium it is calcium MMT. The high layer charge, the very fine particle size, the thin flakes, the high cation exchange capacity (CEC), and the high surface area result in the physical and chemical properties that determine the many industrial applications (Mallakpour and Dinari, 2011). The high layer charge and large surface area of sodium MMT when dispersed in water result in very high viscosity and thixotropy.

A common use of sodium MMT is the preparation of organoclays. The exchangeable sodium ions are replaced with organic cationic molecules such as alkylamines and produce a hydrophobic surface. Another use of sodium MMT is to make pillared clays, which can be tailored for specific catalyst and absorbent uses (Chiu et al., 2008). Interlamellar reactions using sodium MMT are restricted to low temperatures because at high temperatures the layers collapse. By intercalating thermally stable cations, the layers are propped apart. By varying the size of the pillars and/or the spacing between pillars, the pore size and space can be adjusted to suit a particular application (Kasama et al., 2004).

"Palygorskite" and "attapulgite" are synonymous terms for the same hydrated magnesium aluminum silicate mineral. Sepiolite is almost structurally and chemically identical to palygorskite, except it has a slightly larger unit cell. The term "fuller's earth" is used to describe clays that have sorptive and bleaching qualities, so both palygorskite and some smectite clays (particularly calcium MMT) are also called fuller's earth. There is considerable substitution of aluminum by magnesium and iron in the octahedral layer, which gives the particles a moderately high layer charge. This layer charge and the high surface area give palygorskite an intermediate CEC normally about 30–40 meq/100 g. The high surface area, the charge on the lattice, and the inverted structure, which leaves parallel channels through the lattice, give palygorskite and sepiolite a high absorption capacity.

Layer charge is a fundamental property of 2:1 phyllosilicates, which stems from substitutions in the structure of minerals (permanent charge) and from ionizable groups on external surfaces of the minerals (variable charge). In some minerals, this charge is balanced by fixed cations (e.g., micas), whereas in others, it is balanced by the exchangeable cations (e.g., smectites and vermiculite). Layer charge affects many properties of smectites, such as swelling, ion exchange capacity, ion exchange selectivity, and rheological properties of bentonites. Smectites contribute most of the CEC of bentonites, and hence knowledge of layer charge may allow estimation of the smectite content of bentonites from measured CEC values. A number of independent studies using different analytical methods have shown that distribution of layer charge in smectites is not homogeneous but may vary considerably. Similar conclusions have been drawn by the use of the alkylammonium method (AAM). However, the validity of charge heterogeneity calculated by the AAM has been challenged. The layer charge of smectites can be estimated using a variety of analytical methods, which include (1) the structural formula (SF) method, (2) the AAM, and (3) the investigation of XRD traces of K-saturated, ethylene glycol-solvated smectites (Christidis and Eberl method). Additional methods, which, however, have not been used systematically, include (1) NH_4^+ saturation and examination by IR spectroscopy, (2) Methylene Blue adsorption and examination by UV spectroscopy, and (3) quaterner alkylammonium ion adsorption and monitoring by zeta potential measurements. The last two of these methods may provide quantitative information about charge heterogeneity of smectites, since the SF method provides information about charge localization (tetrahedral versus octahedral). It has long been known that there is significant discrepancy between the layer charge determined by the SF method and the AAM. The magnitude of that discrepancy increases systematically with layer charge and may reach 40% for clays with high layer charge. Also it has been argued that local (nanometer scale) rearrangement of interlayer alkylammonium cations will cause gradual monolayer–bilayer transitions, even in smectites with a homogeneous layer charge, and cause the AAM to underestimate the layer charge. By contrast, it is suggested that the SF method may yield too high a layer charge for smectites and that it cannot be used to obtain reasonable CEC due to a permanent charge (CEC_{perm}) (Christidis, 2008; Kaufhold et al., 2002).

Clays are hydrophilic; hence the first step in their modification involves interactions in the aqueous medium (Lin et al., 2004; Mansoori et al., 2010). The small and highly mobile water molecules easily diffuse into interlamellar species, causing a lateral expansion of the clay crystals. The process is diffusion controlled; hence to swell platelets that are twice as large requires four times longer. The rate depends on many factors, and the time required to reach equilibrium swelling varies from minutes to days. Since water monolayer thickness is about 0.28 nm, the first step involves expansion of *d* (001) from 0.96 nm to about 1.25 nm. Further expansion up to complete exfoliation can be accomplished under effective control of conditions, especially the ionic strength and counterion type. MMT exposed to water vapor expands its basal spacing in steps (which correspond to one to four molecular layers of water) to *d* (001) = 1.25–2.0 nm. The origin of water absorption by MMT is, on the one hand, hydration of the counterions in the interlamellar galleries and, on the other, hydrogen bonding to the clay surface. The work to remove the last monolayer of H_2O is about 0.1 J/m^2 at the equivalent pressure of 400 MPa. Dispersing Na-MMT in distilled water causes the ions that are associated with the clay surface to diffuse out. The osmotic pressure pulls the ions away from the clay surface, whereas the electrostatic charge tends to hold them near the surface. Eventually, a steady state is achieved and an electrical double layer is formed that keeps the clay platelets apart. The counterions in the double layer are fairly mobile and can readily be exchanged. The process constant depends on the ionic concentration in the double layer and in the supernatant solution. The double layer is formed by adsorption of solvated ions. Their concentration and the associated electric potential decrease with the distance from the platelet surface; thus, the thickness of the double layer depends on the counterion concentration and valency. Clay platelets can be fully exfoliated in aqueous media, but on standing, since the surface and edges have different charge, the edge-to-face (EF) interactions may lead to the formation of a "house of cards" structure. If enough clay is present, all the water will be tied, resulting in gel formation. Shearing may disrupt the structure, dramatically reducing the viscosity. The gelation is reversible, as evidenced by the thixotropic effects. Another type of structure that may be formed is a tactoid—a parallel alignment of clay platelets, up to 10 nm distant from each other, forming high-clay-concentration regions in suspensions. In some systems, there is a sharp transition between the gel and tactoid phase (Utracki, 2004).

There are different ways to modify 2:1 clay minerals: (1) adsorption; (2) ion exchange with inorganic cations and cationic complexes; (3) ion exchange with organic cations; (4) binding of inorganic and organic anions, mainly at the edges; (5) grafting of organic compounds; (6) reaction with acids; (7) pillaring by different types of poly(hydroxo metal) cations; (8) interlamellar or intraparticle and interparticle polymerization; (9) dehydroxylation and calcination; (10) delamination and reaggregation of smectitic clay minerals; and (11) physical treatments such as lyophilization, ultrasound, and plasma (Bergaya and Lagaly, 2001).

Ion exchange with alkylammonium ions is a well-known method to make clay minerals and clays dispersible in organic solvents and to render them compatible with hydrophobic materials in compounding processes. Exchanging the interlayer cations of natural and synthetic clay minerals (hectorites, fluoromicas, taeniolite, and alkali silicates) by particular organic cations provides a simple method to prepare new types of organic–inorganic hybrid materials (Le Pluart et al., 2002). Intercalation of large organic molecules or cations and macromolecules, in particular proteins, often requires a preceding expansion of the interlayer space (Gemeay et al., 2002; Bergaya et al., 2006).

2.3.1 Adsorption at the Clay–Liquid Interface and Chemical Grafting

There are two main approaches for the surface modification of clay minerals with polymers. One is physical adsorption, and the second is chemical grafting of functional polymers to the surfaces of the clay minerals. Surface modification by polymers is found to be one of the most effective methods, as the surface properties can be widely changed by a variety of functional polymers. The physical adsorption, controlled by thermodynamic criteria, can alter the nature of the clay mineral surfaces and improve their surface physical and chemical properties. The advantage of physical attachment is that the structure of the clay mineral is not altered. The main potential disadvantage is that the forces between the adsorbed molecules and the clay mineral might be weak (Liu, 2007).

The adsorption of surfactants at the solid–liquid interface is strongly influenced by a number of factors: (1) **the nature of the structural groups on the solid surface**, whether the surface contains highly charged sites or essentially nonpolar groupings, and the nature of the atoms of which these sites or groupings are constituted; (2) **the molecular structure of the surfactant being adsorbed** (the adsorbate), whether it is ionic or nonionic and whether the hydrophobic group is long or short, straight chain or branched, aliphatic or aromatic; and (3) **the environment of the aqueous phase**—its pH, its electrolyte content, the presence of any additives such as short-chain polar solutes (alcohol, urea, etc.), and its temperature. Also, these factors determine the mechanism by which adsorption occurs, and the efficiency and effectiveness of adsorption (Dekany et al., 1996; Rosen, 2004; Wang et al., 1999).

2.3.2 Mechanisms of Adsorption and Aggregation

Previous studies have shown three different molecular environments for surfactants within organoclays: (1) **surfactant cations intercalated into the clay interlayers through cation exchange** and **bound to surface sites via electrostatic interaction**; (2) **surfactant (cations and/or molecules) physically adsorbed on the external surface of clay**; and (3) **surfactant molecules located within the clay interlayer** (He et al., 2006; Xu and Boyd, 1995).

There are a number of mechanisms by which surface-active solutes may adsorb onto solid surfaces from an aqueous solution. In general, adsorption of surfactants involves single ions rather than micelles (Rosen, 2004). The adsorption of surfactants onto hydrophilic surfaces from aqueous solutions has been studied intensively for more than half a century. A number of different models for surfactant adsorption have been proposed over the years, relating to different combinations of surfactant and surface. Figure 2.11 shows the probable morphologies for the adsorption of a soluble surfactant onto a hydrophilic substrate (Tyrode et al., 2008). The models are not mutually exclusive but involve probabilities between the two limiting cases. One limit corresponds to surfactant molecules in which the headgroups interact only weakly with the surface. The standard free energy of binding is insufficient to compensate for the loss of translational entropy upon adsorption and the surface coverage remains negligible until the concentration is very near the **critical micelle concentration** (cmc). At a certain concentration, labeled the **critical surface aggregation concentration** (csac), aggregates adsorb to the surface. These aggregates termed "admicelles" bear a resemblance to **micelles** formed in the bulk phase at higher concentrations. The isotherm rises very steeply before leveling out around the cmc. At no point in the isotherm is a monolayer of the surfactant formed. Surfactants typically adsorb to hydrophobic surfaces at concentrations two orders of magnitude below the cmc. Consequently, if a monolayer of the surfactants is formed, the hydrophobic surface that emerges would immediately constitute a second layer of surfactants adsorbed with opposite orientation so that the unfavorable hydrophobic interactions at the interface can be eliminated. This limit can be illustrated by the adsorption onto silica of nonionic surfactants, such as the poly(ethylene glycol) (PEG) and alkyl ethers with short ethylene oxide (EO) groups. The second limiting case is that of a **self-assembled monolayer** (SAM)—the SAM limit. In a SAM, the interaction between the headgroup and the surface, augmented by van der Waals interactions between closely packed chains, is strong enough to drive monolayer formation at concentration orders of magnitude below the cmc. A second layer does not form at these dilute concentrations because elimination of water–hydrocarbon contacts does not compensate for the loss of translational entropy upon adsorption. As the surfactant concentration is increased further, first isolated molecules adsorb to the hydrophobic surface of the SAM with the surface coverage gradually rising until it approaches an asymmetric bilayer at the cmc. Other surfactant/substrate combinations lie between these two extremes with the adsorption isotherm, depending on the relative adsorption energies of the surfactants to the hydrophilic surface and to a hydrophobic surface and on the strength of interactions between chains. In the celebrated **four-region model** of Somasundaran and Fuerstenau (1966), molecules first adsorb to isolated sites on the surface. As the concentration increases, chain–chain interactions lead to the formation of surface clusters, termed "hemi-micelles," at concentrations below those at which the surfactant adsorbs to a hydrophobic surface. In contrast to the admicelles formed with nonionic surfactants, hemi-micelles are noncentrosymmetric and expose a hydrophobic surface to the aqueous solution. At higher concentrations still, a second layer of surfactant with opposite orientation condenses on top of the hemi-micelles. Lateral growth of these aggregates leads to a bilayer or closely packed admicellar structure at the cmc. In an alternative model, adsorption to a hydrophobic surface occurs at concentrations below that of hemi-micelle formation. The adsorption isotherm proceeds directly from a sparse monolayer to an admicellar structure.

The general consensus is that adsorption is initially driven by electrostatic attraction between the cationic surfactant and SiO^- groups on the surface up to the isoelectric point of the interface. The adsorbed molecules are in a single layer and may or may not assemble into hemi-micelles, depending on factors such as surface charge density and electrolyte concentration. At a higher surfactant concentration (around the csac) hydrophobic interactions lead to the adsorption of additional surfactant molecules oriented preferentially with their headgroups away from the surface. The adsorbed surfactant may form uniform bilayers but more commonly forms admicelles that resemble flattened versions of spherical or cylindrical micelles found in bulk solutions. Some important details of the adsorption process remain unclear, however. One of the difficulties in distinguishing experimentally between the various possible adsorption models is that one needs to measure not only the adsorbed amount but also whether the adsorbed molecules are in the form of a bilayer or a monolayer. The sign of the surface charge can, in principle, be determined from the measurements of zeta potential; for example, in the case of cetyltrimethylammonium bromide (CTAB) adsorption on silica, the sign of the surface charge switches from negative to positive as the adsorption progresses.

Two principal models are being suggested to describe the adsorption of ionic surfactants onto hydrophilic surfaces: the **two-step model** and the **four-region model**. The general shape of the adsorption isotherm as well as the proposed molecular description of the different adsorption stages are depicted in Fig. 2.12 for these two models.

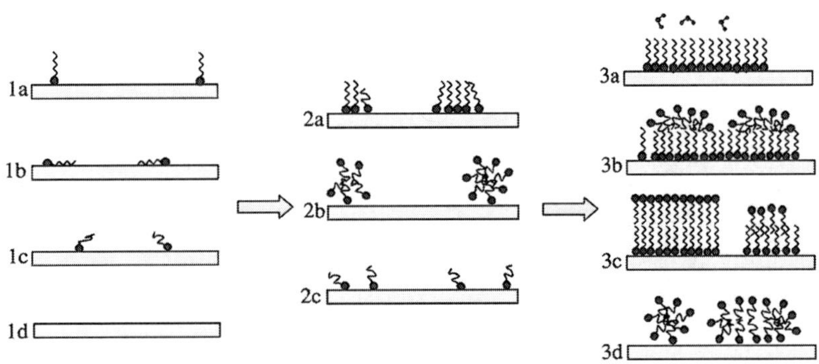

Figure 2.11 Probable morphologies for the adsorption of a soluble surfactant onto a hydrophilic substrate (1d). At very low coverage, the hydrocarbon chains of the adsorbed surfactant may lie perpendicular to the surface (1a), may lie parallel to the surface (1b), or be randomly oriented (1c). As the coverage increases, the molecules may be randomly distributed in a single layer (Langmuir behavior, 2c) or interactions between surfactant molecules may lead to the formation of hemi-micelles (2a) or admicelles (2b). At high coverage, a range of structures is conceivable: monolayer (3a), hemi-micelles on a monolayer (3b), bilayer (3c), or admicelles (3d). Reprinted from Tyrode et al., Copyright 2008, with permission from American Chemical Society.

In both models, the adsorption isotherm is determined by interplay of electrostatic and hydrophobic interactions. At the lowest concentrations (region I in both models), the surfactant adsorbs via electrostatic interactions to oppositely charged surface sites. At the highest concentrations (around the cmc), hydrophobic interactions lead to the formation of aggregates or admicelles at the solid–water interface (region IV in both models). The two models differ in the relative importance of electrostatic and hydrophobic interactions at intermediate concentrations. In the two-step model, there is only ever a low coverage of (isolated) molecules bound electrostatically to the substrate, which then nucleate the formation of admicelles in the steeply rising part of the isotherm. In the four-region model, there is stronger adsorption at low concentrations, leading to the formation of hemi-micelles before the attachment of a second layer through hydrophobic interactions. These models are not mutually exclusive; rather, they describe systems lying at different points along the continuous spectrum of possibilities between the SAM limit and the weak adsorption limit. The CTAB adsorption isotherm closely resembles the double plateau curve of the two-step model in Fig. 2.12. In the lowest surface density range (region I), interactions between surfactant molecules are negligible and the adsorbed amount is expected to increase linearly with concentration (**Henry range**). No difference in the mean orientation of the hydrocarbon chains of the surfactant was detected between low and high coverages, ruling out the possibility of finding flat-lying molecules at low surface densities. The first plateau (region II) corresponds to the formation of a CTAB monolayer or any other asymmetric aggregate, including hemi-micelles. This provides strong evidence that hydrophobic interactions generate symmetric aggregates early on in the adsorption process. No significant changes occur in the mean orientation of the surfactant hydrocarbon chains or the trimethylammonium headgroups in this region. In the following concentration region (region III), the adsorbed amount steeply rises. In region I, surfactant molecules do not lie flat with their alkyl chains on the surface. Hydrophobic interactions come into play from the beginning of region II, as suggested by the four-region model, though hemi-micelles do not form in this or subsequent regions (only centrosymmetric aggregates seem to form). Initial aggregates nucleate at specific sites of the silica surface, possibly where electrostatic interactions anchored the first surfactant molecules at the beginning of the adsorption process (Tyrode et al., 2008).

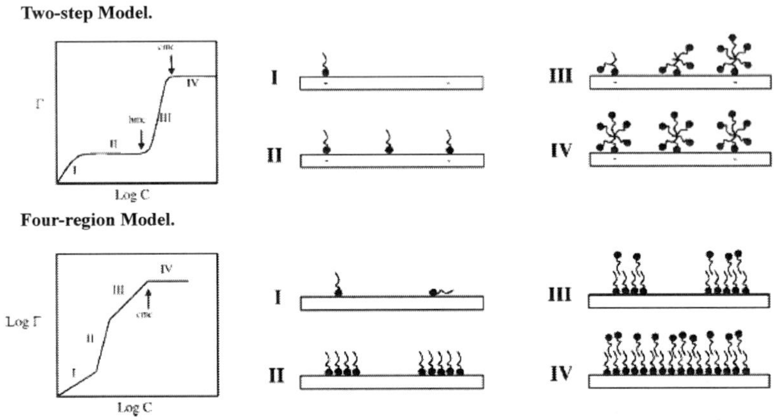

Figure 2.12 Schematic representation of the adsorption isotherm and proposed molecular model for the two-step and four-region adsorption models. Reprinted from Tyrode et al., Copyright 2008, with permission from American Chemical Society.

A theoretical model that uses numerical self-consistent field calculations to study the effect on morphological behavior of varying the surfactant–matrix enthalpic interaction, surfactant coverage, and surfactant length has been proposed (Balazs et al., 1998). The model indicated that longer surfactants (clay organic modifiers) promote intercalation of matrix molecules in the modified clay galleries by providing a reduction in the entropic penalty to the intercalating polymer (Chrissopoulou and Anastasiadis, 2011). However, very high degrees of surfactant coverage on the silicate surface make intercalation and exfoliation unfavorable.

Kurian et al. (2004) described a new scheme termed *mixed coverage*, wherein the silicate is modified with a mixture of PS-based surfactants of different lengths, to create a silicate surface grafted with a bimodal brush. Texturing of the silicate surface was expected to allow intercalation of matrix molecules in the galleries by providing a favorable entropic gain to the intercalating polymer through interaction with the longer grafted molecules, and reduced enthalpic interaction with the clay surface. However, XRD data indicated that the samples remained in a phase-separated morphology, with no indication of intercalation or exfoliation of the layered silicate in the PS matrix. To explain this result, it was suggested that, despite the bimodal nature the brush is still effectively dense (only 17% of the exchangeable cations were replaced by the longer surfactant and the remaining 83% by the shorter surfactant), making the long surfactant fraction ineffective in fostering wetting and intercalation. Figure 2.13 shows a schematic of this scenario. In Fig. 2.13a, a long surfactant is illustrated at low coverage over a surface that is repulsive to the polymer matrix (PS homopolymer). In this case, the surfactant will stretch away from the surface until the energy required to stretch is no longer less than that of the repulsive enthalpic interactions. This is the scenario for typical surfactant-modified clay at very low coverage. In the bimodal brush, the shorter surfactant at full coverage (83%) effectively replaces the silicate surface. As illustrated in Fig. 2.13b, the surface can now be thought of consisting the dense brush. When a small fraction of those surfactants are replaced with longer surfactants, the situation may become something like that illustrated in Fig. 2.13c. The length of the longer surfactant is still important, but the conformation of the longer surfactant must also be considered. If the mixed surfactants used to create the textured surface do not differ much in their lengths, one has effectively a short surfactant at high coverage. As the molecular weight difference between the two surfactants increases, the surfactant present in lesser amount becomes effectively longer, but still it is possible that it does not actually assist in the wetting and intercalation processes, because there is now no enthalpic repulsion causing the surfactant to stretch away from the silicate surface. This would result in the scenario illustrated in Fig. 2.13c, where the longer surfactant now remains close to the brush of chemically similar molecules and does not aid intercalation by interacting with matrix polymer molecules. These clearly demonstrate that when tailoring the silicate characteristics that favor intercalation or exfoliation, one has to simultaneously consider different factors, for example, the surfactant chain length and packing density.

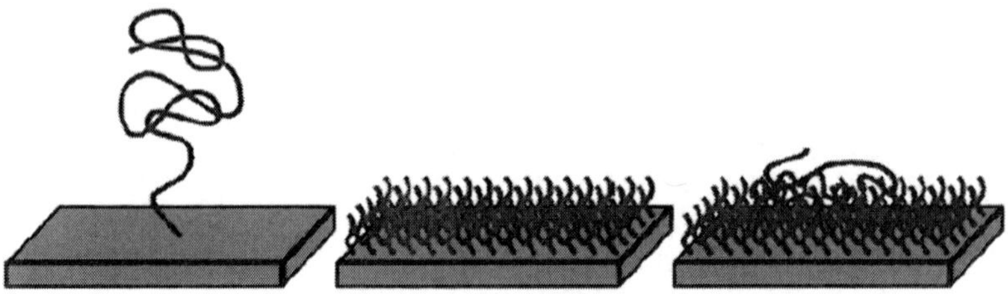

Figure 2.13 Illustrations for a bimodal surfactant brush of one possible outcome on longer surfactant conformation of the densely packed brush of a shorter surfactant. Reprinted from Kurian et al., Copyright 2004, with permission from John Wiley and Sons.

Essentially, surfactant adsorption occurs through various mechanisms or their combination such as ion exchange, ion pairing, acid–base interaction, hydrogen bonding, polarization of π electrons, dispersion interactions and hydrophobic bonding (Gürses et al., 2014; Kooli et al., 2009). Adsorption mechanisms will be explained in detail over some representative adsorption systems in the following sections.

2.3.2.1 Ion exchange

Ion exchange involves replacement of counterions adsorbed onto the substrate from the solution by similarly charged surfactant ions. For example, the surface modification of clay layers can be achieved through a cation exchange process by the replacement of sodium and calcium cations present in the interlayer space or clay galleries by alkylammonium or alkylphosphonium (onium) cations (Fig. 2.14a,b) (Fischer, 2003). In addition to the surface modification and increasing the hydrophobicity of clay layers, the insertion of alkylammonium or alkylphosphonium cations into the galleries causes to some degree of increasing in the interlayer spacing (Ahmad et al., 2009; Chigwada et al., 2006; Le Baron et al., 1999). Also the alkylammonium or alkylphosphonium cations can provide functional groups, which interact with polymer chains and, therefore, increase the interfacial interactions. The ion exchange process is carried out in aqueous solution with the hydrated interlayer cations. The affinity of monovalent sodium cations to hydration promotes the ion exchange process and increases the efficiency of organic modification of clay layers. Clays containing divalent countercations such as calcium cannot be easily hydrated, and therefore their replacement by ion exchange is not efficient. Therefore, in these clays, the divalent cations should be ion-exchanged with easily exchangeable sodium cations prior to modification by onium cations (Ray and Okamoto, 2003). The efficiency of organic modification by the ion exchange process in the increasing of basal spacing depends also on the surface charge of clay layers. The surface charge density of clay layers depends on the clay nature and its preparation (growth) conditions. When the charge density of clay layer is high such as about 1 equiv/mol for mica, the electrostatic forces between layers are also high. Therefore, the interlayer cations do not hydrate or swell, and consequently the ion exchange process cannot be carried out. The medium charge densities (0.5–0.8 equiv/mol) such as in vermiculite can make a high potential for partial swelling and hydration of cations (Zhu et al., 2008). However, the divalent cations in the interlayer spacing of clays such as vermiculite make the swelling process difficult due to the lower affinity of divalent cations for hydration. The lower charge densities (0.25–0.5 equiv/mol) such as formed in MMT result in the weak electrostatic forces between layers. Also the presence of sodium cations, with high affinity for hydration and exchanging, between the MMT layers makes this kind of clays more suitable for swelling and organic modification. The structure of the organoclays and their basal spacing depend on the molecular size of organic onium and its configuration in the interlayer spacing after modification. The organic tail of onium laid up parallel to the clay layer can be obtained with lower charge densities. By increasing the charge density of clay layers, bilayer or trilayer structures with increasing interlayer spacing can be formed. Finally, with the higher charge densities, more interlayer spacing is obtained by the formation of paraffin structure. Also increase in temperature can lead to structures with more basal spacing by increasing the mobility of onium molecules into the galleries. However, at a given temperature, two parameters determine the equilibrium layer spacing: the charge density of layers, the packing and number of onium chains into the interlayer space, and the size or chain length of organic tail in the onium compound (Olad, 2011).

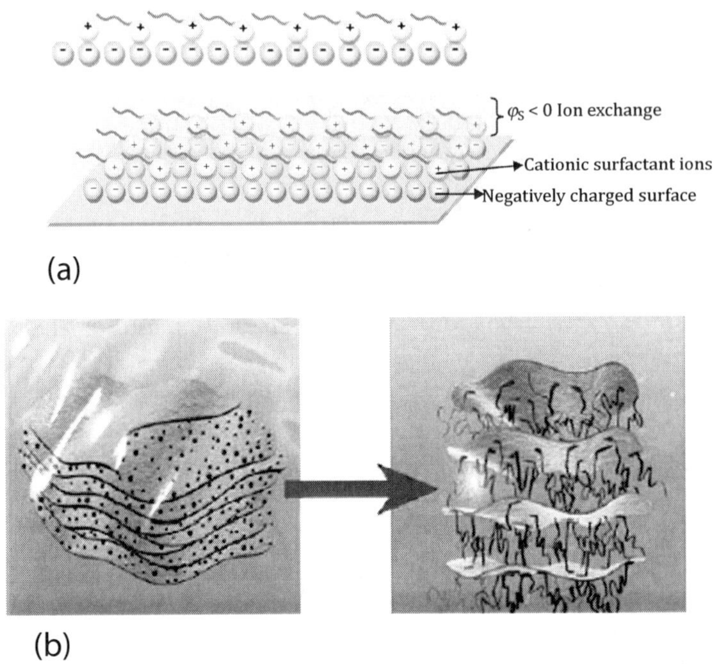

Figure 2.14 (a) Adsorption of alkylammonium cations onto a negatively charged surface by ion exchange mechanism. (b) Schematic presentation of an ion exchange interaction. The inorganic, relatively small (sodium) ions are exchanged against more voluminous organic onium cations. Reprinted from Fischer, Copyright 2003, with permission from Elsevier.

It should be clearly understood that the rigid arrangement of the hydrophobic groups depicted in Fig. 2.14a is only for convenience in illustrating the mechanisms of adsorption. In reality, the hydrophobic groups may assume all conformations, including the interweaving of hydrophobic chains of adjacent molecules, consistent with their surface areas per molecule and the relative orientation of their hydrophobic and hydrophilic groups with respect to the interface (Rosen, 2004). Exchange of the alkali counterions with cationic–organic surfactants, such as ammonium and phosphonium in layered silicates, has been shown schematically in Fig. 2.14a,b, respectively. Alkylammonium ions are mostly used, although other “onium” salts can be used. The ion exchange interaction has two consequences: first, the gap between the single sheets is widened, enabling polymer chains to move in between them, and second, the surface properties of each single sheet are changed from being hydrophilic to hydrophobic (Fischer, 2003).

The intercalation of organic surfactants occurs mostly by ion exchange, which breaks down the interfacial energy barrier due to the lack of affinity between the polymer and the hydrophilic clay surface (Mòczò and Pukànszky, 2008). This also depends on the nature of clay and surfactant as well as on the amount of surfactant present compared with CEC (Mandalia and Bergaya, 2006). In fact, hydrophilicity also increases with a small added amount of surfactant. By addition of surfactant in a slight excess of CEC, the cationic headgroup of alkylammonium ions resides near the external surfaces, while the aliphatic tail remains in the interparticle pores, rendering the clay more hydrophobic and less hydrophilic. At loadings below 1 CEC, the observed basal distances of the organoclays are smaller than 3 nm, and above the CEC, these distances are less than 4 nm. An optimum confined interlayer space of the organoclay that can be correlated with the entropic energy barrier is about 2 nm (Annabi-Bergaya, 2008).

2.3.2.2 Ion pairing

Adsorption of surfactant ions from solution onto oppositely charged sites unoccupied by counterions corresponds to this mechanism. Adsorption of a cationic surfactant such as CTAB onto the negatively charged surface of clay is schematized in Fig. 2.15.

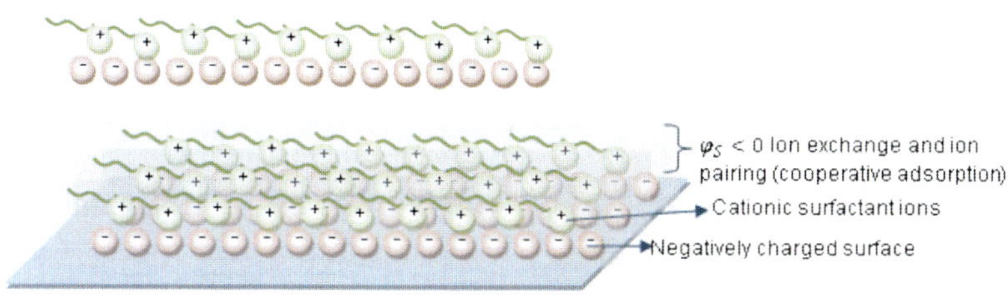

Figure 2.15 Ion pairing.

2.3.2.3 Acid–base interaction and either hydrogen bonding between substrate and adsorbate or Lewis acid–Lewis base reaction

Clay suspensions are dispersed by the addition of a small amount of a weak acid such as tannate, which is adsorbed at the edge surfaces of the clay particles by complexing with the exposed octahedral aluminum ions. Consequently, the edge charge is reversed and a negative double layer is created by which edge-to-face (EF) and edge-to-edge (EE) association is prevented. The anion adsorption on the clay edges is absent on the faces of the clay plates (Zbik and Horn, 2003). As a result, the basal spacing does not increase after such an adsorption. In aqueous medium, the structural –OH groups in MMT form hydrogen bondings with water molecules. The strength of these interactions is expected to be higher than those between water molecules and =Si=O surface groups. Anions are adsorbed on the edges of the clay particles and organic cations on the anionic inner surfaces of the clay platelets. This is evident from the high adsorption capacity of the clay as well as the resulting expansion of the *d* (001) spacing. There is a strong preference for the less mobile organic cations. Often they are adsorbed quantitatively until all the exchange positions are exhausted. The ammonium groups become ionically attached to the clay surface, while the hydrocarbon chains interact with the clay surface and displace the adsorbed water molecules. When the chains are too long to lie flat in the available space, they may tilt, crowding within the interlamellar space. Depending on the conditions, the hydrocarbons may crystallize; hence the spatial organization of the hydrocarbon segments is given by the appropriate crystalline cell unit. When the hydrocarbon chains cover the clay surface, it precipitates from the aqueous suspensions. However, the formed hydrophobic clay can be homogeneously dispersed in an organic medium. Often the concentration of the organic cations exceeds the amount equivalent to the clay CEC. The orientation of the interlayer cations, and hence the *d* (001) spacing, is usually determined by the projected surface area of the cation (in a given orientation), provided that it does not exceed the available surface area per surface anion. Aromatic cations assume either a parallel or an upright position between the layers, depending on the available space. When the modifier molecules are large and bulky, the steric effect precludes total cation exchange. The average distance between the silicilic anions on the MMT surface is about 1.2 nm. However, when the concentration of the organic cations is high, their adsorption may exceed the amount equivalent to the clay CEC. For example, a quaternary ammonium compound having long hydrocarbon chains can be absorbed in the amount equivalent of two-and-a-half times the CEC. The excess enters the interlamellar galleries in a head-to-tail configuration, profiting from the hydrocarbon/hydrocarbon chain solubility and the tendency to crystallize. Under these circumstances, the excess ammonium compounds form a layer with the cationic groups facing the water phase, forming a diffused electric double layer that prevents the particles from precipitation. Thus, the resulting organoclay is hydrophilic. In general, organic molecules can intercalate the galleries of the clays by interacting with the clay surface via ion–dipole, dipole–dipole, and hydrogen bonding interaction mechanisms (Tjong, 2006).

Hydrogen bonding between phenolic resin chains and SiOH groups is schematized in Fig. 2.16a.

A schematic illustration showing the formation of hydrogen bondings in PA6– pristine MMT is depicted in Fig. 2.16b (Maiti and Okamoto, 2003). In this case, the organic modifier consisting of two alkyl tails shields more silicate surface than one alkyl tail, thereby precluding desirable interactions between the polyamide and the clay surface.

2.3.2.4 Adsorption by polarization of π electrons

The process occurs when the adsorbate contains electron-rich aromatic nuclei, and the solid adsorbent has strongly positive sites. Attraction between electron-rich aromatic nuclei of the adsorbate and positive sites on the substrate results in adsorption. Figure 2.17 shows the polarization of π electrons between linear phenolic resin chain and organoclay surface.

2.3.2.5 Adsorption by dispersion forces

Adsorption occurs via London–van der Waals dispersion forces acting between adsorbent and adsorbate molecules. Adsorption by this mechanism generally increases with an increase in the molar mass of the adsorbate. So this mechanism important not only as an independent mechanism but also as a supplementary mechanism in all other types. For example, it accounts in part for the pronounced ability of surfactant ions to displace equally charged simple inorganic ions from solid substrates by an ion exchange mechanism. It is generally accepted that adsorption of hydrophilic, short-chain quaternary ammonium cations on smectites occurs

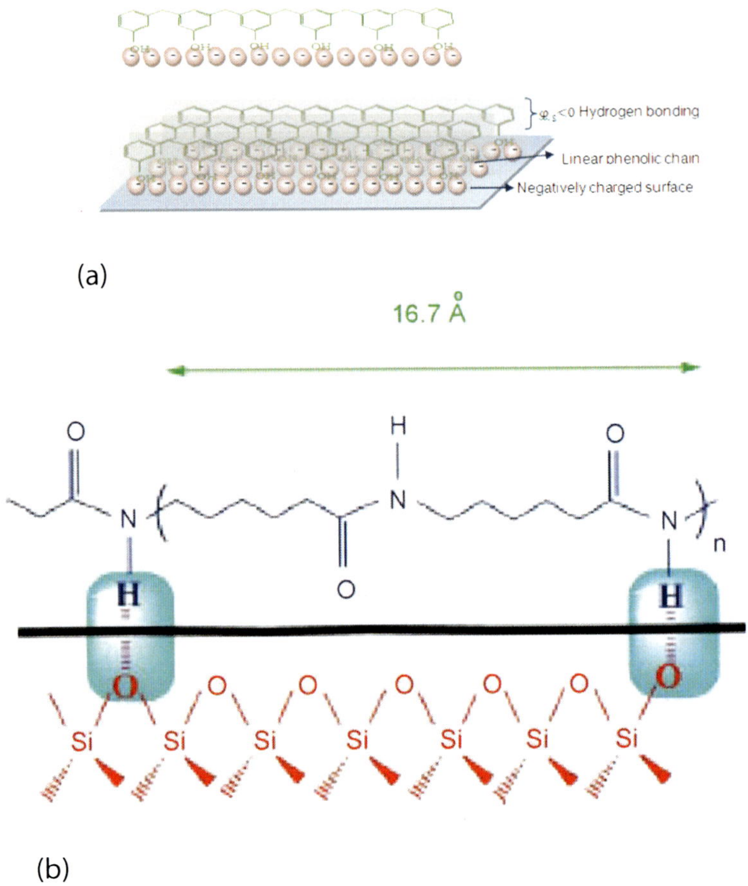

Figure 2.16 (a) Hydrogen bonding. (b) Schematic illustration showing the formation of hydrogen bonds in PA6–pristine MMT. Reprinted from Maiti and Okamoto, Copyright 2003, with permission from John Wiley and Sons.

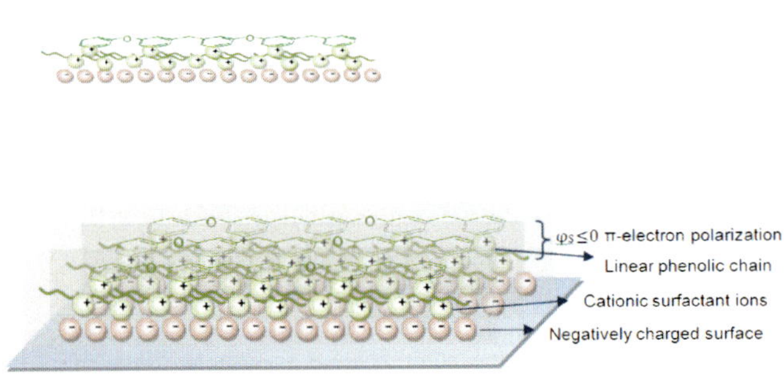

Figure 2.17 Adsorption by polarization of π electrons.

according to the ion exchange mechanism. The extent of adsorption of such cations cannot thus exceed the critical exchange capacity of the mineral. Furthermore, as expected, an increase in ionic strength of the aqueous phase in equilibrium with the organoclay leads to a reduction in the adsorption of the organic cations on its surface. On the other hand, their long-chain homologues appear to undergo adsorption to an extent, which increases with the growth of the alkyl chain and approaches the doubled value of CEC. Van der Waals hydrophobic interactions are suggested to operate in such cases and lead to a bilayer of interdigitated alkyl chains with a positive charge exposed toward the bulk of the solution (Zhang et al., 2012). This process is accompanied by a reversal in charge of the surface and, in consequence, by a lowering of the rate of increase of counterion activity with the growth of total concentration of the surfactant in the bulk of the solution. Such a mechanism for adsorption of long-chain ammonium salts onto a negatively charged surface, which can lead to cooperative adsorption and hemi-micelle formation, has been schematized in Fig. 2.18.

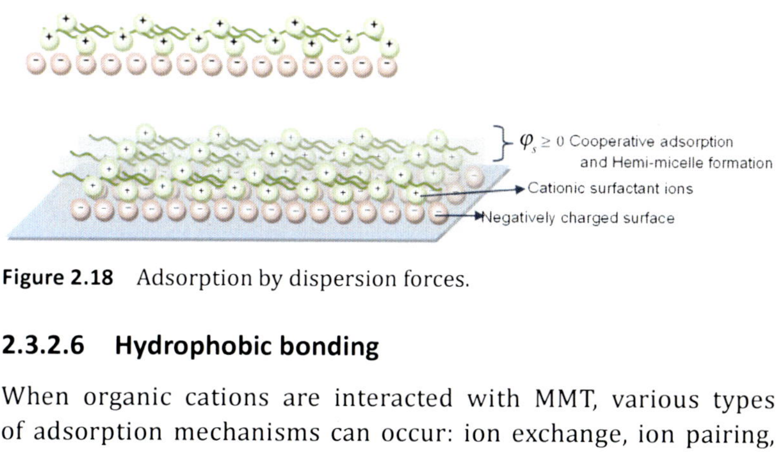

Figure 2.18 Adsorption by dispersion forces.

2.3.2.6 Hydrophobic bonding

When organic cations are interacted with MMT, various types of adsorption mechanisms can occur: ion exchange, ion pairing, hemi-micelle formation, and hydrophobic bonding. When the dose of organic cation is less than or equal to the CEC value of the MMT, all the organic cations are adsorbed by the MMT by ion exchange. When the quantity of the organic cations is greater than the CEC of the MMT, those already existing in the layers of MMT will further adsorb organic compounds in the solution by hydrophobic bonding (Ke and Stroeve, 2005). It seems plausible to assume that contribution of the second mechanism of adsorption of cationics, driven by van der Waals interactions, may lead to a gradual increase in adsorption with the growth of the alkyl chain only when the effects of steric hindrance are weaker compared to those that result from hydrophobic bonding between chains. Such a situation may occur in the case of cationics with a small headgroup such as alkylammonium, alkyltrimethylammonium, or alkylpyridinium salts. One may expect also that as the alkyl chains become long, they may assume gauche conformations and bend away from the surface of the adsorbent, thus avoiding the constraints due to limitations of area available for adsorption. The bent-away hydrocarbon chains of two neighboring cations may stick to themselves hydrophobically, creating a microenvironment for anchoring an excess of the cations, thus leading to the formation of bilayers or admicelles (Kwolek et al., 2003).

Hydrophobic bonding occurs when the combination of mutual attraction between hydrophobic groups of the surfactant molecules and their tendency to escape from an aqueous environment becomes large enough to permit them to adsorb onto the solid adsorbent by aggregating their chains (Fig. 2.19) (Gürses et al., 2009). These surfactant aggregates, termed "admicelles" or "hemi-micelles" by various investigators, were assumed to be more or less flat. In aqueous systems, the formation of structures that will minimize exposure of hydrophobic groups to the aqueous phase depends upon the interactions between the surfactant molecules and the solid surface. Orientation of the adsorbed surfactant molecules with their hydrophobic groups predominantly away from the solid substrate will make the surface more hydrophobic than before surfactant adsorption; orientation with their hydrophilic groups predominantly away from the surface will make it more hydrophilic than before adsorption of the surfactant.

The complex structure of MMT-based organoclays can be partially attributed to the pronounced structural and energetic heterogeneity in pristine MMT clays. Clay minerals always contain several types of adsorption sites with different interaction energies. The energies associated with the basal and lateral surfaces in clay layers can differ greatly from each other and therefore result in different adsorption properties. The face energy alone on the basal surface may also vary at different locations. As a result, when MMT clay is intercalated by small molecules such as water, the distribution of layer thickness can be quite nonuniform. If the long-chain cationic surfactants are used to complex with the clay surface, the nonuniformity in the distribution of layer thickness may be even more severe. As a result, the amount of adsorbed surfactant may significantly exceed the ion exchange capacity. The excessive surfactant may be located either in the vicinity of the clay surface or between the adsorbed organic layers. Conformation of the hydrocarbon chains in the surfactant depends on the grafting density, the type of ion exchange process, and the thermal process utilized during the preparation of organoclays. As the surfactant content approaches the overall ion exchange capacity of the clay, the hydrocarbon tails belonging to surfactant molecules adsorbed at opposing sides of the silicate gallery can interpenetrate. Obviously, the extent of the penetration will affect the thickness of the organic layers. As a result, a few types of distinct layers with different thicknesses may coexist in organoclays, which is the characteristic for the so-called **interstratified systems** (Gelfer et al., 2004).

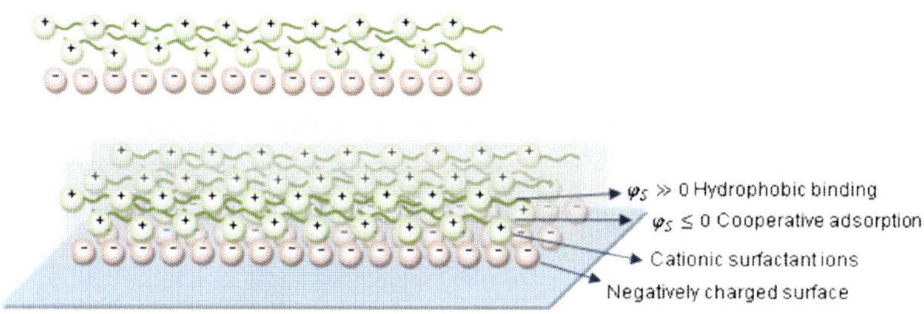

Figure 2.19 Hydrophobic bonding.

To reach adequate modification of an MMT sample, it is necessary to know its CEC and the amount of organic surfactant that can enter between the layers, thus ensuring complete modification of the MMT. This method, suggested by Lagaly (1994), is now used as a routine analytical technique to estimate the interlayer CEC. One important route for the modification is determination of the amount of surfactant adsorbed via ion exchange and also the amount of surfactant adsorbed together with its counterion (Vazquez et al., 2008).

Depending on the functionality, packing density, and length of the organic modifiers, the organomodified layered silicates (OMLSs;

organosilicates or organoclays) may be engineered to optimize their compatibility with a given polymer. It is worth noticing that, on the basis of the CEC of the clay, the content of the surfactant is usually about 35–45 wt.%. Actually, one way to measure the clay CEC is by determining the amount of alkylammonium salt retained by the organoclays.

That is, dried clays that have been subjected to organomodification along with a sample of the corresponding untreated clay are ignited at 1000°C. From the differences in the loss on ignition of the sample and blank and the molar mass of the alkylammonium salt, the milliequivalents of the organic substance retained by the clays are calculated and those values are taken as their CEC. Alternatively, chemical analysis of the clay can also be applied for CEC determination. CEC value determines the number of cationic surfactants that can be intercalated into the galleries by ion exchange. In general, the longer the surfactant chain length, and the higher the charge density of the clay, the further apart the clay layers will be forced. Many methods exist to measure the CEC. Typical CEC values of 2:1 phyllosilicates are listed in Table 2.2 (Tjong, 2006).

Table 2.2 Chemical formula and cation exchange capacity of 2:1 phyllosilicates

Silicate	Formula	CEC (mequiv/100 g)
Montmorillonite	$M_x(Al_{4-x}Mg_x)Si_8O_{20}(OH)_4$	92.6–120
Hectorite	$M_x(Mg_{6-x}Li_x)Si_8O_{20}(OH)_4$	120
Saponite	$M_xMg_6(Si_{8-x}Al_x)Si_8O_{20}(OH)_4$	86.6
Vermiculite	$(Mg, Fe, Al)_3[(Al, Si)_4O_{10}]$ $(OH)_2M_x{\times}nH_2O$	150

Source: Reprinted from Tjong, Copyright 2006, with permission from Elsevier.
Note: M represents the exchangeable cation, and *x* is the layer charge.

The organic cations lower the surface energy of the silicate surface and improve wetting with the polymer matrix. Moreover, the long organic chains of such surfactants, with positively charged ends, are tethered to the surface of the negatively charged silicate layers, resulting in an increase in the gallery height. It then becomes possible for organic species (i.e., polymers or prepolymers) to diffuse between the layers and eventually separate them.

Sometimes, the alkylammonium cations may provide functional groups that could significantly affect the polymerization of monomers (Pavlidou and Papaspyrides, 2008). The microchemical environment in the galleries is, therefore, appropriate for the intercalation of polymer molecules. Conclusively, the surface modification both increases the basal spacing of clays and serves as a compatibilizer between the hydrophilic clay and the hydrophobic polymer. The excess negative charge of layered silicates and their capability to exchange ions can be quantified by a specific property known as the CEC and expressed in milliequivalents per gram. This property is highly dependent on the nature of the isomorphous substitutions in the tetrahedral and octahedral layers and therefore on the nature of the soil where the clay was formed. This explains, for example, why MMTs from different origins show differences in CEC, ranging from approximately 0.9–1.2 meq/g.

On the other hand, there are a number of simple ways of determining the predominant orientation of the surfactant molecules adsorbed on the solid. If the solid can be obtained in the form of a smooth, nonporous planar film or plate, then the measurement of the contact angle of a drop of water placed on the solid surface before and after surfactant adsorption can be used. The greater the contact angle, the greater the hydrophobicity of the surface. If the solid can be obtained in the form of finely divided particles and if surfactant adsorption has made it more hydrophilic, it will disperse better in water after surfactant adsorption than before; if it has made it more hydrophobic, then the particles will either float to the surface when shaken with water or settle out of the water more rapidly than before surfactant adsorption. Alternatively, the particles may be shaken with a mixture of equal volumes of water and a nonpolar solvent (e.g., hexane). If the particles have become more hydrophobic by surfactant adsorption, they will disperse better in the nonpolar phase than before surfactant adsorption; if they have become more hydrophilic, they will disperse better in the aqueous phase than before surfactant adsorption.

Burmistr et al. (2005) used polymeric quaternary ammonium salts to prepare organobentonites. The polymeric quaternary ammonium salts were prepared by reaction between *p*-xylylenedichlorides and diamines. The organoclay showed a basal spacing of 1.67 nm.

Costa Filho et al. (2005) prepared bentonite–ionene complexes based on the adsorption of tetramethyl-[x, y]-ionenes, $x = 2$ and $y = 4$, 9, 10, or 12, and an ionene of epichloridrin-dimethylamine (EPI-DMA) kind. The ionenes were adsorbed on clay mineral surfaces above the ion exchange capacity.

Zhang et al. (2005, 2006b) prepared an organoclay based on an ammonium salt that was synthesized by the reaction of triethylamine and vinylbenzyl chloride and lauryl acrylate copolymer. The organoclay was highly ordered and showed a basal spacing of 3.8 nm.

Zhang et al. (2006a), following the same methodology, prepared an organoclay based on three components: styrene, lauryl acrylate, and vinylbenzyl chloride. In this case, the ammonium salt was synthesized by the reaction of triethylamine and styrene, vinylbenzyl chloride, and lauryl acrylate terpolymer. The organoclay showed a basal spacing of 3.7 nm.

Chigwada et al. (2006b) introduced a larger substituent 4-acetybiphenil on the ammonium cation and obtained a phenylacetophenone dimethyl hexadecylammonium salt, which was used to prepare an organo-MMT (OMMT) that showed a basal spacing of 2.4 nm.

Chigwada et al. (2006a) introduced quinoline and pyridine groups on the hexadecylammonium cation and obtained quinolinium and pyridinium salts that were used to prepare OMMT that showed basal spacings of 1.8 nm and 2.0 nm, respectively. Small changes in the modifier, such as replacing pyridine by quinoline, resulted in significant changes in the organoclays' properties.

Zheng et al. (2006) modified MMT with 2-methylacryloxyethyl dimethyl hexadecylammonium bromide.

Shen (2001) used nonionic surfactants to prepare organobentonites. Nonionic surfactants that are linear alcohol ethoxylates provide higher adsorption capacity and higher chemical stability to bentonite than cationic surfactants.

Yao et al. (2002) reported the use of crown-ethers and cryptand to modify potassium and sodium MMTs. The basal spacings of the crown-ethers/potassium MMT increased to 1.8–1.9 nm and those of sodium MMT to 1.5– 1.7 nm.

The use of imidazole salts to prepare organophilic MMT was also reported by Awad et al. (2004).

Akyüz and Akyüz (2003) studied the adsorption of 2-aminopyrimidine (2APM) on sepiolite and MMT. 2APM was intercalated in MMT that showed a basal spacing characteristic of the monolayer arrangement, while in the sepiolite 2APM molecules were adsorbed on the surface.

According to Chaiko (2003), the adsorption of polyethylene-block-PEG copolymers onto Na^+MMT can facilitate the formation of organoclay–polyethylene nanocomposites. Water soluble polymers, such as poly(ethyleneoxide) (PEO), can be intercalated in organoclays in solution or from the melt. In both approaches, the driving force behind PEO adsorption is entropic and involves at least partial replacement of the water of hydration associated with the exchangeable cations of the clay mineral.

Chaiko (2003) prepared new nanocomposites based on the treatment of Na^+-MMT, Na^+ and Ca^{2+} saponites, and hectorite with PEG and also reported a new mechanism that permits formation of two-dimensional PEO phase. Under acidic conditions, polymer adsorption occurs through an ion exchange process mediated by oxonium cation formation. The protonated PEG complex is stabilized by the formation of the five-membered rings through hydrogen bonding with the terminal ether oxygen. The basal spacing increases with increasing polymer molar mass.

Sánchez-Solís et al. (2004) intercalated maleic anhydride and pentaerythritol in MMT to prepare polyethylene terephthalate–MMT nanocomposites. The basal spacing of the sodium MMT was preserved in the maleic anhydride modified MMT but increased to 1.45 nm in the pentaerythritol modified MMT.

Yu et al. (2004) prepared an OMMT by cation exchange with alkylphosponium cations of (4-carboxybutyl)-triphenylphosphonium bromide. Surface modification by polymers is found to be one of the most effective methods, as the surface properties can be widely changed by a variety of functional polymers.

Grafting of functional polymers to the surface of clay minerals can improve the interaction. The first approach used, named the **one-step grafting method** (Fig. 2.16), consists of the condensation of functionalized polymers with reactive groups of a solid substrate. This method does not give highly dense polymer brushes because chemisorption of the first fraction of chains hinders the diffusion of the following chains to the surface for further attachments. Another approach, named the **two-step grafting method** (Fig. 2.20), has been considered to obtain better densities. In this technique, a monolayer of polymerizable (macromonomer) or initiator (macroinitiator) molecules is covalently attached to a solid surface. After activation, the chains grow from the interface; then the only limit to propagation is the diffusion of monomers to the active species (Liu, 2007).

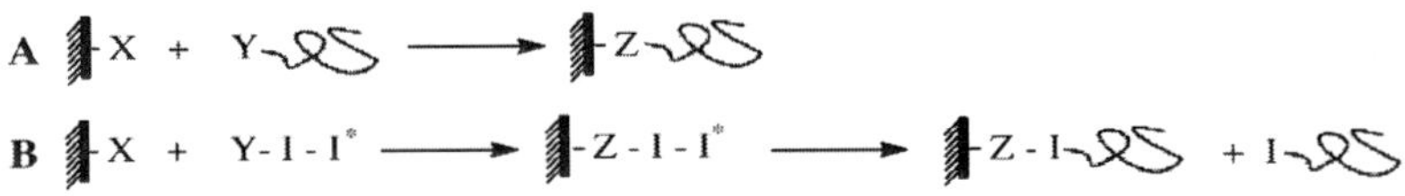

Figure 2.20 General mechanism for one-step (A) and two-step (B) grafting reactions. Reprinted from Liu, Copyright 2007, with permission from Elsevier.

He et al. (2005) described the grafting process of a synthetic fluorohectorite and MMT with (3-aminopropyl)-triethoxysilane ($H_2N\ (CH_2)_3Si(OC_2H_5)_3$). After grafting, the basal spacings increased to 1.45 nm and 1.77 nm and indicated that the 3-aminopropyl triethoxysilane was intercalated. The silane molecules adopted a monolayer arrangement in the synthetic fluorohectorite and a bilayer arrangement in MMT. The study demonstrates that the silylation reaction and the interlayer structure of the grafted products strongly depend on the clay minerals.

Herrera et al. (2005) reported the grafting of laponite with monofunctional γ-metacryloxypropyl dimethyl methoxysilane and trifunctional γ-methacryloxypropyl trimethoxysilane coupling agents. Fourier transform infrared (FTIR) analysis gave evidence of covalent attachment of silane molecules to the hydroxyl groups. The trifunctional silane formed monolayer structure coverage at low concentrations and a multilayer structure at higher concentrations, while the monofunctional silane formed a monolayer.

Wang et al. (2006) modified MMT with γ-aminopropyl trimethoxysilane to be used in the preparation of epoxy nanocomposites.

Yoshimoto et al. (2005) used four aniline salts with different counterions, referred to as AnF, AnCl, AnBr, and AnI, respectively, to modify MMT. Both AnF and AnCl, anilinium cations, and neutral molecules were intercalated at loadings levels beyond the CEC, and the basal spacing expanded up to 2.47 nm. The authors explained that intercalated species form a rigidly packed pseudotrimolecular structure, in which excess neutral molecules are sandwiched by van der Waals interactions between anilinium cations, which are tightly attracted to the upper and lower silicate surfaces by electrostatic interactions. For AnBr and AnI, the basal spacings were lower, 1.46–1.48 nm, and further expansion with higher amounts of the salts was not observed.

Patel et al. (2007) reported the preparation of OMMT with the intercalation of quaternary phosphonium cations. Clays and clay minerals can also be modified with biomolecules, like proteins, enzymes, amino acids, and peptides. There are many examples of modification of clay minerals with biomolecules. Causserand et al. (1997) studied interactions during the adsorption of bovine serum albumin on MMT and also the mechanism that controls the ultrafiltration of this mixture by a polyethersulfone membrane.

Naidja et al. (1997) investigated that tyrosinase immobilized on MMT coated with aluminum hydroxide was less sensitive to temperature and retained its activity for a longer period of time.

Kalra et al. (2003) investigated the adsorption of simple peptides of glycine/alanine (diclycine, triglycine, tetraglycine, and glycylanie) onto Ca^{2+}- and Mg^{2+}-exchanged MMT. The adsorption of the peptides followed the Langmuir type of adsorption, that is, monolayer formation of adsorbates on the surface of the clay mineral. Ca^{2+} MMT exhibited relatively better adsorption as compared to Mg^{2+} MMT. The clay minerals might have played a significant role in the prebiotic formation of proteins via adsorption of simple bio-oligomers on their surface. Smectite modified with aminopropyl was used to anchor biomolecules by covalent bonding and were applied to immobilization of glucose oxidase. Smectite modified with propyltrimethylammonium exhibited favorable electrostatic interactions with negatively charged enzymes and were applied for polyphenol oxidase anchoring.

Gopinath and Sugunan (2007) immobilized α-amylase, glucoamylase, and inverstase on acid-activated MMT via adsorption and grafting. The NH_2 groups interact with the acid sites of clay mineral. Grafting between the enzyme and the glutaraldehyde spacers occurs via condensation of NH_2 groups of enzyme and CHO groups of glutaraldehyde. After immobilization of enzymes, the basal spacing of MMT increased, but the whole enzyme molecules were not intercalated. The protein backbone is situated at the periphery of the clay mineral particles, while the side chains enter in the layers. Surface area and pore volume decreased. The morphology of the particles was changed. ^{27}Al nuclear magnetic resonance (NMR) analysis showed that the tetrahedral Al species were involved in the adsorption process, whereas the octahedral sites were influenced by grafting.

Öztürk et al. (2007) modified bentonite with histidine. Covalent bonds were formed between the amino groups of histidine and the Al^{3+} cations of octahedral sheet of the clay mineral.

Recently, the possibility of grafting some silane molecules on the surface of the clay mineral layers has been studied (Herrera et al., 2004; Park et al., 2004b). In fact, reactive OH groups are present in the clay minerals at the edges and at the structural defects (Lee and Jang, 1996; van Olphen, 1997; He et al., 2005). Therefore, the functionalization of the clay mineral can take place at three possible sites: at the interlayer space, at the external surface (Herrera et al., 2004), and at the edges (Park et al., 2004b). The silanization at the interlayer and at the edges can increase the distance between the layers (Isoda and Kuroda, 2000; Park et al., 2004a). In some cases, the silanization reaction has also been conducted onto organophilic clay minerals to enhance their compatibility with the polymer: these twice-functionalized organoclay minerals are described by Chen and Yoon (2005a, b, c) and Park and Jana (2003).

Some OH groups are present at the edges of the MMT layers, as well as at the structural defects. These OH groups are active sites that can be used to obtain a functionalization of the clay mineral platelets surface (Ek et al., 2004; He et al., 2005). The reaction is the same used in many studies reported in the literature about the functionalization of silica particles (Luechinger et al., 2005) and their compatibilization with organic matrices: an alkoxysilane is typically employed for this purpose. In a study by Di Gianni et al. (2008), glycidylpropyl-triethoxysilane (GPTS) was used so that while the trialkoxysilane moiety can react with the active sites on the clay mineral surface, the epoxy ring can take part in the UV curing of the epoxy resin (CE) employed to build up the polymeric matrix. The functionalization of the clay mineral by means of GPTS is made in two steps: hydrolysis of the silane compound and condensation between the silanols or the alkoxysilane groups.

Organic cations used as clay intercalants and their abbreviations have been tabulated in Table 2.3 (Utracki, 2004).

Table 2.3 Organic cations used as clay intercalants

Abbreviations	Modifiers
2M2DA	Dimethyl didodecyl ammonium chloride
2M2DDA	Dimethyl dilauryl ammonium or dimethyl didodecyl ammonium chloride
2M2HTA	Dimethyl dihydrogenated tallow ammonium chloride (Arquad 2HT)
2M2ODA	*N,N*-Dimethyl-*N,N*-dioctadecyl ammonium bromide or chloride
2M2SA	Dimethyl distearyl ammonium chloride
2M2TA	Dimethyl ditallow ammonium chloride
2MBHTA	Dimethyl benzyl hydrogenated tallow ammonium chloride
2MBODA	Dimethyl benzyl octadecyl- or dimethyl benzyl stearylammonium chloride
2MBSA	Dimethyl benzyl stearylammonium chloride
2MHDI	1,2-Dimethyl-3-*N*-hexadecyl imidazolium
2MHDODA	Dimethyl hexadecyl-octadecyl amine
2MHTL8	Dimethyl hydrogenated tallow 2-ethylhexyl ammonium methyl sulfate
2MODA	Dimethyl octadecyl amine
2MVBDDA	Dimethyl vinylbenzyl dodecyl amine
2S2M	Distearyl dimethyl amine = 2M2ODA
3BHDP	Tributyl-hexadecyl-phosphonium bromide
3MDA	Trimethyl dodecyl amine
3MDDA	Trimethyl dodecyl ammonium bromide
3MHDA	Trimethyl hexadecyl ammonium bromide
3MODA	Trimethyl octadecyl ammonium chloride
4MA	Tetramethyl ammonium chloride
AC12A	α,ω-Dodecyl amino acid (also ADA)
ADA	ω-Amino dodecyl (or lauric) acid, or 12-amino dodecyl acid

(*Continued*)

Table 2.3 (*Continued*)

Abbreviations	Modifiers
AHA	ω-Amino-hexanoic acid
AHDA	ω-Amino-hexadecanoic acid
Allyl-16	*N,N*-Dimethyl-*n*-hexadecyl allyl ammonium chloride
AODA	ω-Amino-octadecanoic acid
APB	1,3-bis(3-Aminophenoxy)benzene
APTS	γ-Aminopropyl-triethoxy silane
ATDA	ω-Amino-tetradecanoic acid
B3HTA	Benzyl trihydrogenated tallow ammonium chloride
Bz-16	*N,N*-Dimethyl benzyl-*n*-hexadecyl ammonium chloride (also 2MBHDA)
CPC	Cetyl pyridinium chloride
DA	Dodecyl amine
DDA	Dodecyl ammonium chloride
DDP	1-Dodecyl-2 pyrrolidone
DIP	Dimethyl isophthalate triphenylphosphonium
HAD	Hexadecyl amine
HDA	Hexadecyl ammonium chloride
HDP	*N*-hexadecyl pyridinium chloride
M2EPPOH	Methyl diethyl polypropylene glycol ammonium chloride
M2HTA	Methyl di[hydrogenated tallow] ammonium
M2S2EtOH	Methyl stearyl bis[2-hydroxyethyl] ammonium
M3HTA	Methyl trihydrogenated tallow ammonium chloride
M3O	Methyl trioctyl ammonium chloride
MC2EtOH	Methyl coco-alkyl di-2-hydroxy ethyl ammonium chloride
MDD2EtOH	Methyl dodecyl di-2-hydroxy ethyl amine
MHA	2-Methacryloyl-oxyethylhexadecyldimethyl ammonium bromide
MOD2EtOH	Methyl octadecyl di-2-hydroxy ethyl amine
MSA	Methyl-styrene ammonium chloride
MT2EtOH	Methyl tallow di-2-hydroxy ethyl ammonium chloride (Ethoquad T12)

Abbreviations	Modifiers
OA	Octyl ammine or ammonium chloride
ODA	Octadecyl amine (C18; also stearyl amine)
OH-16	*N,N*-Dimethyl-*n*-hexadecyl-(4-hydroxymethylbenzyl) ammonium
OM-2	*N*-[4-(4′-Amino-phenoxy)]phenyl phthalimi
OM-l	*N*-[4-(4′-Amino-phenyl)]phenyl phthalimide
P18	Stearyl-tributyl phosphonium bromide
POE-DPA	Polyoxyethylene decyl propyl amine ethyl acrylate
QD1	2-(*N*-Methyl-*N,N*-diethyl ammonium iodide)
QD4	2-(*N*-Butyl-*N,N*-diethyl ammonium bromide) ethyl acrylate
S2EtOH	Stearyl di-2-hydroxy ethyl amine = MOD2EtOH
SA	Stearyl amine (C18; also ODA)
TADA	Tallow diamine chloride
TDA	Tetradecyl ammonium chloride
TPP	Tetraphenyl phosphonium
VB-16	*N,N*-Dimethyl-*n*-hexadecyl-(*p*-vinylbenzyl) ammonium chloride
VDAC	Vinylbenzyl dimethyl dodecyl ammonium chloride

The intercalation mechanisms of the GPTS have been monitored by the XRD patterns, FTIR spectroscopy, and thermogravimetric analysis (TGA). On the basis of the results obtained from the analysis of the modified MMT, it is suggested that many different reactions simultaneously occur between the GPTS and the clay mineral (Scheme 2.1).

MMT can be modified by the condensation reaction (a). Besides the grafting of a single GPTS molecule through reaction of any functionality with the clay mineral OH groups, condensation between couples of grafted GPTS can also occur. GPTS molecules linked to different platelets through the OH groups at the edges can react according to reaction (b). Reaction (c) can occur between GPTS molecules grafted in the same interlayer space; it can also take place between the grafted GPTS and some intercalated molecules. Finally, we cannot exclude that GPTS molecules or oligomers are just intercalated between the platelets (Di Gianni et al., 2008; He et al., 2005; Schanmugharaj et al., 2006).

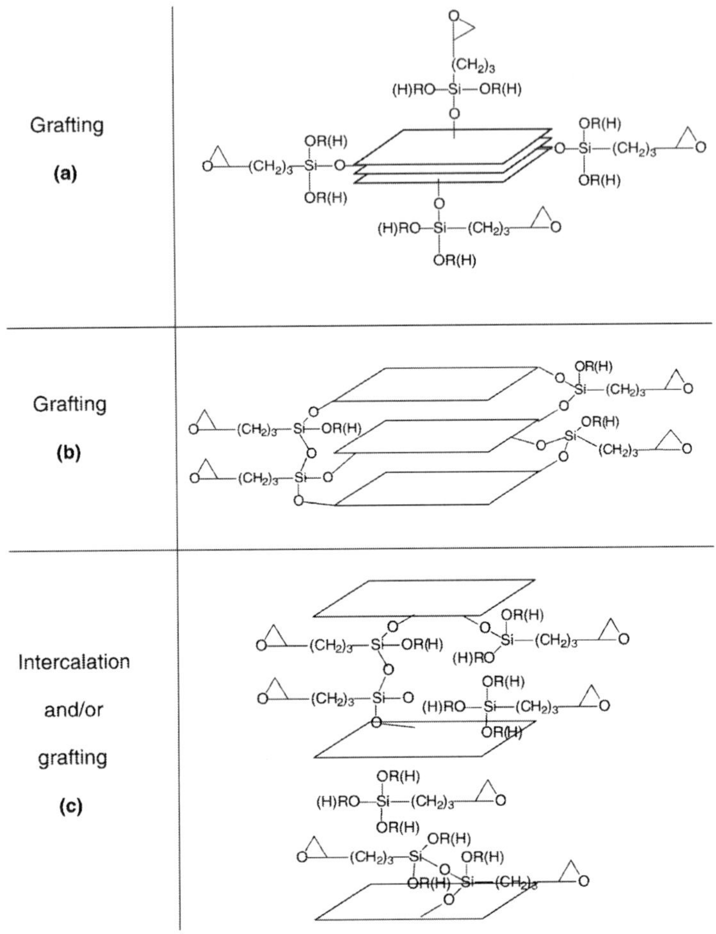

Scheme 2.1 Different possible interactions between the hydrolyzed GPTS and the pristine MMT. Reprinted from Di Gianni et al., Copyright 2008, with permission from Elsevier.

One method that has seen increased attention is surface treatment via ion–dipole bonding of organic molecules, oligomers, or polymers to the exchangeable cation on the clay surface. Utilizing this method, a unique self-assembly of certain organic molecules has been discovered in which the molecules form rigid posts around each cation on the surface. The bonding trends and controlling factors in the self-assembly of the ion–dipole bonding molecules have been described in detail by Beall and Goss (2004). In general, it appears that the headgroup of the molecule is one of the principal factors controlling self-assembly, but the chain length of the alkyl group also plays a role.

One method for surface-modifying smectite clays is the use of organic molecules that contain groups that possess partial negative charges and thus can ion–dipole bond to the exchangeable cation on the surface. This bonding phenomenon was first reported for various glycols and has become a standard method for identification of smectite clays. Extensive studies of alcohols and amines and their complexes with smectic clays have been conducted. These studies demonstrated that under saturated conditions, the longer-chained alcohols and amines can form tight packet structures with fairly large basal spacings. The studies also report extensively on the phase behavior of these compounds as a function of temperature. In these studies, full saturation of the surface was important in stabilizing the phases that exhibit large basal spacings. In these complexes, the alkyl chain interactions and headgroup interactions with the clay surface were the most important bonding factors affecting their stability in undersaturated conditions; the ion–dipole interaction between the polar headgroup becomes a very important factor. The ion–dipole bonding approach in undersaturated systems has been studied extensively in polymer–clay nanocomposites. The main requirement for the attachment of a molecule with a cation by ion–dipole interaction is that the molecule contains groups within the molecule that carry partial negative charges. It has been shown that a large variety of organics will undergo this sort of bonding, including organics containing carbonyls, hydroxyls, and ether groups. In most cases, these molecules will form interdigitated complexes similar to those formed with quaternary ammonium compounds. These organocomplexes will typically exhibit basal d-spacings of approximately 2 nm in the XRD pattern. One very interesting exception to this is the complexes formed between *n*-dodecyl pyrrolidones (DDP). In these complexes, when DDP is added in at least a 2:1 ratio to the exchangeable cation, a self-assembly of DDP occurs around the cation to form a rigid post. This post becomes more ordered and rigid when a 3:1 ratio is employed (Beall and Goss, 2004; Beall and Tsipursky, 1999).

Polyester hydrochlorides are a new type of polycations, prepared from alkyl derivatives of diethanol amine and organic acid chlorides.

They can be used for MMT modification by cation exchange. The length of the organic acid chain did not significantly influence the basal spacing of MMT, but the length of the alkyl group attached to the nitrogen atom increased the basal spacing. For quantitative cation exchange, the molar ratio of cations in MMT and hydrochloride must be at least 1:3, indicating that only a part of polycation chains adheres to MMT, while the remaining segments dangle out into the solution. Hydrophobicity and organic content of OMMT increased with increasing exchange ratio, reaching approximately constant levels for ratios above 1:3 (Huskić et al. 2009).

2.3.3 Organoclay Structures

Surface properties of natural clays can be modified by simply ion exchange with organic cations. By introducing cationic surfactant molecules into the interlamellar space through ion exchange, the properties of clay minerals are enhanced to those of organoclays. The intercalation of a cationic surfactant between the clay layers changes the surface properties from highly hydrophilic to increasingly hydrophobic. In addition, modification of the swelling clay with a cationic surfactant results in an increase in the basal spacing of the layer and exposure of new sorption sites of clays. Organoclays are classified into two different groups, adsorptive clays and organophilic clays, by the structure of quaternary ammonium cations (QACs) and the mechanism of adsorption behavior. Adsorptive clays are synthesized with QACs containing short-chain alkyl groups and/or a benzyl group, such as benzyltriethylammonium (BTEA) or tetramethylammonium (TMA)/trimethylphenylammonium (TMPA). Sorption on this type of organoclay is characterized by Langmuir-type isotherms, which are commonly associated with specific sorption sites. On the other hand, if the QACs contain a long-chain alkyl group, such as hexadecyltrimethylammonium (HDTMA) and didodecyldimethylammonium (DODMA), this group of organoclays is called organophilic clay (Xu and Boyd, 1996). The sorption of this group is characterized by linear isotherms over a wide range of solute concentrations. The different sorption behavior of the two organoclays (adsorptive clays and organophilic clays) is further described in their environmental applications. The modified clay derived from long-chain alkylammonium ions creates an organic partition medium through the conglomeration of the flexible alkyl chains, whereas the clay from small quaternary ammonium ions does not form partition phases. Instead surface adsorption sites for nonionic solute uptake exist as discrete entities on the clay layers because of the small-alkyl-chain ions (Lo et al., 1998; Smith et al., 1990). The size of the alkyl chain length and the charge density of the clay layer are responsible for the arrangement of intercalated organic cations in the organoclays. In the classic model, the organic cations only form monolayers in the adsorptive clays intercalated with short-chain alkyl ammonium ions. In contrast, bilayers, a pseudotrimolecular arrangement, or paraffin complexes may be found in the organophilic clays intercalated with long surfactant cations. The arrangement of surfactant alkyl chains can also be incorporated with the conformational changes of C–C bonds (kicks). It can be assumed that three-layer structures of kinked alkyl chains (pseudotrimolecular layers) are formed by kicks, where some chain ends are shifted above one another. As the highly charged clay minerals intercalate with alkylammonium ions, more paraffin-type structures can be formed (Lagaly, 1986a).

After the synthesis of organoclays by intercalating organic surfactants into the layers of the clays through ion exchange, various modern characterization tools, which include determination of properties such as structure, expansion capacity, layer charge, pore size, crystalline size, charge distribution, and pore distribution, are available for both naturally occurring clays and synthesized organoclays. The structural properties of the organoclays are highly dependent on the structure and molecular environment of the organic surfactants. The essential instrumentations used to determine the characteristics of clays and their modified organic derivatives are XRD, TGA, and electron microscopy such as TEM and SEM. Brunauer–Emmett–Teller (BET) is also used as a specific surface area measurement of organoclays.

XRD is one of the most important techniques to determine the structural geometry, texture, and also identify impurities in layer silicates present in clays. Generally, XRD can offer basal spacing information on the organoclays used. The results of XRD can also provide information on molecular structure configuration of the surfactant intercalated, including a lateral monolayer, lateral bilayer, paraffin monolayer, paraffin bilayer, and pseudotrilayer. The change of basal spacings often indicates that organic cation surfactants are intercalated within the interlayer of organoclays and *d* (0 0 1) spacings of ion-exchanged natural clays with the same intercalants always compare favorably (Carrado, 2000). However, more evidence of organic uptake is essential for the supposition obtained from the change of basal spacings as XRD does not provide detailed information about the local conformation and phase state of the intercalated surfactants (Zhu et al., 2005). Hence, this is done by TGA and FTIR. The thermal stability of the organoclays can be determined by the use of TGA (Hedley et al., 2007). Typical TGA curves run in an oxygen atmosphere to optimize organic degradation, and the weight loss of a sample can be observed as a function of temperature (dynamic heating) or time (isothermal heating). A peak at low temperature (between 60°C and 70°C) is caused by the loss of water from the surface and interlayers, while the high-temperature peak (between 700°C and 750°C) is due to the dehydroxylation of the clay layers. It is often convenient to resolve thermal reactions by plotting the derivative of the TGA curve. Differential thermogravimetry (DTG) is a preferred method to determine the inflection point of mass loss. Differential scanning calorimetry (DSC) can be employed to describe the interlayer structure of organoclays by measuring the amount of energy adsorbed or released at constant temperature. However, differential thermal analysis (DTA) is the primary tool for the investigation of the thermal properties and reactions of minerals and organoclays (Dultz et al., 2005; Li and Ishida, 2002). It is based on the heat differences associated with both the physical and chemical changes of a sample compared to a thermally inert reference material. Using DTA, one can measure all reactions and processes that involve a change in energy during heating and/or cooling of a sample, including dehydration, dehydroxylation, structural decomposition, and transformation (Park et al., 2011; Plante et al., 2009).

Detailed information about the interlayer structure and phase state of alkylammonium in the silicate interlayer was first obtained using FTIR (Vaia et al., 1994). The results revealed that the frequency shift of the CH_2 stretching and scissoring vibrations was a function of interlayer packing density, chain length, and temperature. In addition, a wide range of molecular environments varying from solid-like to liquid-like were found for the first time. Recently, the conformation of alkylammonium confined in the clay interlayer space was extensively studied by techniques of Raman spectroscopy (He et al., 2004), NMR (Cases et al., 1992; Kubies et al., 2002), and electrokinetic measurements (Gürses et al., 2014). These results suggest that the conformation ordering of the intercalated surfactants strongly depends on the loading amount of surfactants and their orientation. The ordered (all-*trans*-) conformation would be predominant when hydrocarbon chains radiated away from the clay surface with increased surfactant concentration (Osman et al., 2002; Venkataraman and Vasudevan, 2000, 2001). More recently, molecular modeling provided some critical information about the structure of organic clay and surfactant–clay and surfactant–surfactant interactions, which was not elucidated by previous experimental results. High-resolution single-pulse ^{13}C magic-angle-spinning nuclear-magnetic-resonance (MAS-NMR) has been proved to be a powerful technique for probing the local environments of atoms in materials (Zeng et al., 2003). However, previous reports paid the most attention to the backbone of the alkyl chain (C2–C14) to elucidate the conformation of the intercalated surfactants because of their strong and distinct resonance peaks, and no attention was paid to the end groups, C1, C15, and C16, which are critical to the conformation and mobility of the intercalated surfactants (Osman et al., 2002; Venkataraman and Vasudevan, 2000; Wang et al., 1996; Zhu et al. 2005).

Originally, Lagaly (1986b) had proposed various arrangements of alkyl chains in organoclays based on XRD analysis. Vaia et al. (1994), on the basis of FTIR spectroscopy studies on a series of primary and quaternary alkyl ammonium cations inside the interlayer space, have proposed interlamellar monolayer, bilayer, or pseudolayer types of arrangements of the chains. Xie et al. (2001a) have studied in detail the thermal degradation behavior of organoclays using TGA, TG/FTIR, and mass spectroscopy to understand the process of thermal degradation and have determined various degradation products (water, CO_2, alkanes, alkenes, linear aldehydes, etc.) from decomposition of intercalated surfactant using the gas chromatography–mass spectrometry (GC-MS) technique. Li and Ishida (2003), from their comparative study of intercalated neutral hexadecylamine and hexadecylammonium ion-modified clay prepared through ion–dipole interaction by direct melt mixing and by cation exchange method, respectively, have looked at conformations of hexadecylamine at different values of the CEC of the organoclay, where they have found a greater degree of ordering for the neutral amine molecules as compared to their alkylammonium salt analogs.

Tiwari et al. (2008) have studied OMMT clays by using ammonium-based organic modifiers that contain one or more dipolar/polar groups such as benzyl, phenyl, hydroxyl, carboxyl, and ether. They characterized basal spacings, interlamellar structure, and thermal stability of these OMMT clays by using wide-angle X-ray diffraction (WAXD) and TGA techniques and reported that the increase in the basal spacing due to organic modification is in good agreement with simple theoretical calculations based on van der Waals volume of the cationic ammonium ions.

Raman spectroscopy is a powerful technique used to probe the microstructural environment of organic compounds within organoclays at various temperatures. By introducing BET gas adsorption theory, the surface area of organoclays can be determined in the presence of N_2 (Carmody et al., 2007). When N_2 gas penetrated the organoclays, the surface area, pore volume, and size distribution of the organoclays are measured, and in the process, neutral N_2 will attach by van der Waals force to form a single layer, and then multilayer adsorption will take place. However, only external surface can be attached with N_2 gas. Electron microscopy analyses such as TEM and SEM are employed to make a direct observation of macrostructural features on organoclays. TEM demonstrates the packing density of organoclays within the clay interlayer space, and SEM micrographs show surface morphology of clay and organoclay samples (He et al., 2006). Besides, ultraviolet-visible (UV-Vis) spectroscopy is applicable to investigate the association between organic cation molecules on the clay surface (Park et al., 2011).

Clay minerals are hydrophilic, and modifying the clay mineral surfaces with cationic surfactants to convert the normally hydrophilic silicate surface to an organophilic surface is a strategic step for various applications. All clay-based applications strongly depend on the surface properties and structure of the organoclays. For synthesis of clay-based mesoporous materials in various organic media, using organoclays as the starting clay material, it is understood that the swelling properties of organoclays will have an important influence on the structure of the resultant. This is because, in this case, the interlayer expanding extent of organoclays in organic media controls the entrance of pillaring agents (e.g., silane and their hydrolysis products) and the template height (He et al., 2006).

The process of intercalation also increases the basal spacing of the clay, and it is known that the basal spacing of organoclays mainly depends on the chain length and the loading of the organocation, which in turn depends upon the ratio of cation to clay in the preparation stage (Lagaly, 1986a; Vaia et al., 1994). Increasing the loading of the organic component in the organoclay results in stepwise increments in the basal spacing. This is usually calculated on the basis of the CEC of the clay and conveniently expressed in CEC multiples (Le Pluart et al., 2002; McLauchlin and Thomas, 2008; Xi et al., 2004).

He et al. (2010) prepared OMMT using MMTs with different cation exchange capacities (CEC) and surfactants with different alkyl chain numbers and chain length. They have reported the following main findings: The basal spacing increased with surfactant loading. The maximum basal spacing also increased with the alkyl chain length of the surfactant. For a given chain length, the maximum basal spacing increased as the chain number increased from one to two. The thermal analysis indicated that the surfactant was located in both the clay mineral interlayer spaces and the interparticle pores with a "house of cards" structure. For a given surfactant, the CEC of the MMT component had little influence on the maximum basal spacing of the OMMT when the same surfactant was used. The level of surfactant loading, however, strongly depended on the CEC.

Yoshimoto et al. (2005) have prepared inorganic–organic intercalation compounds comprising MMT and aniline salts with different counteranions by a mechanochemical processing. The intercalation process and the formed structure of intercalation compounds have been investigated via XRD analysis. It has been found that the amounts of intercalated species were very likely dependent on the types of counteranions and increased with decreasing size of counteranions during the mechanochemical processing. Very interestingly, much larger interlayer expansions of 1.51 nm were observed for aniline hydrofluoride (AnF)– and aniline hydrochloride (AnCl)–MMT systems in higher intercalate-loading levels, suggesting that neutral guest molecules are also introduced within the interlayer regions, together with anilinium cations by van der Waals interactions. They were determined that judging from the larger interlayer expansions and the size of guest molecules, intercalated species were expected to prefer a trimolecular layer arrangement with their aromatic rings perpendicular to the silicate sheets. In contrast, for aniline hydrobromide (AnBr)–MMT, the interlayer expansion was ca. 0.52 nm, which reveals that only anilinium cations are introduced by ion exchange, and they probably adopt a vertical orientation in the interlayers. It is inferred that aniline hydro-iodide (AnI)–MMT compounds have a heterogeneous structure containing both anilinium and sodium cations in the interlayers. Different intercalation behaviors during the mechanochemical processing strongly suggest the smaller the size of counteranions, the more guest molecules can be intercalated into the confined clay interlayers in highly ordered arrangements. On the basis of XRD results, the arrangements of the intercalated species in the clay interlayers are proposed and are schematically depicted in Fig. 2.21. In AnF, AnCl, and AnBr systems, sodium cations in the clay interlayers are replaced by anilinium cations quantitatively through the cation exchange interaction at loading levels below the 1CEC. At the 1CEC, the anionic sites of MMT surfaces are occupied with anilinium cations completely and the basal spacings expand to ca. 1.48 nm. Under these conditions, intercalates, that is, anilinium cations, adopt a perpendicular arrangement, in which the positively charged nitrogen atoms of anilinium cations are tightly attracted to the anionic sites of MMT surfaces (Fig. 2.21a,b). The electrostatic attractions between anilinium cations and the negatively charged clay surfaces give rise to this arrangement. In contrast, the broadness of (001) plane and the existence of excess AnI crystal in the XRD pattern even at 1CEC loading suggest that anilinium cations are partly introduced into the clay interlayers in AnI system. If the replacement of sodium cations by AnI occurs only at the edges of clay platelets, the intercalation complex obtained has heterogeneous structures, including both anilinium and sodium cations in the interlayers, as shown in Fig 2.21c.

In AnBr– and AnI–MMT systems, the higher loading beyond the CEC did not cause any further change in basal spacing, and these results indicate no change in the arrangement of intercalated species in the interlayers, presumably owing to the deposition of the excess intercalate crystals outside the clay layers (Fig. 2.21e,f). However, the larger basal spacing of 2.47 nm observed upon further intercalation of AnF or AnCl beyond the CEC reveals the excess intercalates introduced into the interlayer spaces as neutral molecules together with counteranions over the CEC. With an increase in the density of the intercalated species in the confined clay regions, their bulkiness and the electrostatic repulsion between the aromatic planes and the negatively charged clay surfaces most likely lead to the expansion of the interlayer spacing. The neutral guest molecules further loaded would be expected to be inserted into the void spaces between anilinium cations that are tightly attracted to the upper and lower silicate surfaces by electrostatic interactions. The excess neutral molecules intercalated are sandwiched between anilinium cations by van der Waals interactions, and consequently intercalated species most probably form a rigidly packed trimolecular layer structure with their aromatic planes perpendicular to the clay surface (Fig. 2.21d). In the case of AnF– and AnCl–MMT intercalation compounds, π–π interactions between phenyl rings of intercalated species may contribute to the arrangement of the intercalated species in the confined clay regions. Also, the intercalates that organized as a trimolecular layer in the interlayer region of MMT when using acetonitrile and nitromethane as the guest molecules has been reported (MacEwan, 1948). In addition, π–π interactions are well known as attractive noncovalent bonding interactions (Hunter and Sanders, 1990).

The orientations of organic chains in clay minerals were initially deduced by Lagaly and Weiss (1969). Depending on the layer charge (interlayer cation density and packing density) of the alkylammonium ions of the clay mineral and the chain length of the organic ion, different arrangements of organic molecules between the layers can be formed. The geometry of the surface and the degree of exchange can also influence. Orientations of alkylammonium ions in the galleries of layered silicates are shown schematically in Fig. 2.22. As illustrated in Fig. 2.22, the organic ions may lie flat on the silicate surface as a monolayer or bilayer, or depending on the packing density and the chain length, an inclined paraffin-type structure, with the chains radiating away from the silicate surface, can be formed. In the pseudotrimolecular layers, some chain ends are shifted above one another, so the spacing is determined by the thickness of three alkyl chains (Bergaya et al., 2006; Lagaly, 1981, 1986; Lagaly and Weiss, 1969; Lagaly et al., 2013).

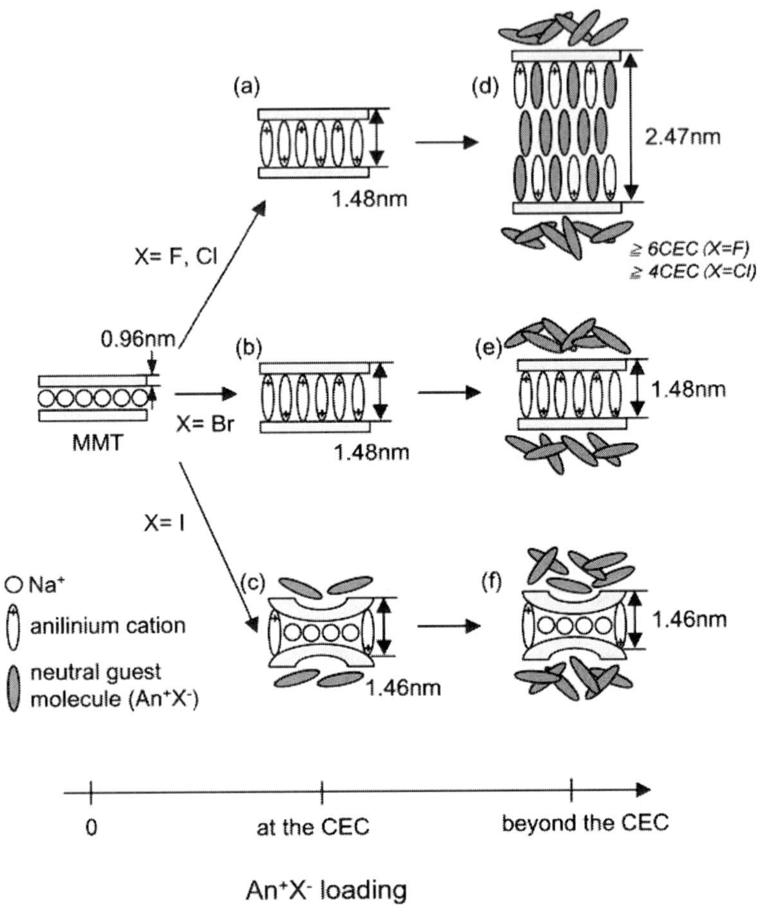

Figure 2.21 Proposed schematic illustration for the arrangements of the intercalated species in the clay interlayers. Reprinted from Yoshimoto et al., Copyright 2005, with permission from Elsevier.

The energies associated with the basal and lateral surfaces in clay layers can be nonuniform. If the long-chain cationic surfactants are used to complex with the clay surface, the nonuniformity in the distribution of layer thickness may be even more severe. The excessive surfactant may be located either in the vicinity of the clay surface or between the adsorbed organic layers.

The basal spacings are, therefore, relatively high (>2 nm) and depend on the amount of intercalated ion pairs and water.

Short-chain alkylammonium ions are lying flat on the surface in a monolayer. Alkylammonium ions that are too long to be packed in monolayers form bilayers. The critical chain length at the monolayer/bilayer transition is related to the layer charge density and provides the best method of layer charge determination of 2:1 clay minerals (Lagaly, 1981).

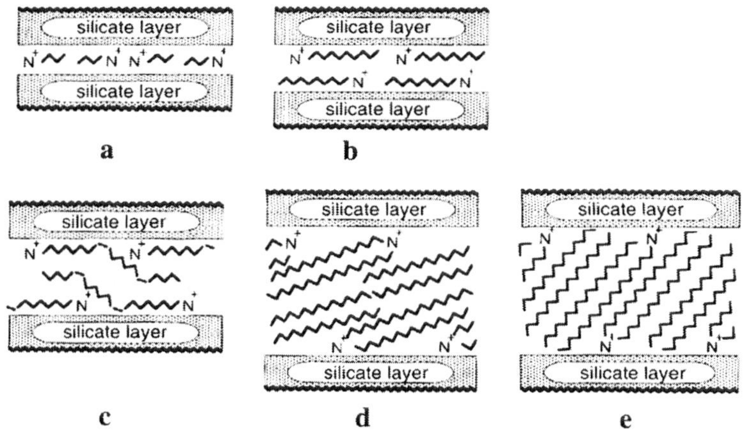

Figure 2.22 Orientations of alkylammonium ions in the galleries of layered silicates: (a) monolayer, (b) bilayers, (c) pseudotrimolecular layers, and (d, e) paraffin-type arrangements of alkylammonium ions with different tilting angles of the alkyl chains. Reprinted from Lagaly et al., Copyright 2013, with permission from Elsevier.

In the case of highly charged smectites (layer charge >0.4 eq/formula unit) or vermiculites, longer alkylammonium ions assume the pseudotrimolecular arrangement. The ammonium groups are in direct contact with the silicate layer so that the structure is bimolecular in regard to the ammonium groups. The alkyl chains, however, assume a trimolecular arrangement by conformational changes from all-*trans*-sequences into kink conformations. Higher layer charges (more highly charged vermiculites, micas) force the alkylammonium ions into the paraffin-type arrangement. Many of the nonsilicatic layered compounds such as several phosphates, arsenates, niobates, and titanates show layer charge densities comparable to vermiculites or even micas, and the alkylammonium ions are often arranged in paraffin-type monolayers or bilayers

(Lagaly and Beneke, 1991). In some cases, the observed basal spacings could not be explained by interlayer alkyl chains in the all-*trans*-conformation, and formation of gauche blocks was assumed.

Quaternary alkylammonium ions such as trimethylammonium ions form structures similar to the primary alkylammonium ions (Kuwaharada et al., 2002; Lagaly et al., 1982; Slade and Gates, 2004). The bulkier dialkyl dimethylammonium or, in industrial applications, ditalloyl dimethylammonium ions arrange in paraffin-type structures. Rotational isomerization of a *trans*-C–C bond near the nitrogen end group into the gauche conformation orients both alkyl chains parallel, which allows a denser packing in mono- or bimolecular films (Favre and Lagaly, 1991; Lagaly and Dèkany, 2005; Schwieger and Lagaly, 2004).

The formation of monolayers, bilayers, and pseudotrimolecular layers of alkylammonium ions in the interlayer spaces of MMTs has been schematized in Fig. 2.23 (de Paiva et al., 2008).

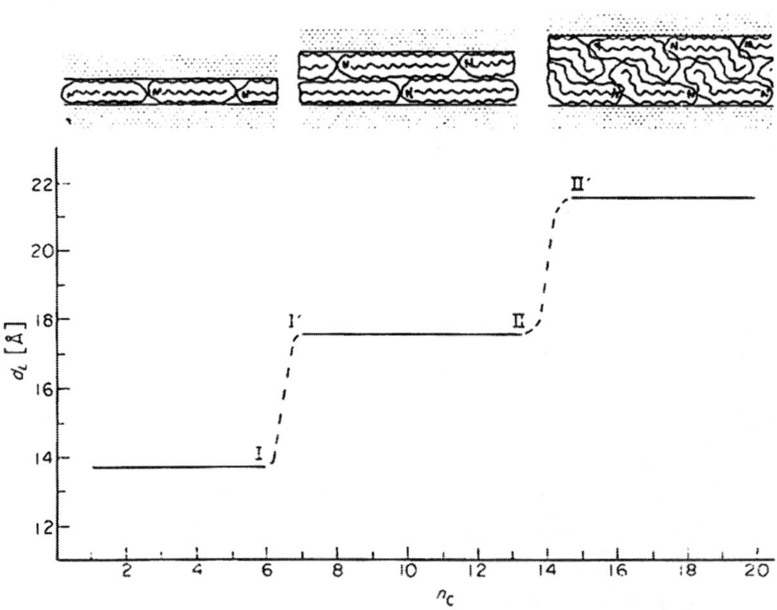

Figure 2.23 Variation of basal spacings of alkylammonium MMTs with chain length (n_C) due to the formation of monolayers, bilayers, and pseudotrimolecular layers of alkylammonium ions in the interlayer spaces. Reprinted from de Paiva et al., Copyright 2008, with permission from Elsevier.

The monolayer structures are characterized by basal spacings of about 1.35 nm. Formation of monolayer structure is impossible if A_c, the area required for the flat-lying cations with *n* carbon atoms in the chain, becomes larger than the equivalent area A_e, the area available to each univalent cation in a monolayer between the two silicate layers. If $A_c > A_e$, the ends of the alkyl chains would protrude beyond the monolayer. Trespassing the limit $A_c = A_e$, bilayers of flat-lying alkylammonium ions are formed with basal spacings of about 1.75 nm. Bilayer structures become unstable if $A_c > 2A_e$. Since three layers of alkylammonium ions are energetically unstable due to the electrical repulsion, the alkyl chains form paraffin-type structures. In the paraffin-type structures, the angles between the chain axes and the silicate layer increase with the layer charge. Therefore, the mean increase in basal spacing with increasing chain length of the cation varies with the charge density (Lagaly and Weiss, 1969). In some cases, the alkyl chains form paraffin-type structures rather than pseudotrimolecular layers. This occurs because the paraffin-type aggregation allows better fit of the ammonium groups to the surface oxygen atoms than it is the case for close-packed chains in pseudotrimolecular layers (Lagaly, 1986a). The microstructure of alkyl chains might be also described in terms of crystalline and amorphous regions like the two-phase model of a polymer, where the packing of chains in amorphous regions is about 10–20% looser than in the crystalline regions (Lagaly et al., 1976). Vaia et al. (1994) have proposed a more realistic description based on FTIR and XRD experiments to probe the interlayer structure and the phase structure of alkylammonium. By monitoring the frequency shifts of the CH_2 stretching and scissoring vibrations as a function of interlayer packing density, chain length, and temperature, they have found that the intercalated molecules exist in states with different degrees of order. The molecular environments vary from solid-like to liquid-like. The presence of a disordered conformation (gauche conformation), liquid-like structure, is more ubiquitous, while an ordered conformation (all-*trans*-conformation) is present only under special circumstances. In general, as the interlayer packing density or the chain length decreases or the temperature increases, the intercalated chains adopt a more disordered liquid-like structure resulting from an increase in the gauche/*trans*-conformer ratio. When the available surface area/molecule is within a certain range, the chains are not completely disordered, but retain some orientational order similar to that in a liquid–crystalline state. Since the chains are free to move around and reorient in the liquid-like state, they can reach equilibrium much faster compared to those in a solid-like state. As the number of carbon atoms increases, the chains adopt a more ordered structure. Hackett et al. (1998) used molecular dynamics simulations to study the orientations of the alkyl chains in organically modified clay mineral. These properties include density profiles, normal forces, chain, and *trans*-gauche conformer ratios and dynamics. The stable arrangement is a monolayer with a basal spacing of 1.32 nm, while bilayer and trilayer correspond to basal spacings of 1.80 nm and 2.27 nm, respectively. The simulations showed a disordered liquid-like arrangement of chains that do not stay flat within a single layer, but instead intertwine with chains in opposite layers. In the bilayer, almost half of the CH_2 groups are in the layer opposite the ammonium, while in the trilayer, CH_2 groups are expected to jump to the middle layer, but relatively few will be found in the layer opposite to the chain end. As density increases, the chain length increases at constant basal spacing, and structures with a minimum deviation of the basal spacing from the monolayer or bilayer are formed. The bulk alkanes form structures with a wide distribution of gallery heights, giving rise to very broad XRD peaks, although the most probable basal spacing is very close to that of a monolayer or bilayer. A way to describe the surfactant chain arrangement is through the number of gauche/*trans*-conformers. The percentage of *trans*-conformers increases sharply near the maximum chain length before the system jumps from a monolayer to a bilayer, because at that point, the surfactant molecules favor all-*trans*-conformations to pack most efficiently and accommodate the highest number of CH_2 and CH_3 groups within the monolayer. Hydrogen bonding of organic molecules to clay minerals can be important, especially for molecules containing OH, NH_2, and NH_3 groups (de Paiva et al., 2008; Moore and Reynolds, 1997).

An important energetic contribution in the adsorption process is related to the reorientation of the alkyl chains. In the more highly charged vermiculites, the chains are in paraffin-type arrangement, that is, they radiate away from the surface and move from the tilted position into the perpendicular orientation. In MMT, the starting situation is different. The alkyl chains are in close contact with the surface oxygen atoms, and rearrangement into the tilted or perpendicular orientation requires a considerable amount of energy, which increases with the chain length.

The energy of reorientation of the alkyl chains is about 5.7 kJ/mol $-CH_2-$ for $n_c = 6–10$, that is, when the chains form monolayers and are strongly fixed between the silicate layers. This energy is reduced to 3.8 kJ/mol $-CH_2-$ when the alkyl chains ($n_c > 10$) form bilayers. Because of the larger distance between the alkyl chains, these values are distinctly smaller than the increment of the heats of sublimation of long-chain compounds (about 9 kJ/mol $-CH_2-$). The paradox that HDTMA smectites adsorb higher amounts of toluene when some ion pairs (HDTMA ions + bromide ions) are intercalated is related to the alkyl chain–surface interaction. Intercalation of some ion pairs increases the alkyl chain density in the interlayer space and causes the chains to move into upright positions so that the energy required for the chain reorientation during toluene adsorption is considerably reduced. As a consequence of the interaction of the CH_2 groups with the surface, the exothermic enthalpy of immersion decreases with increasing chain length and can even change to endothermic. Entropy changes are, therefore, decisive. The adsorption of the liquid molecules on the alkyl chains and the silicate surface and their organization in the cavities are accompanied by a loss of entropy in comparison to the bulk liquid. A possible consequence is that only small amounts of liquid are adsorbed, and the layer separation increases only slightly or remains unchanged. This is observed for alkylammonium MMTs in water. In contrast, the stronger interaction of the organic liquid molecules with the alkyl chains causes the chains to move into the upright position so that the enhanced conformational entropy of the liquid molecules and especially of the alkyl chains can compensate the entropy loss of the adsorbed liquid molecules. This effect is strengthened in binary liquid adsorption. Besides the liquid molecules that are strongly organized around the alkyl chains and on the bar silicate surface (composing the adsorption phase), an additional amount of liquid is taken up. These liquid molecules do not belong to the adsorption phase (i.e., they have bulk-like properties) and diminish the entropy loss of the system because they increase the conformational freedom of the alkyl chains (which are probably not in the all-*trans*-conformation) and also the mobility of the liquid molecules. The entropy loss due to the organization of the water molecules near the alkyl chains and the surface (hydrophobic interaction) cannot be compensated by the two processes, and only some water molecules are taken up by the micropores between the flat-lying alkyl chains. The strong dependence of swelling in water on the layer charge is, therefore, related to the arrangement of the alkyl chains in the dried clay minerals. A considerable increase in the basal spacing occurs only if the energy for the rearrangement of the alkyl chains is low, that is, when the alkyl chains radiate away from the surface (vermiculite) and do not lie flat in tight contact with the surface oxygen atoms as in MMT (Lagaly and Dékany, 2005).

The quaternary alkylammonium salts are cationic surfactants and the most used organic compounds to prepare organoclays. A further advantage is that the large amount of organic material (30–40%) reduces the density of the dispersed particles (Favre and Lagaly, 1991).

Currently, there is a significant amount of research on modification of clay minerals with several kinds of quaternary alkylammonium salts (bromides or chlorides) in laboratory scale. In the case of organoclays for polymer nanocomposites, the kind of quaternary alkylammonium salt influences the affinity between the clay mineral and the polymer (Perrin-Sarazin et al., 2005). For apolar polymers such as polypropylene and polyethylene, the clay minerals are modified with dialkyl dimethylammonium halides, while for polar polymers such as polyamide, the clay minerals are modified with alkyl benzyl dimethylammonium halides or alkyl hydroxyethylammonium halides. Modifications of clay minerals also have been performed with quaternary alkylammonium salts associated with other organic compounds (Sibold et al., 2007).

Liu and Wu (2001) developed a new class of organoclay, which had a larger basal spacing than the organoclays only modified by alkylammonium. First, sodium MMT was modified with HDTMA bromide, resulting in an organoclay with a basal spacing of 1.96 nm. Then, this organoclay was co-intercalated with epoxypropyl methacrylate, increasing the basal spacing to 2.98 nm. Zhang et al. (2004a, 2004b) intercalated octadecylammonium ions and maleic anhydride in MMT. This organoclay was also used to prepare polypropylene nanocomposites.

Kwolek et al. (2003) employed a homologue series of alkyl benzyl dimethylammonium bromides, $[C_6H_5CH_2N(CH_3)_2R]Br$, to obtain OMMT. According to the basal spacings, the short chains are suggested to lie flatly on the surface of MMT, whereas the longer chains form a coiled system of hydrocarbon chains. Tang et al. (2003) also prepared an organophilic MMT with octadecylammonium and HDTMA salts in aqueous solution. Yilmaz and Yapar (2004) modified bentonite with tetradecyl trimethylammonium bromide and HDTMA bromide in amounts equivalent to 25%, 50%, and 100% of CEC of the clay mineral. The increase in the basal spacings characterized a bilayer arrangement, for the two salts at 100% of the CEC, and a monolayer arrangement for tetradecyl trimethylammonium bromide at 25% of the CEC. In all the other cases, random interstratifications of monolayer and bilayer structures were observed.

Lee and Lee (2004) treated bentonite with 1-hexadecylamine, 1-octadecylamine, and cetyl trimethylammonium and octadecyl triethylammonium bromides. All organoclays were delaminated in an epoxy matrix.

Kozak and Domba (2004) used a group of (alkyloxymethyl) dimethyl dodecylammonium chlorides with the general formula $[C_{12}H_{25}N^+(CH_3)_2CH_2OR]Cl^-$ to modify sodium MMT. Someya and Shibata (2004), with the objective of preparing new nanocomposites based on vinylester resin, modified MMTs with octadecylammonium (ODA), bis(2-hydroxyethyl) laurylammonium (BHL-M), diethyl [2(methacryloyloxyl)ethyl] ammonium (DEM-M), and bis(2-hydroxyethyl) lauryl (vinylbenzyl) ammonium chlorides (BHLV-M), respectively.

The basal spacings of the OMMT prepared with the intercalation of some quaternary ammonium and quaternary phosphonium cations are shown in Table 2.4 (de Paiva et al., 2008).

Investigation of the electrokinetic behavior of organoclay particles provides some new insights into the conformation and dynamic properties of the intercalated alkylammonium ions and the conformation and phase state of organic surfactant in MMT interlayer. Gürses et al. (2014) have examined in detail the interactions between clay platelets and CTA^+ ions in aqueous dispersions of Na-MMT by using electrokinetic parameters such as electrical conductivity and zeta potential and by considering XRD patterns and *d*-spacing values of the organoclay samples as a function of both surfactant concentration (CTAB) and clay amount.

Table 2.4 Basal spacings of the OMMT prepared with the intercalation of some quaternary ammonium and quaternary phosphonium cations

Quaternary phosphonium salt	*d* (nm)
Tetrabutyl phosphonium bromide	1.40
Tributyl hexadecylphosphonium bromide	2.32
Tributyl tetradecylphosphonium chloride	2.19
Tetraphenyl phosphonium bromide	1.76
Methyl triphenylphosphonium bromide	1.76
Ethyl triphenylphosphonium bromide	1.76
Propyl triphenylphosphonium bromide	1.76
(Butoxymethyl)dodecyl dimethylammonium chloride	1.45
Dodecyl dimethyl(hexyloxymethyl) ammonium chloride	1.47
Dodecyl dimethyl(octyloxymethyl) ammonium chloride	1.49
(Decyloxymethyl)dodecyl dimethylammonium chloride	1.53
Dodecyl dimethyl(dodecyloxymethyl) ammonium chloride	1.56
(Benzyloxymethyl)dodecyl dimethylammonium chloride	1.41
Octadecylammonium chloride	2.18
Bis(2-hydroxyethyl)laurylammonium chloride	1.81
Diethyl[2(methacryloyloxyl)ethyl] ammonium chloride	1.36
Bis(2hydroxyethyl)lauryl(vinylbenzyl) ammonium chloride	2.37

Source: Reproduced by updating from de Paiva et al., Copyright 2008, with permission from Elsevier.

They found that the curves obtained for the equilibrium adsorption isotherms at the other three solid/liquid ratios, except for 0.3 g/100 mL, exhibit two plateaus (Fig. 2.24).

This was attributed to the change in the mechanism of adsorption of CTA^+ ions onto the interlayer region of clay with increasing equilibrium concentration of CTAB. On the other hand, Fig. 2.25 reveals the variation of zeta potentials of clay particles and the electrical conductivities in aqueous dispersions with CTAB initial concentration as a function of the solid/liquid ratio. Also, to calculate the d-spacing of CTA^+ ion-adsorbed clay platelets, XRD diffractograms for all samples have been used. The variation of zeta potentials and d-spacing values with CTAB initial concentration as a function of the solid/liquid ratio is shown in Fig. 2.26. From Fig. 2.25 it can be seen that zeta potentials of the particles increased with the increase of CTAB initial concentration, parallel to the increase of electrical conductivities in aqueous dispersions. At higher CTAB concentrations, the higher positive zeta potential values of particles and a broad variation at their standard deviations were observed. It is remarkable that at the first three solid/liquid ratios, a **stress region** to be related with hydrophobic binding appeared.

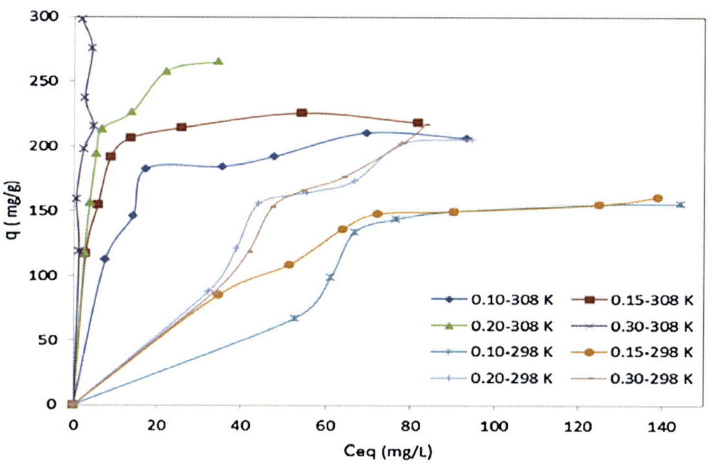

Figure 2.24 Variation of the amount adsorbed (q) versus the equilibrium CTAB concentration (C_{eq}). Reprinted from Gürses et al., Copyright 2014, with permission from Elsevier.

Figure 2.26 clearly shows that the variation of the zeta potential values of clay particles and the calculated d-spacing values with increased initial CTAB concentration exhibit a parallel trend.

The adsorption mechanisms that predominate at certain ranges of CTAB concentration have been schematized by considering the d-spacing values and the zeta potential values of particles in Scheme 2.2. This scheme represents that the absorption of CTA^+ ions onto the clay surface occurs via ion exchange, ion pairing, hemi-micelle formation, or cooperative adsorption mechanisms up to the zeta potential value of 0 mV (plateau 1). The electrical conductivities

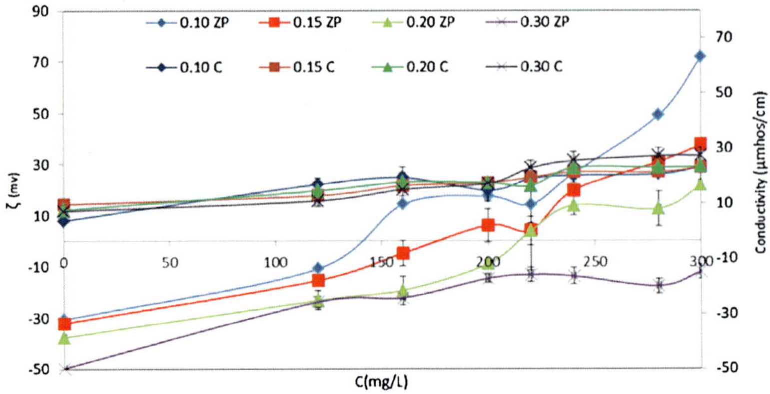

Figure 2.25 The variation of the zeta potentials and electrical conductivity values versus CTAB initial concentration as a function of the solid/liquid ratio. Reprinted from Gürses et al., Copyright 2014, with permission from Elsevier.

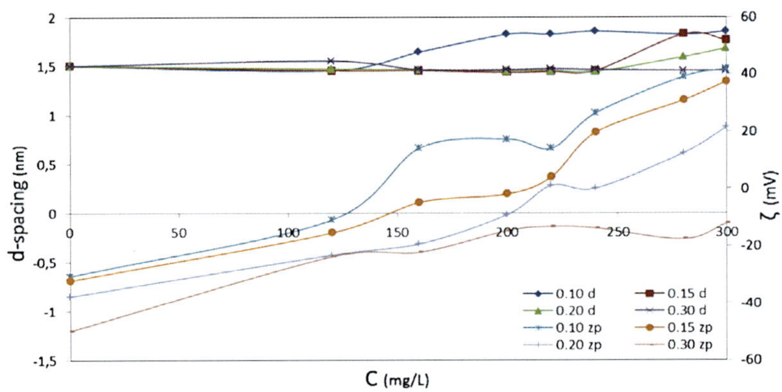

Figure 2.26 The variation of the zeta potentials and *d*-spacing values with CTAB initial concentration as a function of the solid/liquid ratio. Reprinted from Gürses et al., Copyright 2014, with permission from Elsevier.

measured in aqeous dispersions also confirm the mechanisms proposed. The much higher positive zeta potential (plateau 2) implies a sharp mechanism change that corresponds to the hydrophobic bonding mechanism. Thus, this region that is observed during the

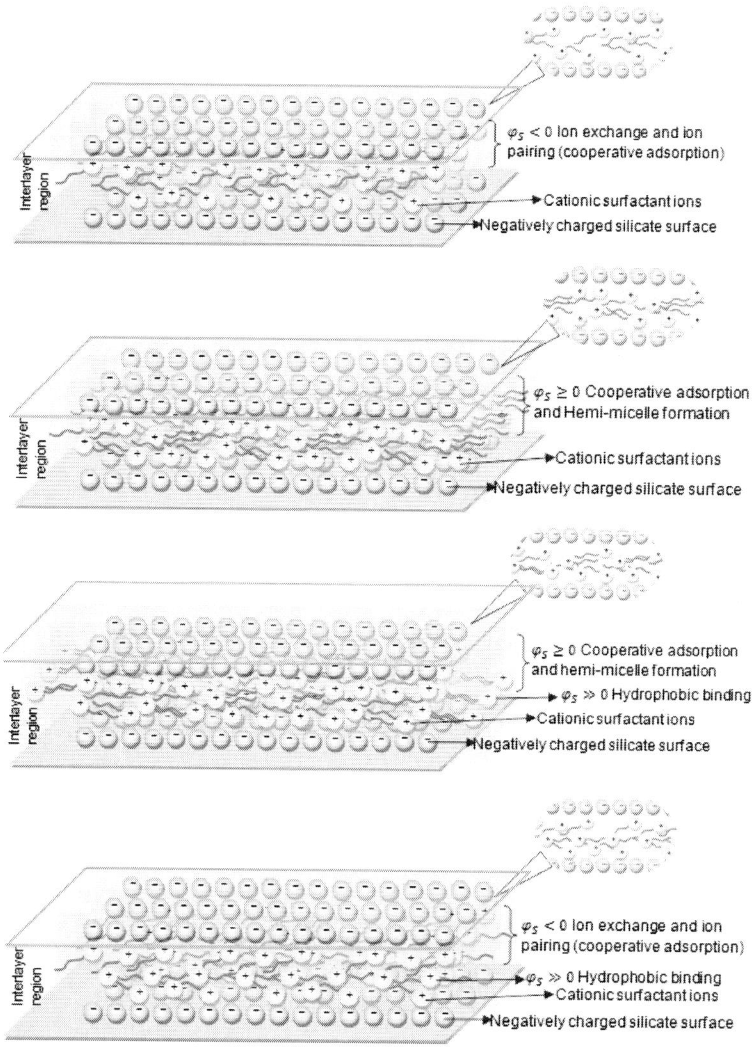

Scheme 2.2 Schematic presentation of the orientation and stacking of CTA^+ ions in interlayer region of the clay.

transition from plateau 1 to plateau 2 may be described as a stress region. Such a transition can be attributed to the adsorption of CTA^+ ions in a flat head–tail orientation as parallel to the decreased interfacial tension because the zeta potential values in this range are close to 0 mV. In this intermediate region, the adsorption of CTA^+ ions virtually depends on interfacial charge neutralization. Therefore, the cohesive interactions, which dominantly occur between the tails of the CTA^+ ions in the stress region, cause the smaller interfacial energy and the larger ionic strength.

It has been suggested that hydrophobic bonding predominates with the combined effect of the tendency of CTA^+ ions to escape from the aqueous phase as a result of the tendency of lowering interfacial energy to the interlayer region of the clay platelets and the increased ionic strength due to the increase of concentrations of CTA^+ ions and other exchangeable ions at the higher CTAB initial concentrations. Accordingly, it is reported that adsorption is initially driven by electrostatic attraction between the cationic surfactant and SiO^- groups on the surface up to the isoelectric point of the interface. The adsorbed molecules are in a single layer and may or may not assemble into hemi-micelles, depending on factors such as surface charge density and electrolyte concentration. At a higher surfactant concentration, hydrophobic interactions lead to the adsorption of additional surfactant molecules oriented preferentially with their headgroups away from the surface (Tyrode et al., 2008). Various types of organic onium cations were used to prepare organoclays. These include the tetramethylammonium cation (TMA^+) and its long-chain alkyltrimethylammonium homologues, trimethylphenylammonium cation ($TMPA^+$), benzyltrimethylammonium cation ($BTMA^+$) and its homologues, and alkylpyridinium cations (Barrer and MacLead, 1955; Brixie and Boyd, 1994; El-Nahhal et al., 2000; Jaynes and Vance, 1999; Kwolek et al., 2003; Nir et al., 1991; Sharmasakar et al., 2000; Zhong Zhang and Sparks, 1993).

It is generally accepted that adsorption of hydrophilic, short-chain quaternary ammonium cations on smectites occurs according to the ion exchange mechanism. The extent of adsorption of such cations cannot thus exceed the critical exchange capacity of the mineral. Furthermore, an increase in ionic strength of the aqueous phase in equilibrium with the organoclay leads to an increase in the adsorption of the organic cations on its surface. Thus, especially their long-chain homologues may exhibit the adsorption to an extent, which increases with growth of the alkyl chain and approaches the doubled value of the critical exchange capacity (Bonczek et al., 2002). Van der Waals hydrophobic interactions are suggested to operate in such cases and lead to a bilayer of interdigitated alkyl chains with a positive charge exposed toward the bulk of the solution. This process is accompanied by a reversal in the charge of the surface and, in consequence, by lowering of the rate of increase in counterion activity with the growth of total concentration of the surfactant in the bulk of the solution (Tahani et al., 1999). Such a mechanism of adsorption of long-chain ammonium salts has been also postulated for silica surfaces (Kwolek et al., 2003).

Weimer et al. (1999) modified an Na-MMT by anchoring an ammonium cation earing a nitroxide moiety known for its ability to mediate the controlled/"living" free-radical polymerization of styrene. The MMT modification with nitroxyl-based organic cations has been schematized in Fig. 2.27.

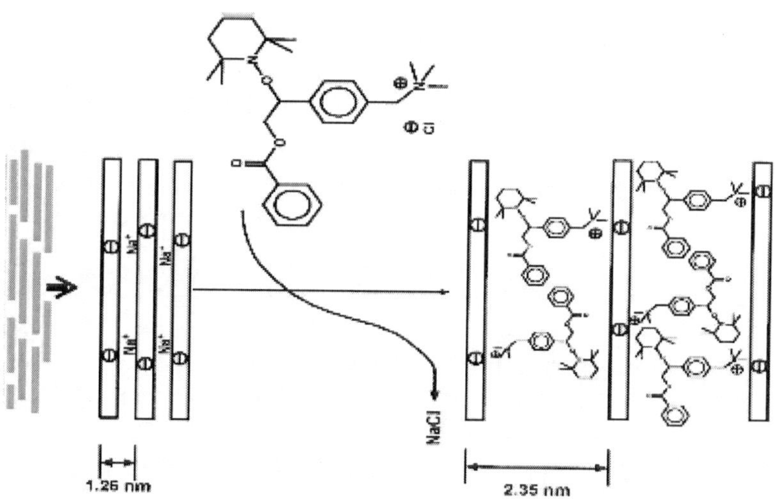

Figure 2.27 Schematic representation of the MMT modification with nitroxyl-based organic cation. Reproduced from Weimer et al., Copyright 1999, with permission from American Chemical Society.

Vaia et al. (1995) have also studied the kinetics of melt intercalation by following the time evolution of XRD diffraction patterns for statically annealed polystyrene/octadecyl ammonium-exchanged fluorohectorite. The change in intensity of the pristine and intercalated diffraction peaks with time was used as a reflection of the kinetics of the polymer intercalation process. The authors found that the nominal silicate particle (agglomerate of ca. 175 μm in diameter) consists of smaller oblong-shaped particles, coined as primary particles forming agglomerates. These primary particles of 1.0–10.0 μm length themselves consist of a compact face-to-face stacking or low-angle intergrowth of individual silicate crystallites (also known as tactoids). These crystallites are built up as a coherent stacking of individual silicate layers. The layers are roughly circular, 0.05–0.50 μm in diameter and ca. 1 nm thick. They are separated by a van der Waals interlayer (or gallery), which contains the alkylammonium cations (≈10 Å thick) in the pristine organosilicate. This overall morphology can be schematized as shown in Fig. 2.28. On the basis of this morphology and TEM observations of partially intercalated composites, it has been demonstrated that the accessibility of the interlayer to the polymer chains depends on the location and orientation of the primary particles within the agglomerates and on the location and orientation of the crystallites within the primary particles, meaning that crystallites near the edge will be more accessible to polymer chains than those near the center. Since the silicate layers are impenetrable, the polymer must enter the gallery from the edges of the crystallites. The authors observed that under their experimental conditions, the polymer penetrates the agglomerate and surrounds the primary particles before the occurrence of substantial intercalation.

The interaction between smectite and tetracycline (TC) in aqueous systems has been examined in detail by Li et al. (2010). They found that TC adsorption onto smectite was a relatively fast process, even though most of the adsorption sites were in the interlayer position involved in intercalation, as confirmed by the expansion of *d* (001) spacing. Intercalation of TC with twisted conformation into the interlayer of smectite with a tilted angle of 45° and with extended conformation has been shown schematically in Fig. 2.29.

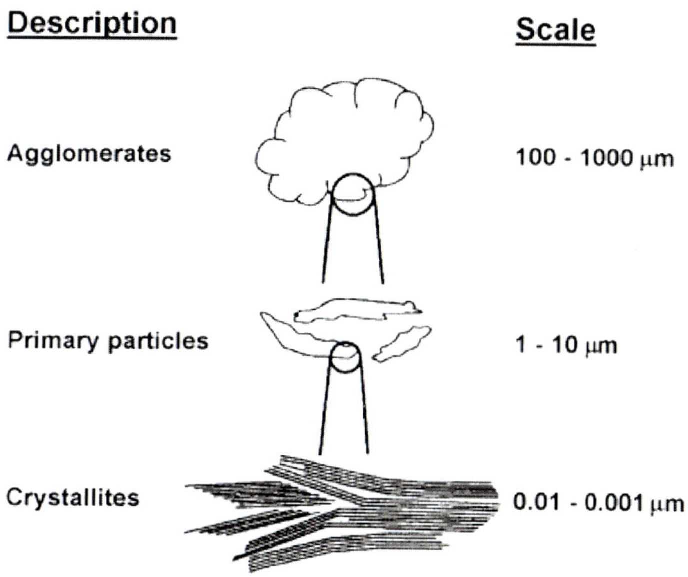

Figure 2.28 Schematic representation of the morphology of organomodified fluorohectorite. Reproduced from Vaia et al., Copyright 1995, with permission from American Chemical Society.

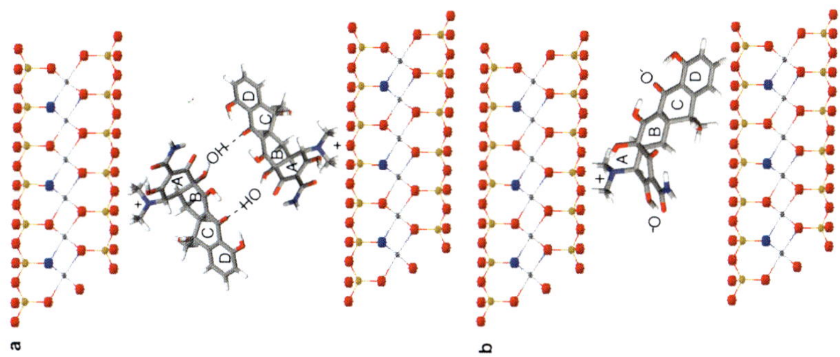

Figure 2.29 Intercalation of TC with twisted conformation into the interlayer of smectite with a tilted angle of 45° (a) and with extended conformation (b). Reproduced from Li et al., Copyright 2010, with permission from Elsevier.

Lee et al. (2005) proposed the ring-opening polymerization of ethylene terephthalate cyclic (ETC) oligomers with organically modified MMT as an alternative approach to the preparation of poly(ethylene terephthalate) (PET)-based nanocomposites.

Intercalation by the ring-opening reaction of cyclic oligomers in between silicate layers has been schemated in Fig. 2.30.

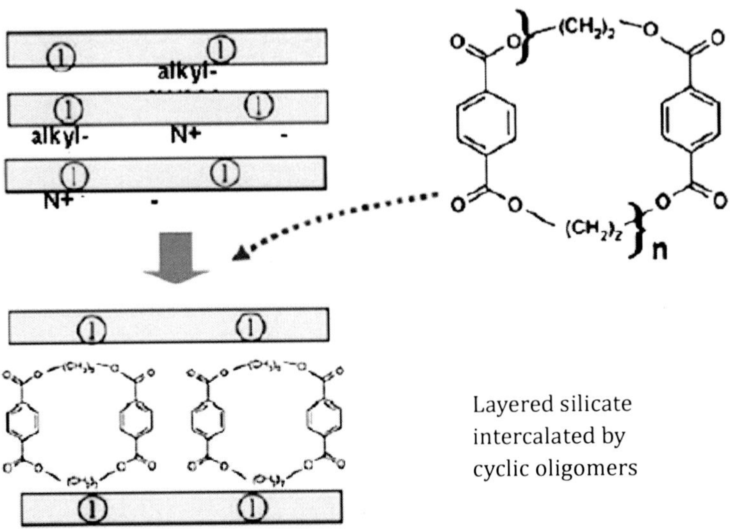

Figure 2.30 Schematic representation of nanocomposite formation by the ring-opening reaction of cyclic oligomers in between silicate layers. Reproduced from Lee et al., Copyright 2005, with permission from Elsevier.

Concerning the conformation of surfactant chains, it is generally accepted that at low interlayer packing densities of the organic modifier, the chains adopt a disordered monolayer arrangement. As the packing density increases, the chains adopt more extended conformations (and thus larger initial gallery heights), ultimately resulting in a solid-like paraffinic arrangement of the chains.

According to Vaia and Giannelis (1997), the optimal structure appears to exhibit a chain arrangement slightly greater than a pseudobilayer. That is, there is an optimum interlayer structure favoring hybrid formation that is intermediate between a disordered monolayer and a solid-like paraffinic arrangement of aliphatic chains (Pavlidou and Papaspyrides, 2008).

In immiscible systems, which typically correspond to the more conventionally filled polymers, the poor physical interaction between the organic and inorganic components leads to poor mechanical and thermal properties. In contrast, strong interactions between the polymer and the layered silicate in polymer-layered silicate (PLS) nanocomposites lead to the organic and inorganic phases being dispersed at the nanometer level (Pollet et al., 2006; Xu et al., 2003).

Alkylammonium or alkylphosphonium cations in the organosilicates lower the surface energy of the inorganic host and improve the wetting characteristics of the polymer matrix and result in a larger interlayer spacing. Additionally, the alkylammonium or alkylphosphonium cations can provide functional groups that can react with the polymer matrix or in some cases initiate the polymerization of monomers to improve the strength of the interface between the inorganic and the polymer matrix (Ray and Okamoto, 2003).

Manias et al. (2001) demonstrated that the polypropylene–MMT nanocomposite formation can be achieved by two ways, that is, either by using neat polypropylene and semifluorinated surfactants for the silicates or by using functionalized polypropylenes and common OMMT. In the first case, a semifluorinated alkyltrichlorosilane (CF_3–$(CF_2)_5$–$(CH_2)_2$–Si–Cl_3) was used to modify MMT, rendering it miscible with neat or unfunctionalized polypropylene, PP.

Wang et al. (2000) used NMR to probe the conformation of alkylammonium surfactant molecules in MMT. They demonstrated the coexistence of order and disordered conformations. Two main resonance peaks are resolved and associated with the backbone of alkyl chains. The resonance at 33 ppm corresponds to the ordered conformation (all-*trans*), and the resonance at 30 ppm corresponds to the disordered conformation (mixture of *trans*- and gauche).

Osman et al. (2004) used IR, NMR XRD, and DSC to study the structure and chain dynamics of mono-, di-, tri-, and tetra-alkylammonium cations of varying lengths (C4, C8, and C18) on MMT platelets. At ambient temperatures, alkylammonium monolayers assembled on MMT adopt a two-dimensional order or a disordered state, depending on the cross-sectional area of the molecules, the area covered by the cation available on the substrate, and the alkyl length. At low temperatures, the alkyl chains preferentially are in all-*trans* conformation. Conformation transformation of the chain takes place with increasing temperature, leading to a disordered phase (liquid-like) in which the chains are in a random conformation. The XRD measurements also showed that the (001) basal spacing of MMT depends upon the number of alkyl chains with different lengths. At a chain length of four carbon atoms (C4), there is nearly no change in the basal-plane spacing of MMT. In the octyl series (C8–4C8), the *d*-spacing tends to increase linearly with increasing number of chains. In the octadecyl series (C18–4C18), the *d*-spacing increases markedly and nonlinearly.

The complex structure of MMT-based organoclays can be partially attributed to the pronounced structural and energetic heterogeneity in pristine MMT clays. On the basis of calorimetric data and adsorption measurements, it has been shown that clay minerals always contain several types of adsorption sites with different interaction energies (Tjong, 2006).

It is known that the variations in the interlayer distances of clay minerals depend upon the nature of the molecules intercalated and their microscopic orientation. Also, the maximum limit of intercalation level depends on the morphologic and chemical characteristics of clay minerals (Yui et al., 2002).

Figure 2.31 shows the morphologies (SEM images) of MMTs (SWy and SAz) modified with HDTMA cations and provides a basis for comparison between the two MMTs (Lee et al., 2005).

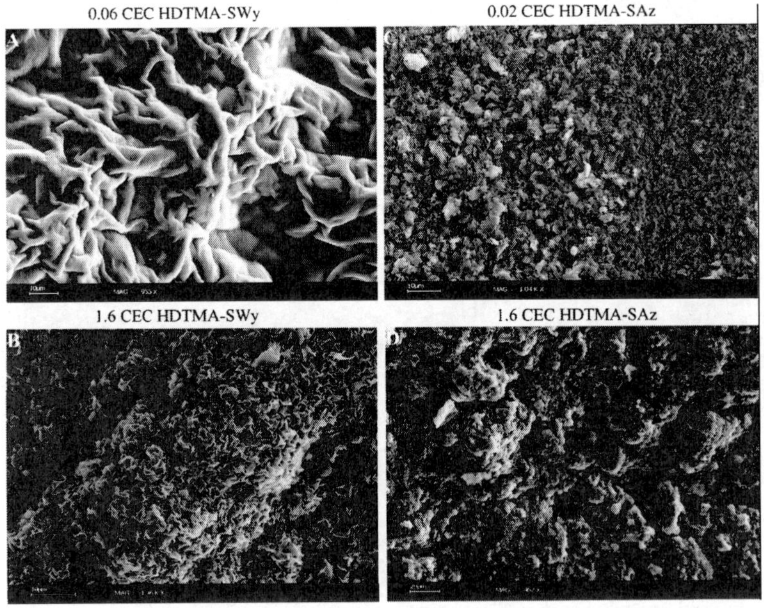

Figure 2.31 SEM images showing morphological changes in MMT particles with the increased HDTMA loading. Reprinted from Lee et al., Copyright 2005, with permission from Elsevier.

0.06 CEC HDTMA-SWy (upper left) exhibits an image consisting the particles adhered each other. In the case of 1.6 CEC HDTMA-SWy (lower left), the morphology progressively varies as a pattern resembling "corn flake". 0.02 CEC HDTMA-SAz (upper right) has grain-like structure, while 1.6 CEC HDTMA-SAz (lower right) shows a more different aggregation.

TEM lattice fringe images of HDTMA-MMTs (SWy and SAz) have been shown in Fig. 2.32 (Lee et al., 2005). Lattice fringes of raw SWy MMT were generally regular and straight (Fig. 2.32A). When HDTMA was added as much as the CEC, the lattice fringes increased up to ~20 Å with a few narrower layers (18 Å) (Fig. 2.32B). There was no significant change in the layer spacing, even at higher HDTMA loading (1.8 CEC) (Fig. 2.32C). 1.2 CEC HDTMA-SAz showed slightly wavy lattice fringes with variable layer spacings (24–27 Å) (Fig. 2.32E), unlike raw SAz (Fig. 2.32D). This reveals the increased interlayer

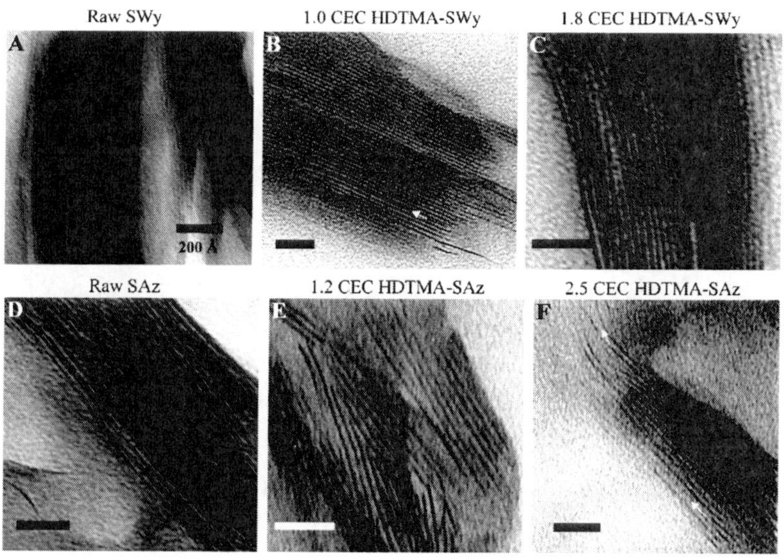

Figure 2.32 TEM lattice fringe images of HDTMA MMTs. (A) Raw SWy MMT (layer spacing ~14 Å). (B) HDTMA loading equivalent to the CEC (~20 Å), except some layers (arrow). (C) No significant increase of the layer spacing (~21 Å) at higher loading. (D) Raw SAz MMT (~15 Å). (E) Layer spacings of 24–27 Å with increasing HDTMA loading. (F) Some highly expanded layers (N40 Å arrows) at high HDTMA loading. Reprinted from Lee et al., Copyright 2005, with permission from Elsevier.

adsorption of HDTMA by SAz MMT; highly expanded layers (>40 Å) appeared at higher HDTMA loading (e.g., 2.5 CEC) (Fig. 2.32F). Considerable lack of a coherent interlayer spacing of HDTMA-SAz may indicate local variation in the density of interlayer charge. It is known that surface charge density is an important factor in determining the adsorbed amount and structure of surfactants onto surfaces (Lagaly, 1994; Manne and Gaub, 1995; Schulz and Warr, 2000). The layer charge distribution in the SWy MMT is heterogeneous, but after intercalation, the layer distances observed by TEM were almost constant because of the regular structure of the intercalated HDTMA. In contrast, the SAz MMT showed slightly different layer spacings with various amounts of alkylammonium ions adsorbed, revealing larger domains of variable charge densities (Lee et al., 2005). Besides, layer charge density, that is, the charge distribution per a unit layer, may be another important factor affecting the arrangement of adsorbed surfactant layers as well as the adsorption of organic compounds from aqueous solution (Li and Bowman, 1997; Sheng and Boyd, 2000).

References

Abollino, O., Giacomino, A., Malandrino, M., Mentasti, E. (2008). Interaction of metal ions with montmorillonite and vermiculite. *Appl. Clay Sci.*, **38**, pp. 227–236.

Ahmad, M. B., Hoidy, W. H., Ibrahim, N. A. B., Al-Mulla, E. A. J. (2009). Modification of montmorillonite by new surfactants. *J. Eng. Appl. Sci.*, **4**(3), pp. 184–188.

Akyüz, S., Akyüz, T. (2003). FT-IR spectroscopic investigations of surface and intercalated 2-aminopyrimidine adsorbed on sepiolite and montmorillonite from Anatolia. *J. Mol. Struct.*, **651–653**, pp. 205–210.

Alexandre, M., Dubois, P. (2000). Polymer-layered silicate nanocomposites: Preparation, properties and uses of a new class of materials. *Mater. Sci. Eng.*, **28**, pp. 1–63.

Annabi-Bergaya, F. (2008). Layered clay minerals: Basic research and innovative composite applications. *Microporous Mesoporous Mater.*, **107**, pp. 141–148.

Auerbach, S. M., Carrado, K. A., Dutta, P. K. (2004). *Handbook of Layered Materials.* USA: Marcel Dekker.

Awad, W. H., Gilman, J. W., Nyden, M., Harris Jr., R. H., Sutto, T. E., Callahan, J., Truvole, P. C., DeLong, H. C., Fox, D. M. (2004). Thermal degradation studies of alkyl-imidazolium salts and their application in nanocomposites. *Thermochim. Acta,* **409**, pp. 3–11.

Balazs, A. C., Singh, C., Zhulina, E. (1998). Modeling the interactions between polymers and clay surfaces through self-consistent field theory. *Macromolecules,* **31**, pp. 8370–8381.

Baldassari, S., Komarneni, S., Mariani, E., Villa, C. (2006). Microwave versus conventional preparation of organoclays from natural and synthetic clays. *Appl. Clay Sci.,* **31**, pp. 134–141.

Barrer, R. M., MacLeod, D. M. (1955). Activation of montmorillonite by ion exchange and sorption complexes of tetra-alkyl ammonium montmorillonites. *Trans. Faraday Soc.,* **51**, pp. 1290–1300.

Beall, G. W., Goss, M. (2004). Self-assembly of organic molecules on montmorillonite. *Appl. Clay Sci.,* **27**, pp. 179–186.

Beall, G. W., Tsipursky, S. J. (1999). Nanocomposites produced utilizing a novel ion–dipole clay surface modification in polymer–clay composites. In Al-Malaika, S., Golovoy, A., Wilkie, C.A. (eds.), *Chemistry and Technology of Polymer Additives*. Blackwell Science, pp. 266–280.

Beall, G. W., Tsipursky, S., Sorokin, A., Goldman, A. (1996). Process for manufacturing intercalates and exfoliates and composite materials containing same. U.S. Patent 5578672, 33 pp.

Bergaya, F., Lagaly, G. (2001). Surface modification of clay minerals. *Appl. Clay Sci.,* **19**, pp. 1–3.

Bergaya, F., Theng, B. K. G., Lagaly, G. (eds.) (2006). *Handbook of Clay Science.* Amsterdam, Netherlands: Elsevier.

Bonczek, J. L., Harris, W. G., Nkedi-Kizza, P. (2002). Monolayer to bilayer transitional arrangements of hexadecyltrimethylammonium cations on Na-montmorillonite. *Clays Clay Miner.,* **50**, pp. 11–17.

Bormans, P. (2003). *Ceramics Are More than Clay Alone: Raw Materials, Products, and Applications.* Cambridge, UK: Cambridge International Science.

Brixie, J. M., Boyd, S. A. (1994). Treatment of contaminated soils with organoclays to reduce leachable pentachlorophenol. *J. Environ. Qual.,* **23**, pp. 1283–1290.

Burmistr, M. V., Sukhyy, K. M., Shilov, V. V., Pissis, P., Spanoudaki, A., Sukha, I. V., Tomilo, V. I., Gomza, Y. P. (2005). Synthesis, structure, thermal and mechanical properties of nanocomposites based on linear polymers and layered silicates modified by polymeric quaternary ammonium salts (ionenes). *Polymer,* **46**, pp. 12226–12232.

Carmody, O., Frost, R., Xi, Y., Kokot, S. (2007). Surface characterisation of selected sorbent materials for common hydrocarbon fuels. *Surface Sci.*, **601**(9), pp. 2066–2076.

Carrado, K. A. (2000). Synthetic organo- and polymer–clays: Preparation, characterization, and materials applications. *Appl. Clay Sci.*, **17**, pp. 1–23.

Cases, J. M., Bèrend, I., Besson, G., François, M., Uriot, J. P., Thomas, F. and Poirier, J. E. (1992). Mechanism of adsorption and desorption of water vapor by homoionic montmorillonite. 1. The sodium-exchanged form. *Langmuir*, **8**, pp. 2730–2739.

Causserand, C., Jover, K., Aimar, P., Meireles, M. (1997). Modification of clay cake permeability by adsorption of protein. *J. Membr. Sci.*, **137**, pp. 31–44.

Cavani, F., Trifiro, F., Vaccari, A. (1991). Hydrotalcite-type anionic clays. Preparation, properties and applications. *Catal. Today*, **11**, pp. 173–301.

Chaiko, D. J. (2003). New poly (ethylene oxide)-clay composites. *Chem. Mater.*, **15**, pp. 1105–1110.

Chen, G. X., Yoon, J. S. (2005a). Clay functionalization and organization for delamination of the silicates tactoids in poly (L-lactide) matrix. *Macromol. Rapid Comm.*, **26**(11), pp. 899–904.

Chen, G. X., Yoon, J. S. (2005b). Thermal stability of poly (L-lactide)/poly (butylene succinate)/clay nanocomposites. *Polym. Degrad. Stab.*, **88**(2), pp. 206–212.

Chen, G. X., Yoon, J. S. (2005c). Nonisothermal crystallization kinetics of poly [(butilene succinate)-co-(butilene adipate)] (PBSA) and twice functionalized organoclay. *J. Polym. Sci. Pol. Phys.*, **43**(7), pp. 817–826.

Chigwada, G., Wang, D. D., Wilkie, C. A. (2006a). Polystyrene nanocomposites based on quinolinium and pyridinium surfactants. *Polym. Degrad. Stab.*, **91**, pp. 848–855.

Chigwada, G., Wang, D. D., Jiang, D. D., Wilkie, C. A. (2006b). Styrenic nanocomposites prepared using a novel biphenyl-containing modified clay. *Polym. Degrad. Stab.*, **91**, pp. 755–762.

Chiu, C. W., Chu, C. C., Cheng, W. T., Lin J. J. (2008). Exfoliation of smectite clays by branched polyamines consisting of multiple ionic sites. *Eur. Polym. J.*, **44**, pp. 628–636.

Choy, J. H., Choi, S. J., Oh, J. M., Park, T. (2007). Clay minerals and layered double hydroxides for novel biological applications. *Appl. Clay Sci.*, **36**, pp.122–132.

Chrissopoulou, K., Anastasiadis, V. (2011). Polyolefin/layered silicate nanocomposites with functional compatibilizers. *Eur. Polym. J.*, **47**, pp. 600–613.

Christidis, G. E. (2008). Validity of the structural formula method for layer charge determination of smectites: A re-evaluation of published data. *Appl. Clay Sci.*, **42**, pp. 1–7.

Cole, K. C. (2008). Use of infrared spectroscopy to characterize clay intercalation and exfoliation in polymer nanocomposites. *Macromolecules*, **41**, pp. 834–843.

Costa Filho, A. P., Gomes, A. S., Lucas, E. S. (2005). Preparação e caracterização de organobentonita modificada com ionenos alifáticos. *Polímeros: Ciência e Tecnologia*, **15**, pp. 212–217.

Darder, M., Colilla, M., Ruiz-Hitzky, E. (2005). Chitosan–clay nanocomposites: Application as electrochemical sensors. *Appl. Clay Sci.*, **28**, pp. 199–208.

de Paiva, L. B., Morales, A. R., Francisco, R., Díaz, V. (2008). Organoclays: Properties, preparation and applications. *Appl. Clay Sci.*, **42**, pp. 8–24.

Dekany, I., Farkas, A., Kiraly, Z., Klumpp, E., Narres, H. D. (1996). Interlamellar adsorption of 1-pentanol from aqueous solution on hydrophobic clay mineral. *Colloids Surf., A*, **119**, pp. 7–13.

Di Gianni, A., Amerio, E., Monticelli, O., Bongiovanni, R. (2008). Preparation of polymer/clay mineral nanocomposites via dispersion of silylated montmorillonite in a UV curable epoxy matrix. *Appl. Clay Sci.*, **42**, pp. 116–124.

Dultz, S., Riebe, B., Bunnenberg, C. (2005). Temperature effects on iodine adsorption on organo-clay minerals: II. Structural effects. *Appl. Clay Sci.*, **28**(1–4), pp. 17–30.

Ek, S., Liskola, E. I., Niinistö, L. (2004). Atomic layer deposition of aminofunctionalized silica surfaces using N-(2-Aminoethyl)-3-aminopropyltrimethoxysilane as a silylating agent. *J. Phys. Chem. B*, **108**(28), pp. 9650–9655.

El-Nahhal, Y., Nir, S., Margulies, L., Rubin, B. (1999). Reduction of photodegradation and volatilization of herbicides in organo-clay formulations. *Appl. Clay Sci.*, **14**, pp. 105–119.

El-Nahhal, Y., Nir, S., Serban, C., Rabinowitz, O., Rubin, B. (2001). Organoclay formulation of acetochlor for reduced movement in soil. *J. Agr. Food Chem.*, **49**, pp. 5364–5371.

Fang, C., Yu, R., Zhang, Y., Hu, J., Zhang, M., Mi, X. (2012). Combined modification of asphalt with polyethylene packaging waste and organophilic montmorillonite. *Polym. Testing*, **31**(2), pp. 276–281.

Favre, H., Lagaly, G. (1991). Organo-bentonites with quaternary alkylammonium ions, *Clay Miner.*, **26**, pp. 19–32.

Fischer, H. (2003). Polymer nanocomposites: From fundamental research to specific applications. *Mater. Sci. Eng. C*, **23**, pp. 763–772.

Galan, E. (1996). Properties and applications of palygorskite–sepiolite clays. *Clay Miner.*, **31**, pp. 443–453.

Gao, F. (2004). Clay/polymer composites: The story. *Mater. Today*, **7**(11), pp. 50–55.

Gelfer, M., Burger, C., Fadeev, A., Sics, I., Chu, B., Hsiao, B. S., Heintz, A., Kojo, K., Hsu, S-L., Si, M., Rafailovich, M. (2004). Thermally induced phase transitions and morphological changes in organoclays. *Langmuir*, **20**, pp. 3746–3758.

Gemeay, A. H., El-Sherbiny, A. S., Zaki, A. B. (2002). Adsorption and kinetic studies of the intercalation of some organic compounds onto NaC-montmorillonite. *J. Colloid Int. Sci.*, **245**, pp. 116–125.

Giannelis, E. P. (1996). Polymer layered silicate nanocomposites. *Adv. Mater.*, **8**, pp. 29–35.

Gopinath, S., Sugunan, S. (2007). Enzymes immobilized on montmorillonite K10: Effect of adsorption and grafting on the surface properties and the enzyme activity. *Appl. Clay Sci.*, **35**, pp. 67–75.

Grim, R. E. (1968). *Clay Mineralogy.* New York, USA: McGraw-Hill.

Gürses, A., Ejder-Korucu, M., Karaca, S. (2011). *Clay-organoclay and Organoclay/polymer Nanocomposites.* USA: Nova Science.

Gürses, A., Günes, K., Mindivan, F., Ejder-Korucu, M., Açıkyıldız, M., Doğar, C. (2014). The investigation of electrokinetic behaviour of micro-particles produced by CTA^+ ions and Na-montmorillonite. *Appl. Surf. Sci.*, **318**, pp. 76–84.

Gürses, A., Karaca, S., Açıkyıldız, M., Ejder-Korucu, M. (2009). Thermodynamics and mechanism of cetyltrimethylammonium adsorption onto clayey soil from aqueous solutions. *Chem. Eng. J.*, **147**, pp. 194–201.

Hackett, E., Manias, E., Giannelis, E. P. (1998). Molecular dynamics simulations of organically modified silicates. *J. Chem. Phys.*, **108**, pp. 7410–7415.

Hanley, H. J. M., Muzny, C. D., Butler, B. D. (1997). Surfactant adsorption on a clay mineral: Application of radiation scattering. *Langmuir*, **13**, pp. 5276–5282.

Hartman, H., Sposito, G., Yang, A., Manne, S., Gould, S. A. C., Hansma, P. K. (1990). Molecular-scale imaging of clay mineral surfaces with the atomic force microscope. *Clays Clay Miner.*, **38**, pp. 337–342.

He, H., Duchet, J., Galy, J., Gerard, J. F. (2005). Grafting of swelling clay materials with 3-aminopropyltriethoxysilane. *J. Colloid Interface Sci.*, **288**, pp.171–176.

He, H., Duchet, J., Galy, J., Gérard, J. F. (2006). Influence of cationic surfactant removal on the thermal stability of organoclays. *J. Colloid Inerface. Sci.*, **295**, pp. 202–208.

He, H., Frost, R. L., Bostrom, T., Yuan, P., Duong, L., Yang, D., Xi, Y., Kloprogge, J. T. (2006). Changes in the morphology of organoclays with $HDTMA^+$ surfactant loading. *Appl. Clay Sci.*, **31**(3–4), pp. 262–271.

He, H., Frost, R. L., Xi, Y. F., Zhu, J. X. (2004). Raman spectroscopic study of organo-montmorillonites. *J. Raman Spectrosc.*, **35**, pp. 316–323.

He, H., Ma, Y., Zhu, J., Yuan, P., Qing, Y. (2010). Organoclays prepared from montmorillonites with different cation exchange capacity and surfactant configuration. *Appl. Clay Sci.*, **48**, pp. 67–72.

Hedley, C. B., Yuan, G., Theng, B. K. G. (2007). Thermal analysis of montmorillonites modified with quaternary phosphonium and ammonium surfactants. *Appl. Clay Sci.*, **35**, pp. 180–188.

Henderson, G. S., Vrdoljak, G. A., Eby, R. K., Wicks, F. J., Rachlin, A. L. (1994). Atomic-force microscopy studies of layer silicate minerals. *Coll. Surf. A*, **87**, pp. 197–212.

Herrera, N. N., Letoffe, J. M., Putaux, J. M., David, L., Bourgeat-Lami, E. (2004). Aqueous dispersions of silane-functionalized laponite clay platelets. A first step toward the elaboration of water-based polymer/clay nanocomposites. *Langmuir*, **20**(5), pp. 1564–1571.

Herrera, N. N., Letoffe, J. M., Reymond, J.-P., Bourgeat-Lami, E. (2005). Silylation of laponite clay particles with monofunctional and trifunctional vinyl alkoxysilanes. *J. Mater. Chem.*, **15**, pp. 863–871.

Holeman, J. N. (1965). *U.S. Department of Agriculture Soil Conservation Service Engineering Division Technical Release No. 28 Geology.*

Huh, J. K., Song, D. I., Jeon, Y. W. (2000). Sorption of phenol and alkylphenols from aqueous solution onto organically modified montmorillonite and applications of dual-mode sorption model. *Sep. Purif. Technol.*, **35**, pp. 243–259.

Hunter, C. A., Sanders, J. K. M. (1990). The nature of pi-pi interactions. *J. Am. Chem. Soc.*, **112**, pp. 5525–5534.

Huskić, M., Žagar, E., Žigon, M., Brnardić, I., Macan, J., Ivanković, M. (2009). Modification of montmorillonite by cationic polyesters. *Appl. Clay Sci.*, **43**, pp. 420–424.

Işık, I. (2007). *Impact modified polyamide-organoclay nanocomposites.* Ph. D. Thesis. Middle East Technical University, Ankara, Turkey.

Isoda, K., Kuroda, K. (2000). Interlamellar grafting of γ-methacryloxypropylsilyl groups on magadiite and copolymerization with methyl methacrylate. *Chem. Mater.*, **12**(6), pp. 1702–1707.

Jaber, M., Miehe-Brendle, J., Michelin, L., Delmotte, L. (2005). Heavy metal retention by organoclays: Synthesis, applications, and retention mechanism. *Chem. Mater.*, **17**, pp. 5275–5281.

Jaynes, W. F., Boyd, S. A. (1991). Clay mineral type and organic compound sorption by hexadecyltrimethylammonium-exchanged clays. *Soil Sci. Soc. Am. J.*, **55**, pp. 43–48.

Jaynes, W. F., Vance, G. F. (1999). Sorption of benzene, toluene, ethylbenzene, and xylene (BTEX) compounds by hectorite clays exchanged with aromatic organic cations. *Clays Clay Miner.*, **47**, pp. 358–365.

Kalra, S., Pant, C. K., Pathak, H. D., Mehata, M. S. (2003). Studies on the adsorption of peptides of glycine/alanine on montmorillonite clay with or without co-ordinated divalent cations. *Colloids Surf. A Physicochem. Eng. Asp.*, **212**, pp. 43–50.

Karube, J. (1998). Hysteresis of the colloidal stability of imogolite. *Clays Clay. Miner.*, **46**, pp.583–585.

Kasama, T., Watanabe, Y., Yamada, H., Murakami, T. (2004). Sorption of phosphates on Al-pillared smectites and mica at acidic to neutral pH. *Appl. Clay Sci.*, **25**, pp. 167–177.

Kaufhold, S., Dohrmann, R., Ufer, K., Meyer, F. M. (2002). Comparison of methods for the quantification of montmorillonite in bentonites. *Appl. Clay Sci.*, **22**, pp. 145–151.

Ke, Y. C., Stroeve, P. (2005). *Polymer-Layered Silicate and Silica Nanocomposites.* Amsterdam, The Netherlands: Elsevier.

Keller, W. D. (1986). *Clay Minerals and the Origin of Life*. Cairns-Smith, A. G., Hartman, H. (eds.). Great Britain: Cambridge University Press.

Kevadiya, B. D., Rajkumar, S., Bajaj, H. C., Chettiar, S. S., Gosai, K., Brahmbhatt, H., Bhatt, S. A., Barvaliya, Y. K., Dave, G. S., Kothari, R. K. (2014). Biodegradable gelatin–ciprofloxacin–montmorillonite composite hydrogels for controlled drug release and wound dressing application. *Colloid Surf. B*, **122**, pp. 175–183.

Kirk-Othmer (1993). *Encyclopedia of Chemical Technology*, 4th Ed., Vol. **24**, pp. 71–117. New York: John Wiley and Sons.

Kooli, F., Liu, Y., Alshahateet, S. F., Messali, M., Bergaya, F. (2009). Reaction of acid activated montmorillonites with hexadecyl trimethylammonium bromide solution. *Appl. Clay Sci.*, **43**, pp. 357–363.

Kozak, M., Domba, L. (2004). Adsorption of the quaternary ammonium salts on montmorillonite. *J. Phys. Chem. Solids*, **65**, pp. 441–445.

Kubies, D., Jèrôme, R., Grandjean, J. (2002). Surfactant molecules intercalated in laponite as studied by ^{13}C and ^{29}Si MAS NMR. *Langmuir*, **18**, pp. 6159–6163.

Kurian, M., Dasgupta, A., Beyer, F. L., Galvin, M. E. (2004). Investigation of the effects of silicate modification on polymer–layered silicate nanocomposite morphology. *J. Polym. Sci. Polym. Phys.*, **42**, pp. 4075–4083.

Kuwaharada, S., Tateyama, H., Nishimura, S., Hirosue, H. (2002). Smectite quasicrystaln aqueous solutions as a function of cationic surfactant concentration. *Clays Clay Miner.*, **50**, pp. 18–24.

Kwolek, T., Hodorowicz, M., Stadnicka, K., Czapkiewicz, J. (2003). Adsorption isotherms of homologous alkyldimethylbenzylammonium bromides on sodium montmorillonite. *J. Colloid Interface. Sci.*, **264**, pp. 14–19.

Lagaly, G. (1981a). Characterization of clays by organic compounds. *Clay Miner.*, **16**, pp. 1–21.

Lagaly, G. (1981b). Inorganic layer compounds. Phenomena of interface reactions with organic compounds. *Naturwiss*, **68**, pp. 82–88.

Lagaly, G. (1986a). Interaction of alkylamines with different types of layered compounds. *Solid State Ionics*, **22**(1), pp. 43–51.

Lagaly, G. (1986b). Smectitic clays as ionic macromolecules. In Wilson, A. D., Prosser, H. J. (eds.), *Development of Ionic Polymers*, Vol. **2**, pp. 77–140. London: Elsevier.

Lagaly, G. (1994). Layer charge determination by alkylammonium ions. In Mermut, A. R. (ed.), *Layer Charge Characteristics of 2:1 Silicate Clay Minerals*, pp. 2 – 46. Boulder, Colorado: The Clay Minerals Society.

Lagaly, G., Beneke, K. (1991). Intercalation and exchange reactions of clay minerals and non-clay layer compounds. *Coll. Polymer Sci.*, **269**, pp. 1198–1211.

Lagaly, G., Dèkany, T. I. (2005). Adsorption on hydrophobized surfaces: Clusters and self-organization. *Adv. Colloid Interface. Sci.*, **114–115**, pp. 189–204.

Lagaly, G., Gonzalez, M. F., Weiss, A. (1976). Problems in layer-charge determination of montmorillonites. *Clay Miner.*, **11**, pp. 173–187.

Lagaly, G., Weiss, A. (1969). Determination of the layer charge in mica-type layer silicates. *Proceedings of the International Clay Conference*, Tokyo, pp. 61–80.

Lagaly, G., Witter, R. (1982). Clustering of liquid molecules on solid surfaces. *Ber. Bunsenges. Phys. Chem.*, **86**, pp. 74–80.

Lagaly, G., Ziesmer, S. (2003). Colloid chemistry of clay minerals: The coagulation of montmorillonite dispersions. *Adv. Colloid Interface. Sci.*, **100–102**, pp. 105–128.

Lagaly, G., Ogawa M., Dékány I. (2013). *Clay Mineral–Organic Interactions*. In Bergaya, F., Lagaly, G. (Eds.). *Developments in Clay Science*. pp. 435–505. Amsterdam, Netherland: Elsevier.

Lan, T., Pinnavaia, T. J. (1994). Clay-reinforced epoxy nanocomposites. *Chem. Mater.*, **6**, pp. 2216–2219.

Le Pluart, L., Duchet, J., Sautereau, H., Gérard J. F. (2002). Surface modifications of montmorillonite for tailored interfaces in nanocomposites. *J. Adhes.*, **78**(7), pp. 645–666.

LeBaron, P. C., Wang, Z., Pinnavaia, T. J. (1999). Polymer-layered silicate nanocomposites: An overview. *Appl. Clay Sci.*, **15**, pp. 11–29.

Lee, D. C., Jang, L. W. (1996). Preparation and characterization of PMMA-clay hybrid composite by emulsion polymerization. *J. Appl. Polym. Sci.*, **61**(7), pp. 1117–1122.

Lee, J. Y., Lee, H. K. (2004). Characterization of organobentonite used for polymer nanocomposites. *Mater. Chem. Phys.*, **85**, pp. 410–415.

Lee, S., Tiwari, M. D. (2012). Organo and inorgano-organo-modified clays in the remediation of aqueous solutions: An overview. *Appl. Clay Sci.*, **59–60**, pp. 84–102.

Lee, S.-S., Ma, Y. T., Rhee, H.-W., Kim, J. (2005). Exfoliation of layered silicate by ring opening reaction of cyclic oligomers in PET–clay nanocomposites. *Polymer*, **46**, pp. 2201–2210.

Lee, S. Y., Cho, W. J., Hahn, P. S., Lee, M., Lee, Y. B., Kim, K. J. (2005). Microstructural changes of reference montmorillonites by cationic surfactants. *Appl. Clay Sci.*, **30**, pp. 174–180.

Li, F., Gao, Z., Li, X., Fang, L. (2014). The effect of *Paecilomyces catenlannulatus* on removal of U (VI) by illite. *J. Environ. Radioactiv.*, **137**, pp. 31–36.

Li, Y., Ishida, H. (2002). A differential scanning calorimetry study of the assembly of hexadecylamine molecules in the nanoscale confined space of silicate galleries. *Chem. Mater.*, **14**(3), pp. 1398–1404.

Li, Y., Ishida, H. (2003). Concentration-dependent conformation of alkyl tail in the nanoconfined space: Hexadecylamine in the silicate galleries. *Langmuir*, **19**, pp. 2479–2484.

Li, Z., Bowman, R. S. (1997). Counterion effects on the sorption of cationic surfactant and chromate on natural clinoptilolite. *Environ. Sci. Technol.*, **31**, pp. 2407–2412.

Li, Z., Chang, P. H., Jean, J. S., Jiang, W. T., Wang, C. J. (2010). Interaction between tetracycline and smectite in aqueous solution. *J. Colloid Interface. Sci.*, **341**, pp. 311–319.

Lin, J. J., Chang, Y. C., Cheng, I. J. (2004). Novel mechanism for layered silicate clay intercalation by poly (propylene oxide)-segmented carboxylic acid. *Macromol. Rapid Commun.*, **25**, pp. 508–512.

Lindgreen, H. (2000). Electrical conduction in layer silicates investigated by combined scanning tunnelling microscopy and atomic force microscopy. *Clay Miner.*, **35**, pp. 643–652.

Lindgreen, H., Garnaes, J., Hansen, P. L., Besenbacher, F., Laegsgaard, E., Stensgaard, I., Gould, S. A. C., Hansma, P. K. (1991). Ultrafine particles of north-sea illite smectite clay-minerals investigated by STM and AFM. *Am. Miner.*, **76**, pp. 1218–1222.

Liu, A. G., Wu, R. C., Eschenazi, E., Papadopoulos, K. (2000). AFM on humic acid adsorption on mica. *Colloids Surf., A*, **174**, pp. 245–252.

Liu, P. (2007). Polymer modified clay minerals: A review. *Appl. Clay Sci.*, **38**, pp. 64–76.

Liu, R., Frosta, R. L., Martens, W. N., Yuan, Y. (2008). Synthesis, characterization of mono, di and tri alkyl surfactant intercalated Wyoming montmorillonite for the removal of phenol from aqueous systems. *J. Colloid Interface Sci.*, **327**, pp. 287–294.

Liu, X., Wu, Q. (2001). PP/clay nanocomposites prepared by grafting–melt intercalation. *Polymer*, **42**, pp. 10013–10019.

Luechinger, M., Prins, R., Pirngruber, G. D. (2005). Functionalization of silica surfaces with mixtures of 3-aminopropyl and methyl groups. *Micropouros Mesopouros Mater.*, **85**(1–2), pp. 111–118.

MacEwan, D. M. C. (1948). Complexes of clays with organic compounds. I. Complex formation between montmorillonite and halloysite and certain organic liquids. *Trans. Faraday Soc.*, **44**, pp. 349–367.

Maiti, P., Okamoto, M. (2003). Crystallization controlled by silicate surfaces in nylon 6–clay nanocomposites. *Macromol. Mater. Eng.*, **288**, pp. 440–445.

Mallakpour, S., Mohammad, D. (2009). Novel nanocomposites based on reactive organoclay of L-tyrosine and amine end-capped poly(amide–imide): Synthesis and characterization. *Appl. Clay Sci.*, **44**, pp. 151–155.

Malucelli, G., Ronchetti, S., Lak, N., Priola, A., Dintcheva, N. T., La Mantia, F. P. (2007). Intercalation effects in LDPE/o-montmorillonites nanocomposites. *Eur. Polym. J.*, **43**, pp.328–335.

Mandalia, T., Bergaya, F. (2006). Organo clay mineral–melted polyolefin nanocomposites. Effect of surfactant/CEC ratio. *J. Phys. Chem. Solids*, **67**, pp. 836–845.

Manias, E., Touny, A., Wu, L., Strawhecker, K., Lu, B., Chung, T. C. (2001). Polypropylene/montmorillonite nanocomposites. Review of the synthetic routes and materials properties. *Chem. Mater.*, **13**, pp. 3516–3523.

Manjanna, J. (2008). Preparation of Fe (II)-montmorillonite by reduction of Fe(III)-montmorillonite with ascorbic acid. *Appl. Clay Sci.*, **42**, pp. 32–38.

Manne, S., Gaub, H. E. (1995). Molecular organization of surfactants at solid–liquid interfaces. *Science*, **270**, pp. 1480–1482.

Mansoori, Y., Atghia, S. V., Zamanloo, M. R., Imanzadeh, G., Sirousazar, M. (2010). Polymer–clay nanocomposites: Free-radical grafting of polyacrylamide onto organophilic montmorillonite. *Eur. Polym. J.*, **46**, pp. 1844–1853.

McLauchlin, A. R., Thomas, N. L. (2008). Preparation and characterization of organoclays based on an amphoteric surfactant. *J. Colloid Interface Sci.*, **321**, pp. 39–43.

Meng, J., Hu, X. (2004). Synthesis and exfoliation of bismaleimide–organoclay nanocomposites, *Polymer*, **45**, pp. 9011–9018.

Meng, X., Du, X., Wang, Z., Bi, W., Tang, T. (2008). The investigation of exfoliation process of organic modified montmorillonite in thermoplastic polyurethane with different molecular weights. *Compos. Sci. Technol.*, **68**, pp. 1815–1821.

Meunier, A. (2005). *Clays.* Berlin, Heidelberg, New York: Springer.

Mòczò, J., Pukànszky, B. (2008). Polymer micro and nanocomposites: Structure, interactions, properties. *J. Ind. Eng. Chem.*, **14**, pp. 535–563.

Moore, D. M., Reynolds, R. C. (1997). *X-ray Diffraction and the Identification and Analysis of Clay Minerals*, New York USA: Oxford University Press. ISBN 0195087135.

Moraru, V. N. M. (2001). Structure formation of alkylammonium montmorillonites in organic media. *Appl. Clay Sci.*, **19**, pp. 11–26.

Mottana, A., Sassi, F. P., Thompson, J. B., Guggenheim, S. (2002). *Micas: Crystal Chemistry and Metamorphic Petrology.* Washington, DC: Mineralogical Society of America.

Möckel, D., Staude, E., Dal-Cin, M., Darcovich, K., Guiver, M. (1998). Tangential flow streaming potential measurements: Hydrodynamic cell characterization and zeta potentials of carboxylated polysulfone membranes. *J. Membrane Sci.,* **145**(2), pp. 211–222.

Murray, H. (2003). Clays in industry. In Dominguez, E. A., Mas, G. R., Cravero, F. (eds.), *A Clay Odyssey. Proceedings of the 12th International Clay Conference 2001*. Elsevier, Amsterdam, pp. 3–14.

Murray, H. H. (2000). Traditional and new applications for kaolin, smectite, and palygorskite: A general overview. *Appl. Clay Sci.,* **17**, pp. 207–221.

Naidja, A., Huang, P. M., Bollag, J.-M. (1997). Activity of tyrosinase immobilized on hydroxyaluminum–montmorillonite complexes. *J. Mol. Catal., A Chem.,* **115**, pp. 305–316.

Newman, A. C. D (1987). *Chemistry of Clays and Clay Minerals.* New York: Wiley-Interscience, pp. 1–128.

Nir, S., Undabeytia, T., Yaron-Marcovich, D., El-Nahhal, Y., Polebusova, T., Serban, C., Rytwo, G., Lagaly, G., Rubin, B. (2000). Optimization of adsorption of hydrophobic herbicides on montmorillonite preadsorbed by monovalent organic cations: Interaction between phenyl rings. *Env. Sci. Tec.,* **34**(7), pp. 1269–1274.

Nishimura, S., Biggs, S., Scales, P. J., Healy, T. W., Tsunematsu, K., Tateyama, T. (1994). Molecular-scale structure of the cation modified muscovite mica basal-plane. *Langmuir,* **10**, pp. 4554–4559.

Occelli, M. L., Drake, B., Gould, S. A. C. (1993). Characterization of pillared montmorillonites with the atomic-force microscope (AFM). *J. Catal.,* **142**, pp. 337–348.

Occelli, M. L., Gould, S. A. C. (2001). Examination of coked surfaces of pillared rectorite catalysts with the atomic force microscope. *J. Catal.,* **198**, pp. 41–46.

Occelli, M. L., Gould, S. A. C., Drake, B. (1994). Atomic-scale imaging of pillared rectorite catalysts with the atomic-force microscope. *Microporous Mater.,* **2**, pp. 205–215.

Occelli, M. L., Gould, S. A. C., Tsai, J. M., Drake, B. (1995). Examination of vanadium contaminated pillared rectorite catalysts with the atomic force microscope. *J. Mol. Catal. A,* **100**, pp. 161–166.

Odin, G. S. (ed.) (1988). Glaucony from the Gulf of Guinea. *In Green Marine Clays*, pp. 225–247. Amsterdam: Elsevier.

Olad, A. (2011). *Polymer/Clay Nanocomposites, Advances in Diverse Industrial Applications of Nanocomposites.* Reddy, B. (Ed.), ISBN: 978-953-307-202-9, InTech, DOI: 10.5772/15657. Available from: http://www.intechopen.com/books/advances-in-diverse-industrial-applications-of-nanocomposites/polymer-clay-nanocomposites.

Osman, M. A., Ploetze, M., Skrabal, P. (2002). Structure and molecular dynamics of alkane monolayers self-assembled on mica platelets. *J. Phys. Chem. B,* **106**, pp. 653–662.

Osman, M. A., Ploetze, M., Skrabal, P. (2004). Structure and properties of alkylammonium monolayers self-assembled on montmorillonite platelets. *J. Phys. Chem. B,* **108**(8), pp. 2580–2588.

Öztürk, N., Tabak, A., Akgöl, S., Denizli, A. (2007). Newly synthesized bentonite–histidine (Bent–Hist) micro-composite affinity sorbents for IgG adsorption. *Colloids Surf. A Physicochem. Eng. Asp.,* **301**, pp. 490–497.

Park, J., Jana, S. C. (2003). Effect of plasticization of epoxy networks by organic modifier on exfoliation of nanoclay. *Macromolecules,* **36**(22), pp. 8391–8397.

Park, K. W., Jeong, S. Y., Kwon, O. Y. (2004a). Interlamellar silylation of H-kenyaite with 3-aminopropyltriethoxysilane. *Appl. Clay Sci.,* **27**(1–2), pp. 21–27.

Park, M., Shim, I. K., Jung, E. Y., Choy, J. H. (2004b). Modification of external surface of laponite by silane grafting. *J. Phy. Chem. Solids,* **65**(2–3), pp. 499–501.

Park, Y., Ayoko, G. A., Frost, R. L. (2011). Application of organoclays for the adsorption of recalcitrant organic molecules from aqueous media. *J. Colloid Interface Sci.,* **354**, pp. 292–305.

Patel, H. A., Somani, R. S., Bajah, H. C., Jasra, R. V. (2007). Preparation and characterization of phosphonium montmorillonite with enhanced thermal stability. *Appl. Clay Sci.,* **35**, pp. 194–200.

Patrick, H. N., Warr, G. G., Manne, S., Askay, I. A. (1999). Surface micellization patterns of quaternary ammonium surfactants on mica. *Langmuir,* **15**, pp. 1685–1692.

Pavlidou, S., Papaspyrides, C. D. (2008). A review on polymer–layered silicate nanocomposites. *Prog. Polym. Sci.,* **33**, pp. 1119–1198.

Perrin-Sarazin, F., Ton-That, M. T., Bureau, M. N., Denault, J. (2005). Micro- and nano-structure in polypropylene/clay nanocomposites. *Polymer,* **46**, pp. 11624–11634.

Plante, A. F., Fernández, J. M., Leifeld, J. (2009). Application of thermal analysis techniques in soil science. *Geoderma,* **153**(1–2), pp. 1–10.

Pollet, E., Delcourt, C., Alexandre, M., Dubois, P. (2006). Transesterification catalysts to improve clay exfoliation in synthetic biodegradable polyester nanocomposites. *Eur. Polym. J.*, **42**, pp. 1330–1341.

Porter, T. L., Eastman, M. P., Hagerman, M. E., Attuso, J. L. Bain, E. D. (1996). Scanning force microscopy and polymerization studies on cast thin films of hectorite and montmorillonite. *J. Vac. Sci. Technol. A*, **14**, pp. 1488–1493.

Ray, S. S., Okamoto, M. (2003). Polymer/layered silicate nanocomposites: A review from preparation to processing. *Prog. Polym. Sci.*, **28**, pp. 1539–1641.

Roelofs, J. C. A. A., Berben, P. H. (2006). Preparation and performance of synthetic organoclays. *App. Clay Sci.*, **33**, pp. 13–20.

Rooj, S., Das, A., Stöckelhuber, K. W., Mukhopadhyay, N., Bhattacharyya, A. R., Jehnichen, D., Heinrich, G. (2012). Pre-intercalation of long chain fatty acid in the interlayer space of layered silicates and preparation of montmorillonite/natural rubber nanocomposites. *App. Clay Sci.*, **67–68**, pp. 50–56.

Rosen, M. J. (2004). *Surfactants and Interfacial Phenomena*. 3rd Ed. New York: Wiley-Interscience.

Ruiz-Hitzky, E., Van Meerbeek, A. (2006). *Clay Mineral- and Organoclay–Polymer Nanocomposite*. In Bergaya, F., Theng, B. K. G., Lagaly, G. (eds.), *Handbook of Clay Science Developments in Clay Science*, pp. 583–621. Amsterdam, The Netherlands: Elsevier.

Sànchez-Martìn, M. J., Dorado, M. C., Del Hoyo, C., Rodrìguez-Cruz, M. S. (2008). Influence of clay mineral structure and surfactant nature on the adsorption capacity of surfactants by clays. *J. Hazard. Mater.*, **150**, pp. 115–123.

Sánchez-Solís, A., Ibarra-Romero, I., Estrada, M. R., Calderas, F., Manero, O. (2004). Mechanical and rheological studies on polyethylene terphtalate–montmorillonite nanocomposites. *Polym. Eng. Sci.*, **44**, pp. 1094–1102.

Schanmugharaj, A. M., Rhee, K. Y., Ryu, S. H. (2006). Influence of dispersing medium on grafting of aminopropyltriethoxysilane in swelling clay materials. *J. Colloid Interf. Sci.*, **298**(2), pp. 854–859.

Schulz, J. C., Warr, G. G. (2000). Adsorbed layer structure of cationic surfactants on clays (mica is not a typical substrate for adsorption studies). *Langmuir*, **16**, pp. 2995–2996.

Schwieger, W., Lagaly, G. (2004). In Auerbach M., Carrado K., Dutta P. K., (eds.), *Handbook of layered materials*. New York, USA: Marcel Dekker Inc. pp. 541–629.

Sharmasakar, S., Jaynes, W. F., Vance, G. F. (2000). BTEX sorption by montmorillonite organo-clays: TMPA, ADAM, HDTMA. *Water Air Soil Pollut.*, **119**, pp. 257–273.

Shaw, D. J. (1992). *Introduction to Colloid and Surface Chemistry*, 4th Ed. Butterworth, London: Heinemann.

Shen, Y. H. (2001). Preparations of organobentonite using nonionic surfactants. *Chemosphere*, **44**, pp. 989–995.

Sheng, G., Boyd, S. A. (2000). Polarity effect on dichlorobenzene sorption by hexadecyltrimethylammonium-clays. *Clays Clay Miner.*, **48**, pp. 43–50.

Sibold, N., Dufour, C., Gourbilleau, F., Metzner, M. N., Lagrève, C., Pluart, L. L., Madec, P. J., Pham, T. N. (2007). Montmorillonite for clay-polymer nanocomposites: Intercalation of tailored compounds based on succinic anhydride, acid and acid salt derivatives: A review. *Appl. Clay Sci.*, **38**, pp. 130–138.

Slade, P. G., Gates, W. P. (2004). The swelling of HDTMA smectites as influenced by their preparation and layer charges. *Appl. Clay Sci.*, **25**, pp. 93–101.

Smith, J. A., Jaffe, P. R., Chiou, C. T. (1990). Effect of ten quaternary ammonium cations on tetrachloromethane sorption to clay from water. *Environ. Sci. Technol.*, **24**(8), pp. 1167–1172.

Somasundaran, P., Fuerstenau, D. W. (1966). Mechanisms of alkyl sulfonate adsorption at the alumina-water interface. *J. Phys. Chem.*, **70**, pp. 90–96.

Someya, Y., Shibata, M. (2004). Morphology, thermal, and mechanical properties of vinylester resin nanocomposites with various organo-modified montmorillonites. *Polym. Eng. Sci.*, **44**, pp. 2041–2046.

Suga, K., Rusling, J. F. (1993). Structural characterization of surfactant and clay-surfactant films of micrometer thickness by FT-IR spectroscopy. *Langmuir*, **9**, pp. 3649–3655.

Tahani, A., Karroua, M., Van Damme, H., Levitz, P., Bergaya, F. (1999). Adsorption of a cationic surfactant on Na–montmorillonite: Inspection of adsorption layer by X-ray and fluorescence spectroscopies. *J. Colloid Interface. Sci.*, **216**, pp. 242–249.

Taroco, H. A., Santos, J. A. F., Domingues, R. Z., Matencio, T. (2011). *Ceramic Materials for Solid Oxide Fuel Cells, Advances* in *Ceramics: Synthesis and Characterization, Processing and Specific Applications*. Sikalidis, C. (ed.). ISBN: 978-953-307-505-1, InTech, DOI:10.5772/18297.

Tiwari, R. R., Khilar, K. C., Natarajan, U. (2008). Synthesis and characterization of novel organo-montmorillonites. *Appl. Clay Sci.*, **38**, pp. 203–208.

Tjong, S. C. (2006). Structural and mechanical properties of polymer nanocomposites. *Mater. Sci. Eng. R,* **53**, pp. 73–197.

Tombàcz, E., Szekeres, M., Baranyi, L., Michèli E. (1998). Surface modification of clay minerals by organic polyions. *Colloids Surf. A,* **141**, pp. 379–384.

Tyrode, E., Rutland, M. W., Bain, C. D. (2008). Adsorption of CTAB on hydrophilic silica studied by linear and nonlinear optical spectroscopy. *J. Am. Chem. Soc.,* **130**, pp. 17434–17445.

Tzavalas, S., Gregoriou, V. G. (2009). Infrared spectroscopy as a tool to monitor the extent of intercalation and exfoliation in polymer clay nanocomposites. *Vib. Spectrosc.,* **51**, pp. 39–43.

Unnikrishnan, L., Mohanty, S., Nayak, S. K., Ali, A. (2011). Preparation and characterization of poly(methyl methacrylate)–clay nanocomposites via melt intercalation: Effect of organoclay on thermal, mechanical and flammability properties. *Mater. Sci. Eng. A,* **528**, pp. 3943–3951.

Uskarcı, T. (2006). *Behaviour of Bentonite Suspensions in Non-Aqueous Media.* Master Thesis. Middle East Technical University, Ankara, Turkey.

Utracki, L. (2004). *Clay-Containing Polymeric Nanocomposites.* 1st Ed. Shawbury, Shrewsbury, Shropshire, UK: Rapra Technology.

Vaia, R. A., Giannelis, E. P. (1997). Lattice of polymer melt intercalation in organically-modified layered silicates. *Macromolecules,* **30**, pp. 7990–7999.

Vaia, R. A., Jandt, K. D., Kramer, E. J., Giannelis, E. P. (1995). Kinetics of polymer melt intercalation. *Macromolecules,* **28**, pp. 8080–8085.

Vaia, R. A., Teukolsky, R. K., Giannelis, E. P. (1994). Interlayer structure and molecular environment of alkylammonium layered silicates. *Chem. Mater.,* **6**, pp. 1017–1022.

van Olphen, H. (1977). *An Introduction to Clay Colloid Chemistry.* New York, USA: Wiley- Interscience.

Vazquez, A., López, M., Kortaberria, G., Martín, L., Mondragon, I. (2008). Modification of montmorillonite with cationic surfactants. Thermal and chemical analysis including CEC determination. *App. Clay Sci.,* **41**, pp. 24–36.

Velde, B. (1992). *Introduction to Clay Minerals.* New York: Chapman & Hall.

Venkataraman, N. V., Vasudevan, S. (2000). Conformation of an alkane chain in confined geometry: Cetyl trimethyl ammonium ion intercalated in layered $CdPS_3$. *J. Phys. Chem. B,* **104**, pp. 11179–11185.

Venkataraman, N. V., Vasudevan, S. (2001). Conformation of methylene chains in an intercalated surfactant bilayer. *J. Phys. Chem. B,* **105**, pp. 1805–1812.

Vrdoljak, G. A., Henderson, G. S. (1994). Specific ion adsorption at the mineral-water interface - cesium adsorption on chlorite. *Coll. Surf. A,* **87**, pp. 187–196.

Vrdoljak, G. A., Henderson, G. S., Fawcett, J. J., Wicks, F. J. (1994). Structural relaxation of the chlorite surface imaged by the atomic-force microscope. *Am. Miner.,* **79**, pp. 107–112.

Wang, J., Han, B., Yan, H., Li, Z., Thomas, R. K. (1999). Adsorption and adsolubilization behaviors of cationic surfactant and hydrophobically modified polymer mixtures on Na-kaolinite. *Langmuir,* **15**, pp. 8207–8211.

Wang, L., Wang, K., Chen, L., He, C., Zhang, Y. (2006). Hydrothermal effects on the thermomechanical properties of high performance epoxy/clay nanocomposites. *Polym. Eng. Sci.,* **46**, pp. 215–221.

Wang, L.-Q., Liu, J., Exarhos, G. J., Bunker, B. C. (1996). Investigation of the structure and dynamics of surfactant molecules in mesophase silicates using solid-state ^{13}C NMR. *Langmuir,* **12**, pp. 2663–2669.

Wang, L.-Q., Liu, J., Exarhos, G. J., Flanigan, K. Y., Bordia R. (2000). Conformation heterogeneity and mobility of surfactant molecules in intercalated clay minerals studied by solid-state NMR. *J. Phys. Chem. B,* **104**, pp. 2810–2816.

Wang, X., Huang, A., Jia, D., Li, Y. (2008). From exfoliation to intercalation: Changes in morphology of HNBR/organoclay nanocomposites. *Eur. Polym. J.,* **44**, pp. 2784–2789.

Weimer, M. W., Chen, H., Giannelis, E. P., Sogah, D. Y. (1999). Direct synthesis of dispersed nanocomposites by in situ living free radical polymerization using a silicate-anchored initiator. *J. Am. Chem. Soc.,* **121**, pp. 1615–1616.

Wicks, F. J., Kjoller, K., Eby, R. K., Hawthorne, F. C., Henderson, G. S. Vrdoljak, G. A. (1993). Imaging the internal atomic-structure of layer silicates using the atomic-force microscope. *Can. Miner.,* **31**, pp. 541–550.

Wing, M. Y., Zhong-Zhen, Y. (2006). *Polymer Nanocomposites*. Cambridge, UK: Woodhead, Maney.

Xi, Y., Ding, Z., He, H., Frost, R. L. (2004). Structure of organoclays—an X-ray diffraction and thermogravimetric analysis study. *J. Colloid Interface Sci.,* **277**, pp. 116–120.

Xiao, X., Hu, J., Charych, D. H., Salmeron, M. (1996). Chain length dependence of the frictional properties of alkylsilane molecules self-assembled on mica studied by atomic force microscopy. *Langmuir,* **12**, pp. 235–237.

Xie, W., Gao, Z., Pan, W. P., Hunter, D., Singh, A., Vaia, R. (2001a). Thermal degradation chemistry of alkyl quaternary ammonium montmorillonite. *Chem. Mater.*, **13**, pp. 2979–2990.

Xu, M., Choi, Y. S., Kim, Y. K., Wang, K. H., Chung, I. J. (2003). Synthesis and characterization of exfoliated poly(styrene-co-methyl methacrylate)/clay nanocomposites via emulsion polymerization with AMPS. *Polymer*, **44**, pp. 6387–6395.

Xu, S., Boyd, S. A. (1995). Alternative model for cationic surfactant adsorption by layer silicates. *Environ. Sci. Tech.*, **29**, pp. 3022–3028.

Xu, S., Boyd, S. A. (1996). Cationic surfactant adsorption by swelling and nonswelling layer silicates. *Langmuir*, **11**, pp. 2508–2514.

Yao, H., Zhu, J., Morgan, A. B., Wilkie, C. A. (2002). Crown ether-modified clays and their polystyrene nanocomposites. *Polym. Eng. Sci.*, **42**, pp. 1808–1814.

Yariv, S., Cross, H. (2002). *Organo-Clay Complexes and Interactions*. New York: Marcel Dekker.

Yoshimoto, S., Osashi, F., Kameyama, T. (2005). X-ray diffraction studies of intercalation compounds prepared from aniline salts and montmorillonite by mechanical processing. *Sol. Sta. Comm.*, **136**, pp. 251–256.

Yu, Y. H., Yeh, J. M., Liou, S. J., Chang, Y. P. (2004). Organo-soluble polyimide (TBAPPODPA)/clay nanocomposite materials with advanced anticorrosive properties prepared from solution dispersion technique. *Acta Mater.*, **52**, pp. 475–486.

Yui, T., Yoshida, H., Tachibana, H., Tryk, D. A., Inoue, H. (2002). Intercalation of polyfluorinated surfactants into clay minerals and the characterization of the hybrid compounds. *Langmuir*, **18**, pp. 891–896.

Zampori, L., Stampino, P. G., Cristiani, C., Dotelli, G., Cazzola, P. (2010). Synthesis of organoclays using non-ionic surfactants: Effect of time, temperature and concentration. *Appl. Clay Sci.*, **48**, pp. 97–102.

Zbik, M., Horn, R. G. (2003). Hydrophobic attraction may contribute to aqueous flocculation of clays. *Colloids Surf. A*, **222**, pp. 323–328.

Zbik, M., Smart, R. St. C. (2002). Dispersion of kaolinite and talc in aqueous solution: Nano-morphology and nano-bubble entrapment. *Mine. Eng.*, **15**(4), pp. 277–286.

Zen, J. M., Senthil Kumar, A. (2004). The prospects of clay mineral electrodes. *Anal. Chem.*, **76**, pp. 205A–211A.

Zeng, Q. H., Yu, A. B., Lu, G. Q., Standish, R. K. (2003). Molecular dynamics simulation of organic-inorganic nanocomposites: Layering behavior and interlayer structure of organoclays'. *Chem. Mater.,* **15**, pp. 4732–4738.

Zhang, J., Jiang, D. D., Wilkie, C. A. (2005). Polyethylene and polypropylene nanocomposites based upon an oligomerically modified clay. *Thermochim. Acta,* **430**, pp. 107–113.

Zhang, J., Jiang, D. D., Wilkie, C. A. (2006a). Polyethylene and polypropylene nanocomposites based on a three component oligomerically-modified clay. *Polym. Degrad. Stab.,* **91**, pp. 641–648.

Zhang, J., Jiang, D. D., Wilkie, C. A. (2006b). Thermal and flame properties of polyethylene and polypropylene nanocomposites based on an oligomerically-modified clay. *Polym. Degrad. Stab.,* **91**, pp. 298–304.

Zhang, Y., Fu, L., Yang, H. (2012). Insights into the physicochemical aspects from natural halloysite to silica nanotubes. *Colloid Surfaces A.,* **414**, pp. 115–119.

Zhang, Y., Zhao, Y., Zhu, Y., Wu, H., Wang, H., Lu, W. (2012). Adsorption of mixed cationic-nonionic surfactant and its effect on bentonite structure. *J. Environ. Sci.,* **24**(8), pp. 1525–1532.

Zhang, Y.-Q., Lee, J.-H., Jang, H.-J., Nah, C.-W. (2004b). Preparing PP/clay nanocomposites using a swelling agent. *Composites: Part B,* **35**, pp. 133–138.

Zhang, Y.-Q., Lee, J.-H., Rhee, J. M., Rhee, K. Y. (2004a). Polypropylene-clay nanocomposites prepared by in situ grafting–intercalating in melt. *Comp. Sci. Technol.,* **64**, pp. 1373–1389.

Zhang, Z. Z., Sparks, D. L., Scrivner, N. C. (1993). Sorption and desorption of quaternary amine cations on clays. *Environ. Sci. Technol.,* **27**, pp. 1625–1631.

Zheng, X., Jiang, D. D., Wang, D., Wilkie, C. A. (2006). Flammability of styrenic polymer clay nanocomposites based on a methyl methacrylate oligomerically-modified clay. *Pol. Degrad. Stab.,* **91**, pp. 289–297.

Zhu, J., He, H., Zhu, L., Wen, X., Deng, F. (2005). Characterization of organic phases in the interlayer of montmorillonite using FTIR and ^{13}C NMR. *J. Colloid Interface Sci.,* **286**, pp. 239–244.

Zhu, R., Zhu, L., Zhu, J., Xu, L. (2008). Structure of cetyltrimethylammonium intercalated hydrobiotite. *Appl. Clay Sci.,* **42**, pp. 224–231.

Chapter 3

Organoclay Synthesis Methods

3.1 Introduction

The interactions between organic matter and clay minerals are among the most widespread interactions in nature. The adsorption of organic material by clay minerals has been widely investigated during the last decade and has been extensively reviewed. The interactions include cation exchange and adsorption of polar and nonpolar molecules. In these interactions, adsorption, in which physical or chemical interactions occur between the mineral and organic materials, is the primary process (Yariv and Cross, 2002). Moreover, such interactions involve a wide variety of interacting mechanisms and energies (Table 3.1) resulting in organic–inorganic materials of variable stability (Auerbach et al., 2004).

Surface modifications of clay minerals have received attention because they allow the creation of new materials and new applications. The main focus of surface modifications of clays is materials science, because organoclays are essential to develop polymer nanocomposites. Nanocomposites constitute one of the most developed areas of nanotechnology. Modified clays are also used in other applications such as adsorbents of organic pollutants in soil, water, and air; rheological control agents; paints; cosmetics; refractory varnish; and thixotropic fluids (de Paiva et al., 2008; Yuehong et al., 2010).

Introduction to Polymer–Clay Nanocomposites
Ahmet Gürses

ISBN 978-981-4613-02-6 (Hardcover), 978-981-4613-03-3 (eBook)
www.panstanford.com

Table 3.1 Mechanisms of clay–organic interactions

Nature of the interactions	Characteristics
Electrostatic	Ion exchange of interlayer cations with organic cations
Van der Waals forces Hydrogen bonding and water bridges Ion dipole and coordination Proton transfer Electron transfer	Adsorption of neutral molecules by interactions with external or internal (intracrystalline region of silicates) surfaces
Covalent bonding	Grafting reactions of organic groups

To modify clays and clay minerals, as mentioned in Chapter 2, several routes have been employed, such as adsorption, ion exchange with inorganic cations and organic cations, binding of inorganic and organic anions (mainly at the edges), grafting of organic compounds, reaction with acids, pillaring by different types of poly(hydroxo metal) cations, intraparticle and interparticle polymerization, dehydroxylation and calcination, delamination and reaggregation of smectites, and lyophilization, ultrasound, and plasma (Bergaya and Lagaly, 2001).

Ion exchange with alkylammonium ions is well known and the preferential method to prepare organoclays. Generally, experimental studies describe the preparation of organoclays at the laboratory scale, with different experimental conditions, clays from several regions and suppliers, and several kinds of organic compounds (de Paiva et al., 2008).

Moreover, the importance of clay–organic interactions arises because such compounds are used in industrial applications, such as additives, fillers, rheological agents, and specific sorbents, and, recently, in technological applications as advanced materials (Ruiz-Hitzky and Meerbeek, 2006). On the other hand, clay–organic interactions are of crucial importance regarding agricultural production, the origin and exploitation of oil resources, and the origin of life. The current relevance and vitality of the field can be inferred from the exponential increase in scientific works appearing during the last few decades. This chapter has been devoted to review and summarize the main mechanisms governing clay–organic interactions, from discrete molecules or cations to large entities, such as polymers. Only a few detailed examples will be presented in trying to group these interactions into several representative systems. In addition, the characterization of organic–inorganic solids derived from various common clays, as well as a brief discussion of current research and trends in the development of new hybrid compounds will be addressed. Consequently, a general overview associated with the application perspectives of these hybrids compounds as advanced materials will be presented. Milestones concerning clay–organic interactions and relevant research achievements are given in Table 3.2.

Table 3.2 Milestones in clay–organic interactions

Date	Milestones	Examples
8th century AD	Clay–dye hybrid compounds	Palygorskite/indigo mixtures (Maya blue)
1911	Catalytic transformation of organic compounds activated by clays	Pinene to camphene over palygorskite
1939–41	Intercalation of organic cations in smectites	Montmorillonite/ aliphatic and aromatic ammonium cations
1944	Intercalation of neutral species in clays	Montmorillonite/ glycerol
1945	Structural features by XRD, one-dimensional Fourier analysis of clay–organic compounds	Montmorillonite/ diamines and glycols
1949	Application of thermal analysis in organoclay characterization	DTA of montmorillonite/ alkylammonium
1954	Selective sorption of hydrocarbons by palygorskite and sepiolite	Palygorskite/*n*-heptane iso-octane
1961	Intercalation of salts in kaolinite	Kaolinite/K-acetate
1961	Intercalation of neutral molecules in kaolinite	Kaolinite/urea
1961	UV-Vis application to study clay–organic systems	Montmorillonite/ benzidine
1961	Polymer–clay intercalation compounds	Montmorillonite/ polyacrylonitrile

(*Continued*)

Table 3.2 (*Continued*)

Date	Milestones	Examples
1965	Orientation of organic molecules in the interlayer space by IR spectroscopy	Montmorillonite and vermiculite/pyridine and pyridinium
1968	Organic derivatives of clays through covalent bonds (grafting)	Vinyl derivatives of chrysotile
1969	π bonds in clay–aromatic compounds	Cu-montmorillonite/ benzene
1974	Organic reactions in the interlayer space of clays Intracrystalline sorption of organic compounds in sepiolite	Vermiculite/L-ornithine peptide formation Sepiolite/hexane
1976	Organic pillared clays	Montmorillonite/ diprotonated triethylenediamine
1980	Application of 13C-NMR to characterize clay–organic systems	Ag-hectorite/benzene
1983	Catalysts based on organometallic–clay complexes	Montmorillonite/ [Rh(PPh3)3]
1984	LMMS application to characterize clay–organic systems	Sepiolite/organosilanes
1985	Photostabilization of co-adsorbed labile bioactive species	Montmorillonite/ methyl green/ bioresmethrin
1989	Microwave activation of organic reactions on clay-adsorbed compounds	Rearrangement of pinacol to pinacolone intercalated in montmorillonite
1990	Ion-conducting polymers–2D intercalated materials	Montmorillonite/PEO
1993	Polymer melt intercalation in organosmectites	Montmorillonite/ polystyrene
1993	Grafting of organic groups in the interlayer space of kaolinite	Methoxy derivatives of kaolinite
1998	Templated synthesis of polymer–clay nanocomposites	Synthetic fluorosmectite/ polyvinylpyrrolidone

Source: Reproduced from Auerbach et al., Copyright 2004, with permission from Taylor and Francis Group.

3.2 Synthesis of Organoclays

The synthesis of organoclays is based on the mechanisms of the interactions between clay minerals and organic compounds. A displacement process occurs when water molecules in the interlayer space of smectites and vermiculites are displaced by polar molecules (Bergaya et al., 2006). Neutral organic compounds can form complexes with the interlayer cations (de Paiva et al., 2008). There are different ways to modify 2:1-type clay minerals: hydrogen bonding; ion–dipole interactions; coordination bonds; acid–base reactions; charge transfer; van der Waals interactions; ion exchange with organic and inorganic ions and cationic complexes; grafting of organic compounds; pillaring with the different types of poly(hydroxy metal) cations; interlamellar or intraparticle polymerization; and the delamination and reaggregation of smectitic clay minerals, with the thermal processes and some physical techniques, such as dehydroxylation, calcination, lyophilization, ultrasound, and plasma (see Chapter 2). The organoclays are generally prepared in solutions by cation exchange and solid-state interaction (de Paiva et al., 2008). However, to increase the effectiveness of these methods, new efforts are conducted. For example, Baldassari et al. (2006) synthesized six organo derivatives with each of the lower-charged fluorophlogopite-type clays by a microwave-assisted procedure. They found that using natural sodium montmorillonite (Na-MMT) or low-charged micas as precursors, complete intercalation can be easily achieved with conventional methods under mild temperature conditions (~60°C), but a microwave-assisted process is more effective than the conventional process when applied to the intercalation of higher-charged clays.

3.2.1 Cation Exchange

Clay minerals consist of small crystalline particles with silica oxygen tetrahedral sheets and aluminum or magnesium octahedral sheets where an aluminum or magnesium ion is octahedrally coordinated to six oxygens or hydroxyls. The octahedral sheet is located between two Si tetrahedral sheets. Organoclays or organomontmorillonites are clays that have been modified with organic surfactants with single and dual cationic surfactants, anionic–cationic surfactants, and nonionic surfactants. Cation exchange has been used for nearly 50 years. This method is based on the displacement with cations in the interlayer region of clay minerals of quaternary alkylammonium cations in aqueous solution (Mandalia and Bergaya, 2006). Generally, this can be done by ion exchange with cationic surfactants, including primary, secondary, tertiary, and quaternary alkylammonium or alkylphosphonium cations. Alkylammonium or alkylphosphonium cations in the organosilicates lower the surface energy of the inorganic host and improve the wetting characteristics of the polymer matrix and result in a larger interlayer spacing. The structure and properties of the resultant organoclays are affected by the type of both surfactant and clay minerals. From the ion exchange, the interlayer spacing between the single sheets is broadened. This enables the adsorption of organic cation chains and changes the surface properties of each single sheet from hydrophilic to hydrophobic or organophilic. The interlayer cation density or packing density of the alkylammonium ions of the clay minerals and the chain length of the organic ion are important factors to determine the arrangement of organic molecules between the layers. The formation of monolayers, bilayers, and pseudotrimolecular layers of alkylammonium ions in the interlayer spaces of montmorillonite was characterized by the basal spacings (Park et al., 2011).

Additionally, the alkylammonium or alkylphosphonium cations can provide functional groups that can react with the polymer matrix or in some cases initiate the polymerization of monomers to improve the strength of the interface between the inorganic and the polymer matrix (Sinha Ray and Okamoto, 2003). Schematic representation of a cation exchange mechanism between the silicate and an alkylammonium salt has been shown in Fig. 3.1.

Mallakpour and Mohammad (2013) prepared organomodified clays by a cation exchange method, which is a displacement of the sodium cations of Cloisite-Na^+ with the protonated amino acids.

Vazquez et al. (2008) prepared an organomontmorillonite according to a standard ion exchange procedure. Cloisite-Na^+ was dried at 110°C for 48 h and washed with distilled water (10 wt.%) with vigorous stirring for 24 h to cause the delamination of the montmorillonite. The amount of surfactant added was about 1.5 cation exchange capacity (CEC). The suspension was stirred at room temperature for 15 min. The montmorillonite dispersion was introduced in a 500 mL flask and vigorously stirred in reflux for 6 h, filtered in vacuum, and washed with water and in some cases subsequently with 50/50 vol.% ethanol/water mixture. Two indicators were used to point out the removal of the anions and the excess of organic cations: on the one hand, a chloride and bromide test with a few drops of 1 M $AgNO_3$ solution up to no precipitate was observed, and on the other hand, the absence of lather in the filtered solution after stirring was observed. Several treatments were tested to improve the amount of intercalated surfactant as follows:

(1) **Water**. After modification, the organomontmorillonite was washed only with distilled water in reflux. The water volume depended on the type of surfactant.
(2) **EtOH/water**. After washing with distilled water, a 50/50 mixture of ethanol/water in reflux was used.
(3) **Sonication**. Sonication was used before the modification to promote delamination of the montmorillonite particles. The time was 6 h, and the organically modified montmorillonite was washed with the EtOH/water method.

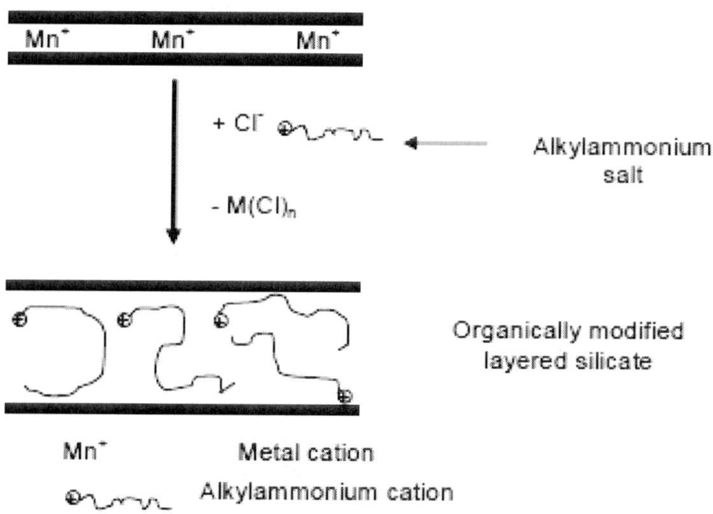

Figure 3.1 Schematic representation of a cation exchange mechanism between a silicate and an alkylammonium salt. Reprinted from Zanetti, Copyright 2000, with permission from John Wiley and Sons.

(4) **Polar activator [PAC]**. After stirring the montmorillonite for 24 h in 500 mL distilled water, a PAC solution (ethanol/water 5/95 vol.%, 40 wt.% related to the montmorillonite) was added to get higher exfoliation of the layers. The prepared organomontmorillonite was washed with ethanol/water. This method was carried out at two temperatures: 90°C and in reflux (110°C) (Awad et al., 2004; He et al., 2005; Litina et al., 2006).

Zhu et al. (2008) prepared surfactant–clay mineral complexes as follows: 5.0 g of montmorillonite or hydrobiotite was dispersed in 100 mL of distilled water, and then desired amounts of cetyltrimethylammonium bromide (CTMAB) from 0.22 mmol/g to 2.16 mmol/g and from 0.11 mmol/g to 2.0 mmol/g were added for the montmorillonite and the hydrobiotite, respectively. The suspensions were stirred at 50°C for 10 h and then aged at 25°C for 1 week. Subsequently, the products were washed by filtration for several times and dried at 80°C for 8 h. The complexes synthesized from the montmorillonite and hydrobiotite were designated as CTMA–Mont and CTMA–Hyd, respectively. CTMA–Mont complexes with different surfactant loading amounts were denoted as xC–Mont (*x* meaning the surfactant loading amount). For example, 21C–Mont represented the CTMA–Mont sample with the surfactant loading amount equal to 0.21 mmol/g, and the same notation was used for CTMA–Hyd complexes. All samples were ground to pass 100 meshes and then characterized after drying.

Burmistr et al. (2005) prepared organobentonite by cationic exchange between the inorganic cations in the bentonite galleries and polymeric quaternary ammonium salts (PQAS) in an aqueous solution. One hundred and ten grams of bentonite was suspended in 10 L of distilled water. The dispersion was stood for 1 h, after which the finely dispersed fraction was separated. This operation was repeated three to four times. After that, an aqueous solution of 5 g Na_2CO_3 was added gradually to exchange all inorganic cations in the bentonite galleries to Na^+ cations. The mixture was agitated vigorously for 15 min at boiling temperature. Then, 20 g of PQAS in 200 mL distilled water was added to the dispersion. The mixture was stirred for 24 h at 40°C and then filtered, washed with hot water, and dried at 60°C in vacuum. The organobentonite thus obtained is highly hydrophobic. Particles potential of the water suspension for pure bentonite is ζ = −32 mV, whereas for organobentonite, ζ = +34 mV.

3.2.2 Solid-State Interaction

Organic molecules can be intercalated in dried clay minerals by solid-state interaction without the use of solvents. The absence of solvents preceding the preparation is environmentally good and makes the process more suitable for industrialization. The intercalation of neutral compounds into dried montmorillonite and vermiculites is not necessarily accompanied by cation movement midway between the silicate layers (outer-surface complexes). The cations can remain in contact with one silicate layer, that is, the oxygen atoms of the silicate surface occupy the coordination sites of the cations.

The first solid-state interaction of clay minerals and ammonium cations was studied by Ogawa et al. (1990). Breakwell et al. (1995) then employed a "dry method," which consists the reaction of the clay with the quaternary alkylammonium salt at 60°C in a high shear-mixer. X-ray diffraction and X-ray photoelectron analyses showed that the "dry process" provided the basal spacings in the range of 1.24–1.27 nm and also the quaternary alkyl ammonium ion exchange in the ratio of 68 to 78%.

Yoshimoto et al. (2005) also used solid-state interactions through a mechanochemical process to intercalate aniline salts with different counterions in montmorillonite. Many solid-state processes are based on the ion–dipole interaction. In the ion–dipole interaction, organic molecules with polar groups are attached to the interlayer cations. The negative parts of the molecules interact with cations on the surface, and water molecules are displaced from the interlayer cations (Beall and Goss, 2004; Merinska et al., 2002; Yoshimoto et al., 2005).

According to Merinska et al. (2002), the ion–dipole treatment to prepare organoclays for polymer nanocomposites is a technique that was discovered in the 1990s by Nanocor. Merinska et al. (2002) intercalated octadecylamine in montmorillonite by ion–dipole interactions.

Beall and Goss (2004) intercalated three different families of organic compounds in montmorillonite: the alcohols butanol, hexanol, octanol, dodecanol, tetradecanol, and octadecanol; the aldehydes butanal, hexanal, octanal, decanal, and dodecanal; and the *n*-alkyl pyrrolidones methylpyrrolidone, octylpryrrolidone, and dodecypyrrolydone. The authors concluded that the ion–dipole interaction of small polar molecules with exchangeable cations on the montmorillonite surface appeared to be controlled principally by the polar group, but the chain length of the alkyl group also played a role.

3.3 Quaternary Alkylammonium Cations Used to Prepare Organoclays

Quaternary alkylammonium salts are cationic surfactants and the most used organic compounds to prepare organoclays. Most recently, there is an important amount of research on modification of clay minerals with several kinds of quaternary alkylammonium salts (bromides or chlorides) at the laboratory scale. In the case of polymer–organoclay nanocomposites, the kind of quaternary alkylammonium salt affects the affinity between the clay mineral and the polymer (Someya and Shibata, 2004). For apolar polymers such as polypropylene (PP) and polyethylene (PE), the clay minerals are modified with dialkyl dimethylammonium halides, while for polar polymers such as polyamide (PA), poly(vinyl alcohol) (PVA), and poly(ethylene oxide) (PEO), the clay minerals are modified with alkyl benzyl dimethylammonium halides or alkyl hydroxyethylammonium halides. Modifications of clay minerals also have been performed with quaternary alkylammonium salts associated with other organic compounds (de Paiva et al., 2008).

Clay interactions with ammonium ions were studied in the early 1930s. The principal method used to modify the interlayer space of smectites was based on cation exchange with alkylammonium ions. For several decades, such a method was thoroughly investigated by numerous researchers (Gieseking, 1939; Lee and Tiwari, 2012). The alkylammonium species used to prepare organoclays were quaternary ammonium compounds containing alkyl, phenyl, benzyl, and pyridyl groups (see Table 3.3).

Table 3.3 Some samples of quaternary ammonium cations frequently used to prepare organoclays

Abbreviation	Quaternary cation	Formula
TMA	Tetramethylammonium	$(CH_3)_4N^+$
TMPA	Trimethyl phenylammonium	$C_6H_5N^+(CH_3)_3$
BTMA	Benzyl trimethylammonium	$C_6H_5CH_2N^+(CH_3)_3$
HDPY	Hexadecylpyridinium	$C_6H_5N^+(C_{16}H_{33})$
BDTDA	Benzyldimethyl tetradecylammonium	$C_6H_5CH_2N^+(C_{14}H_{29})(CH_3)_2$
HDTMA	Hexadecyltrimethylammonium	$C_{16}H_{33}N^+(CH_3)_3$
DODMA	Dioctadecyldimethylammonium	$(C_{18}H_{37})_2N^+(CH_3)_2$

Wang et al. (2001) modified Na-MMT with dodecylamine, hexadecylamine, and octadecylamine. They reported that the alkylammonium chain length may change the degree of interaction between clay and PE.

Liu and Wu (2001) developed a new class of organoclay that had a larger basal spacing than the organoclays only modified by alkylammonium. First, Na-MMT was modified with hexadecyltrimethylammonium bromide (HDTMAB), resulting in an organoclay with a basal spacing of 1.96 nm. Then, this organoclay was co-intercalated with epoxypropyl methacrylate, increasing the basal spacing to 2.98 nm. This organoclay was used to prepare PP nanocomposites.

Klapyta et al. (2001) prepared organoclays by cation exchange with dodecyl- and octadecyl trimethylammonium bromide. The concentration of ammonium salts varied from 0.25 to 2.0 CEC of the silicate.

Lee and Kim (2002) modified smectite with HDTMA cations equivalent to 1.0 and 2.5 times the CEC converts to HDTMA-smectite. Smectite is modified with the HDTMA loading of 1.0 and 2.5 times the CEC, and the *d*-value of clay increases to 2.1 nm and 3.94 nm.

Tang et al. (2003) prepared an organophilic montmorillonite with octadecylammonium and HDTMA salts in aqueous solution to be used in PP nanocomposites.

Lee and Lee (2004) treated bentonite with 1-hexadecylamine, 1-octadecylamine, and cetyltrimethylammonium and octadecyltriethylammonium bromides. All organoclays were delaminated in an epoxy matrix. Young's modulus and tensile strength of the nanocomposite increases with an increase in the organoclay content.

Xi et al. (2004) investigated X-ray diffraction to determine the changes in the surface properties of montmorillonitic clay through the changes in the basal spacings of montmorillonite (SWy-2) and surfactant-intercalated organoclays. The concentrations of surfactant (octadecyltrimethylammonium [$ODTMA^+$]) used are equivalent to 0.2 CEC, 0.4 CEC, 0.6 CEC, 0.8 CEC, 1.0 CEC, 1.5 CEC, 2.0 CEC, 3.0 CEC, and 4.0 CEC of the montmorillonite. Variation in the *d*-spacing was found to be a steep function of surfactant concentration.

Li et al. (2005) modified Na-MMT with CTMAB and aminoun decanoic acid (AUA). After the cation exchange with CTMAB and AUA, the *d* (001) spacing was expanded to 1.84 nm and 1.72 nm. This organoclay was also used to prepare exfoliated polystyrene/clay nanocomposites.

Xi et al. (2005) investigated FTIR spectra of the ODTMA bromide intercalated into an Na-MMT. The concentrations of $ODTMA^+$ used are equivalent to 0.2 CEC, 0.4 CEC, 0.6 CEC, 0.8 CEC, 1.0 CEC, 1.5 CEC, 2.0 CEC, 3.0 CEC, and 4.0 CEC of the montmorillonite. They found that the spectra of the ODTMA-intercalated montmorillonite are very different from the pure ODTMA bromide. A change in both the wavenumber and the intensity of the bands occurs as the CEC increases.

Zheng et al. (2006) worked the thermal stability and compatibility of organoclays by using novel surfactants such as stibonium and tropylium. These compounds provide to the clay minerals good thermal stability and show very few degradation below 300°C.

He et al. (2007) modified montmorillonite with different HDTMAB concentrations. This study demonstrates that the surfactant distribution within the organoclays depends strongly on the surfactant loadings. With an increase in surfactant concentration, the basal spacings of the resultant organoclays increase in the following order: (Na-Mt) → 1.24 nm (0.5 CECMt) → 1.48 nm (0.7 CECMt) → 1.78 nm (1.0 CECMt) → 1.95 nm (1.5 CECMt) → 2.23 nm (2.0 CECMt) → 3.61 nm (2.5 CECMt) → 3.84 nm.

Zhu and Zhu (2008) modified montmorillonite with CTMAB. Four organoclays were prepared by intercalating different amounts of the cationic surfactant $CTMA^+$ into montmorillonite. In this work, the thermodynamics of sorption of hydrophobic organic compounds (HOCs) on organoclays were investigated.

Borisover et al. (2010) examined how the thermal treatment of organoclays can affect the sorptive properties of organoclay sorbents. The organoclay sorbents were prepared from Na-MMT by *n*-HDTMA cations.

Nowadays, surface modification of clay minerals has become increasingly important for improving the practical applications of clays and clay minerals. The most important new application of organoclays is the preparation of various types of nanocomposites.

Table 3.4 presents an inclusive review of the works involving organoclays, together with the method of synthesis, organic agents, and basal spacings.

Table 3.4 Works involving organoclays: the methods of synthesis, organic agents, and basal spacings

Clay	Modifier	Methodology	Reference
1994			
Sodium montmorillonite Lithium fluorohectorite	Dioctadecyl-dimethyl-ammonium bromide Primary amines ($C_nH_{2n+1}NH_2$, $n = 6, 9–14$ and 18)	Cation exchange in water/ethanol	Vaia et al. (1994)
Bentonites: Verde Lodo, Róseo-bofe, Chocolate (Campina Grande, PB, Brazil) Green, brown, and mixture (Vitória da Conquista, BA, Brazil) Bentonite Volclay-SPN (Wyoming, USA)	Alkyl benzyl dimethyl-ammonium chloride	Cation exchange in aqueous solution	Diaz (1994)

(*Continued*)

Table 3.4 (*Continued*)

1995			
Wyoming bentonite	Dehydro-genated tallow dimethyl-ammonium chloride	1. "Dry" process using a high-shear mixer (solid-state reaction)	Breakwell et al. (1995)
		2. Cation exchange in aqueous solution	
1996			
Synthetic micas: fluorotaeniolite and fluorohectorite (Topy Industries)	Alkyl trimethyl-ammonium chlorides	Cation exchange in aqueous solution	Tamura and Nakazawa (1996)
2000			
Sodium montmorillonite Kunipia F (Kunimine Industries)	Octadecyl-ammonium chloride	Cation exchange in aqueous solution	Hasegawa et al. (2000)
2001			
Sodium montmorillonite Kunipia F (Kunimine Industries)	Dodecyltri-methyl-ammonium bromide	Cation exchange in aqueous solution	Klapyta et al. (2001)
Synthetic micas: Na-fluoro-tetrasilicic-mica and Li-fluorotaeniolite (Topy Industries)	Octadecyltri-methyl-ammonium bromide		
Sodium montmorillonite	Hexadecyltri-methyl-ammonium Epoxypropyl methacrylate	Cation exchange in aqueous solution (ammonium salt) + co-intercalation of epoxypropyl methacrylate	Liu and Wu (2001)

Pyzhevsky sodium montmorillonite (Ukraine)	Several alkyl-ammonium salts	Cation exchange in organic liquids	Moraru (2001)
Natural bentonite from the Cauca Valley (Colombia)	Tetramethyl-ammonium bromide Hexadecyltri-methyl-ammonium bromide Benzyl dimethyl hexadecyl-ammonium chloride Alkyl benzyl dimethyl-ammonium chloride	Cation exchange	Rodríguez-Sarmiento and Pinzón-Bello (2001)
Bentonite (Fisher Science)	Brij 78 Poly-ethylene glycol ether Dodecyl pyridinium chloride	Adsorption of surfactants in aqueous solution (proceeding similar to cation exchange)	Shen (2001)
2002			
Sodium montmorillonite (Wyoming, USA) Calcium montmorillonite (Texas, USA)	Tetramethyl-ammonium chloride	1. Acid activation and ion exchange in aqueous solution 2. Ion exchange in aqueous solution and acid activation	Moronta et al. (2002)
Montmorillonite K-10 (Aldrich)	Dodecyl-ammonium chloride	Cation exchange in aqueous solution	Park et al. (2002)
Sodium hectorite (Hector, California)	Dioctadecyl-dimethyl-ammonium chloride	Cation exchange in aqueous solution	Vougaris and Petridis (2002)

(Continued)

Table 3.4 *(Continued)*

Sodium montmorillonite Cloisite (Southern Clay) Potassium montmorillonite (obtained from the sodium form)	Crown ethers Cryptand	Similar to ion exchange: 1. Water/acetone (sodium and potassium clay minerals) 2. Iodide solution (sodium clay minerals)	Yao et al. (2002)
2003			
Natural montmorillonite from Cankiri (Anatolia, Turkey) Sepiolite from Mihalliccik-Eskisehir (Turkey)	2-amino pyrimidyne	Adsorption of 2-amino pyrimidine in ethyl alcohol solution	Akyüz and Akyüz (2003)
Sodium montmorillonite (Tolsa S.A., Spain)	Octadecyl-ammonium chloride	Cation exchange in aqueous solution	Arroyo et al. (2003)
Sodium montmorillonite (Wyoming, USA)	Dioctadecyl-dimethyl-ammonium chloride	Cation exchange	Bongiovanni et al. (2003)
Montmorillonite	Imidazolium salts		Bottino et al. (2003)
Sodium montmorillonite (Cloisite-Na) Sodium and calcium saponites (Arizona, USA) Hectorite (Bentone MA, Elementis)	Poly(ethylene-glycol)	Cation exchange	Chaiko (2003)
Sodium montmorillonite Kunipia F (CO-OP Chemical Co., Japan)	Hexadecyl-amine salt	Cation exchange in aqueous solution	Chang et al. (2003)

Fluorohectorite (CO-OP Chemical Co., Japan)	Octadecyl-ammonium chloride	Cation exchange in aqueous solution	Gorassi et al. (2003)
Bentonite (Sigma Chemical Co.)	A series of alkyl benzyl dimethyl-ammonium bromide	Cation exchange in aqueous solution	Kwolek et al. (2003)
Montmorillonite (Keyan Company)	Hexadecyltri-methyl-ammonium bromide Octadecyl-ammonium bromide	1. Cation exchange (2 salts) 2. Solubilization of hexadecyltrimethyl-ammonium in alcohol, wetting of montmorillonite in alcohol, and mixing of them in a blender at high speed and drying 3. Mixing of hexadecyltrimethyl-ammonium bromide and montmorillonite in a mortar (solid-state reaction)	Tang et al. (2003)
2004			
Bentonite of Resadiye (Tokat,Turkey)	Tetraethyl-ammonium iodide Tetrabutyl-ammonium bromide	Cation exchange	Akçay et al. (2004)
Sodium montmorillonite Cloisite (Southern Clay)	Alkyl-imidazolium salts	Cation exchange in water/ethanol (50/50, v/v)	Awad et al. (2004)
Sodium montmorillonite Cloisite (Southern Clay)	Butanol Hexanol Octanol	Ion–dipole technique (solid-state reaction)	Beall and Goss (2004)

(*Continued*)

Table 3.4 (*Continued*)

	Decanol Tetradecanol Octadecanol		
	Butanal Hexanal		
	Octanal Decanal Dodecanal		
	Methylpyrro- lidone Octylpyrro- lidone Dodecylpyrro- lidone		
Montmorillonite (Hebei, China)	Hexadecyltri-methyl-ammonium bromide	Cation exchange in aqueous solution	Hongping et al. (2004)
Sodium montmorillonite Kunipia F (Kunimine Industries)	Octadecyl-amine	Cation exchange	Kim et al. (2004)
Bentonite from Kampo (Korea)	Hexadecyl-amine-1 octadecyl-amine-1 Cetyltrimethyl-ammonium bromide Octadecyl-triethyl-ammonium bromide	Cation exchange in water/ethanol (4/1, v/v)	Lee and Lee (2004)
Sodium montmorillonite Somasif® ME100 (CO-OP Chemical Co., Japan)	A series of (alkyloxy-methyl) dimethyl-dodecyl Ammonium chlorides	Cation exchange in aqueous solution	Kozak and Domba (2004)

Sodium montmorillonite (Wyoming, USA)	*N*-cetylpyridinium chloride	Cation exchange	Pozsgay et al. (2004)
Sodium montmorillonite (Nanocor)	Maleic anhydride (MA) Pentaerythritol Decacylammoniumchloride Dodecylammoniumchloride Tetradecylammoniumchloride Octadecylammoniumchloride	MA dissolution and stirring in anhydrous acetone pentaerythritol-aqueous solution (proceeding similar to cation exchange) Ammonium salts cation exchange in aqueous solution	Sánchez Solís et al. (2004)
Calcium montmorillonite (Nanjing Tangshan Clay Factory, China)	Octadecylammonium chloride	Cation exchange in aqueous solution	Shishan et al. (2004)
Sodium montmorillonite Kunipia F (Kunimine Industries)	Octadecylammonium chloride Bis (2-hydroxyethyl) laurylammonium chloride Diethyl[2-(methacryloyoxyl) ethyl] ammonium chloride Bis(2-hydroxyethyl) lauryl (vinylbenzyl) ammonium chloride	Cation exchange in aqueous solution	Someya and Shibata (2004)

(*Continued*)

Table 3.4 (*Continued*)

Sodium montmorillonite (Ke Yan)	Hexadecyltri methyl-ammonium bromide	Cation exchange in aqueous solution	Wang et al. (2004)
Sodium montmorillonite (Wyoming, USA)	Octadecyl-trimethyl-ammonium bromide	Cation exchange in aqueous solution	Xi et al. (2004)
Bentonite from Resadiye (Tokat, Turkey)	Tetradecyl-trimethyl-ammonium bromide Hexadecyl-trimethyl-ammonium bromide	Cation exchange in aqueous solution	Yilmaz and Yapar (2004)
Sodium montmorillonite (Industrial Tech. Res. Institute)	(4-carboxy-butyl) -triphenyl-phospho-nium-bromide	Cation exchange	Yu et al. (2004)
Sodium montmorillonite (Kunimine Industries)	Octadecyl-ammonium chloride Maleic anhydride	Cation exchange in aqueous solution + cointercalation of maleic anydride	Zhang et al. (2004a,b)
2005			
Bentonite (Dash-Bent Trading Company)	Polymeric quaternary ammonium salt	Cation exchange in aqueous solution	Burmistr et al. (2005)
Sodium bentonite (Bentonita, Brazil)	Tetramethyl-[*x*,*y*] -ionenes (ammonium polyquater-naries)	Cation exchange in aqueous solution	Costa Filho et al. (2005)
Synthethic fluorohectorite	3-aminopropyl triethoxysilane	Grafting reaction in water/ethanol (25/75, v/v)	He et al. (2005)

Laponite (synthetic hectorite from Rockwood Additives Ltd.)	γ-metacryloxypropyl dimethyl methoxysilane γ-methacryloxypropyl trimethoxy silane	Grafting	Herrera et al. (2005)
Sodium montmorillonite (Wyoming, USA) Calcium montmorillontie (Arizona, USA)	Hexadeyltrimethylammonium bromide	Cation exchange in aqueous solution	Lee et al. (2005)
Bentonite (Sigma-Aldrich)	Hexadecyltrimethylammonium bromide	Cation exchange in aqueous solution	Majdan et al. (2005)
Bentonite Brasgel PA (Bentonita, Brazil)	Alkyl dimethyl benzylammonium chloride, cetyltrimethylammonium bromide	Cation exchange in aqueous solution	Pinto et al. (2005)
Sodium montmorillonite (Zhangjiakou Qinghe Chemical Factory, Hebei, China)	Octadecyltrimethylammonium chloride	Cation exchange in aqueous solution	Qin et al. (2005)
Natural bentonite (Bentonita, Brazil)	Cetyltrimethylammonium bromide	Cation exchange in aqueous solution	Ramos Filho et al. (2005)
Sodium montmorillonite (Wyoming, USA)	Cetylpyridiniumchloride monohydrate	Cation exchange	Százdi et al. (2005)

(*Continued*)

Table 3.4 (*Continued*)

Bentonite (Sigma-Aldrich) Bentonite Light Green Lages (Paraíba, Brazil) Alkyl dimethyl benzyl-ammonium chloride	Hexadecyl-trimethyl-ammonium chloride	Cation exchange in aqueous solution	Vianna (2005)
Sodium montmorillonite (Wyoming, USA)	Octadecyl-trimethyl-ammonium bromide	Cation exchange in aqueous solution	Xi et al. (2005)
Bentonite of Resadiye (Tokat, Turkey)	Hexadecyl-trimethyl-ammonium bromide	Cation exchange in aqueous solution	Yapar et al. (2005)
Sodium montmorillonite Kunipia F (Kunimine Industries)	Aniline salts	Solid-state reaction	Yoshimoto et al. (2005)
Sodium montmorillonite Cloisite (Southern Clay)	Ammonium salt based on triethylaminte and copolymer of vinylbenzyl chloride and lauryl acrylate	Dispersion of montmorillonite in a solution of tetrahydrofuran, THF	Zhang et al. (2005)
Montmorillonite (Hebei, China).	Cetyltrimethyl-ammonium bromide	Cation exchange in aqueous solution	Zhu et al. (2005)
2006			
Bentonite of Resadiye (Tokat, Turkey)	Tetraethyl-ammonium iodide Tetrabutyl-ammonium bromide	Cation exchange	Akçay (2006)

Bentonite Brasgel PA (Bentonita, Brazil)	Alkyl dimethyl benzyl ammonium chloride Dialkyl dimethyl-ammonium chloride Cetyl trimethyl-ammonium chloride/bromide	Cation exchange in aqueous solution	Araújo et al. (2006)
Sodium montmorillonite Cloisite (Southern Clay)	Quinolinium and pyridinium salts	Cation exchange in THF/water solution	Chigwada et al. (2006a)
Sodium montmorillonite Cloisite (Southern Clay)	Phenylaceto-phenone Dimethyl hexadecyl-ammonium salt	Cation exchange in THF/water solution	Chigwada et al. (2006b)
Montmorillonite K-10 (Aldrich)	Octadecyl-amine	Cation exchange in aqueous solution	Contreras et al. (2006)
Montmorillonite Volclay 325 (Wyoming, USA)	Hexadecyltri-methyl-ammonium bromide	Cation exchange in aqueous solution + adsorption of 1,3,4-thiadizole-2,5-dithiol	Filho and Carmo (2006)
Montmorillonite (IMI)	Dialkyl dimethyl-ammonium-chloride Octadecyl-amine	Not presented	Li et al. (2006)
Natural bentonites from San Juan Province (Argentine)	Hexadecyl-trimethyl-ammonium bromide	Cation exchange in aqueous solution	Volzone et al. (2006)

(*Continued*)

Table 3.4 (*Continued*)

Sodium montmorillonite Cloisite (Southern Clay)	Ammonium salt based on triethylamine and terpolymer of vinilbenzyl chloride, styrene, and lauryl acrylate	Dispersion of montmorillonite in a solution of THF/water	Zhang et al. (2006a)
Sodium montmorillonite Cloisite (Southern Clay)	Ammonium salt based on triethylaminte and copolymer of vinylbenzyl chloride and lauryl acrylate	Dispersion of montmorillonite in a solution of THF/water	Zhang et al. (2006b)
Sodium montmorillonite Cloisite (Southern Clay)	2-methyacrylo-xyloxyethyl dimethyl hexadecyl -ammonium bromide	Dispersion of clay in a solution of THF/water	Zheng et al. (2006)
2007			
Sodium montmorillonite Cloisite (Southern Clay)	Dimethyl-dihydro-genated Tallow ammonium chloride	Cation exchange in aqueous solution	Filippi et al. (2007)
Attapulgite	Octodecyl trimethyl ammonium chloride (OTMAC)	Exchage/aqueous/sonication	Huang et al. (2007)
Bentonite (Akli Mines, Barmer, Rajasthan, India)	Quaternary phosphonium salts	Cation exchange in aqueous solution	Patel et al. (2007)
Bentonite	Hexadecyltri-methyl-ammonium bromide	Cation exchange in aqueous solution	Richards and Bouazza, (2007)

Sodium montmorillonite Cloisite (Southern Clay)	Dimethyl bis (hydrogenated -tallow) ammonium	Ion exchange reaction	Yoo et al. (2007)
Montmorillonite-Na+	Hexadecyltri-methyl-ammonium bromide	Exchange/aqueous	Yan et al. (2007)
Bentonite	Al pillared and TMCS and OTS modified	Intercalation in aqueous medium, grafting of organic compound in pyridine, N_2 atmosphere	Zhu et al. (2007)
2008			
Sodium montmorillonite Cloisite (Southern Clay)	Glycidyl-propyl-triethoxy-sylane	Dispersion of clay in ethanol/water	Di Gianni et al. (2008)
Sepiolite	Dodecyl-trimethyl-ammonium bromide (DTMA)	Cation exchange in aqueous solution	Gok et al. (2008)
Na-bentonite (Wyoming, Columbia)	Berberine, metolachlor, diquat, and difenzoquat	Organic cation reaction	Rytwo et al. (2008)
Smectite	3-mercapto-propyl-trimethoxy silane (MPTMS)	Grafted/toluene medium/argon atmosphere	Tonle et al. (2008)
Montmorillonite clay vesicle	Didodecyl-dimethyl-ammonium bromide (DDAB)	Cation exchange in aqueous solution	Undabeytia et al. (2008)

(*Continued*)

Table 3.4 (*Continued*)

Sodium montmorillonite	Hexyltri-phenyl-phosphonium bromide	Cationic exchange reaction	Wang et al. (2008)
Montmorillonite	Octyltri-methyl-ammonium-bromide (OTAB) Dodecyl-trimethyl-ammonium bromide (DTAB) Cetyltri-methyl-ammonium bromide Stearyltri-methyl-ammonium bromide (STAB)	Cation exchange in aqueous solution	Wang and Wang (2008)
2009			
Montmorillonite	Tetrabutyl-ammonium (TBA)	Cation exchange in aqueous solution	Akcay et al. (2009)
Bentonite Kaolinite	Hexadecyltri-methyl-ammonium bromide Phenyltri-methyl-ammonium-bromide (PTMA)	Adsorption in aqueous medium/ stirring	Alkaram et al. (2009)
Bentonite	Al and Cr Intercalated and CTMAB modified	Aqueous medium	Bouberka et al. (2009)

Bentonite	Poly(epi-choloro-hydrin dimethyl-amine) (EPIDMA)	Aqueous/ complexation	Kang et al. (2009)
Bentonite	Trimethyl-ammonium bromide (HDTMA) Benzyltriethyl-ammonium chloride (BTEA)	Cation exchange in aqueous solution	Marsal et al. (2009)
Montmorillonite	2-(3-(2-amino-ethylthio) propylthio) ethanamine (AEPE)	Grafted/ acetonitrile/N_2 atmosphere	Phothiton-timongkol et al. (2009)
Bentonite	Polydiallydi-methyl-ammonium (PDADMA)	Cation exchange in aqueous solution	Shen et al. (2009)
Montmorillonite	Tetramethyl-ammonium (TMA)	Cation exchange in aqueous solution	Vidal and Volzone (2009)
2010			
Bentonite	Hexadecyltri-methyl-ammonium bromide	Cation exchange in aqueous solution	Jović -Jovičić et al. (2010)
Bentonite Zeolite	Cetyltrimethyl-ammonium bromide Hexadecyltri-methyl-ammonium bromide	Cation exchange in aqueous solution	Koswojo et al. (2010)
Sodium montmorillonite Cloisite (Southern Clay)	Methyl tallow bis-2-hydroxyethyl, quaternary ammonium	Not presented	Nemato-llahi et al. (2010)

(*Continued*)

Table 3.4 (*Continued*)

Bentonite	Hexadecyltri-methyl-ammonium	Cation exchange in aqueous solution	Xi et al. (2010)
Montmorillonite	Hexadecyltri-methyl-ammonium bromide	Cation exchange in aqueous solution	Zhou et al. (2010)
2011			
Sodium montmorillonite Cloisite (Southern Clay)	Ditallow dimethyl alkyl-ammonium chloride Ditallow dimethyl ammonium chloride (DTDMA)	Cation exchange in water/ethanol	McLauchlin et al. (2011)
Montmorillonite (Wyoming, SWY-1)	Dodecyltri-methyl-ammonium bromide (DoDTMA) Octadecyltri-methyl-ammonium chloride Hexadecyltri-methyl-ammonium bromide	Cation exchange in aqueous solution	Kim et al. (2011)
Montmorillonite	Tetradecyltri-methyl-ammonium bromide	Ion exchange reactions	Park et al. (2011)

Sodium montmorillonites Cloisite (Southern Clay)	Benzyldi-methyldodecyl ammonium chloride Dimethyl-dioctadecyl ammonium bromide Cetyltrimethyl-ammonium bromide Octadecyltri-methyl-ammonium bromide	Cation exchange in aqueous solution	Seliem et al. (2011)
Montmorillonite	Polydiallyldi-methyl-ammonium (PDADMA) Cetyltri-methyl-ammonium bromide	Cation exchange in aqueous solution	Zhu et al. (2011)
2012			
Sodium montmorillonite Cloisite (Southern Clay)	Methyl tallow bis-2-hydroxyethyl, quaternary ammonium	Not presented	Belver et al. (2012)
Montmorillonite Cloisite (Southern Clay)	Dimethyl hydrogenated tallow benzyl ammonium (DMHTB)	Cation exchange in aqueous solution	Elban et al. (2012)
Illite–smectite	Hexadecyltri-methyl-ammonium bromide	Cation exchange in aqueous solution	Gammoudi et al. (2012)

(*Continued*)

Table 3.4 *(Continued)*

Bentonite	Hexadecyltri-methyl-ammonium bromide	Cation exchange in aqueous solution	Mirmo-hamad-sadeghi et al. (2012)
Bentonite	1-Hexadecyl-pyridinium bromide	Not presented	Zhang et al. (2012)
2013			
Montmorillonite	Ethyl hexadecyl dimethyl ammonium (EHDDMA)	Cation exchange in aqueous solution	Beltrán et al. (2013)
Sodium montmorillonite (Jebel Sbih, Tunisia)	Hexadecyltri-methyl-ammonium bromide	Cation exchange in aqueous solution	Dammak et al. (2013)
Montmorillonite	Cetyltrimethyl-ammonium bromide	Cation exchange in aqueous solution	Fatimah and Huda (2013)
Sodium montmorillonite	Tetradecyl-trimethyl-ammonium bromide (TTAB), cetyltri-methyl-ammonium bromide, teartri-monium bromide (STAB)	Cation exchange in aqueous solution	Hu et al. (2013)
Laterite of Haripur Bentonite of Swabi Khyber, Pakhtunkhwa	Hexadecyltri-methyl-ammonium chloride Tetradecyltri-methyl-ammonium bromide	Cation exchange reaction of lattice ions	Nafees et al. (2013)

Sodium montmorillonite Cloisite (Southern Clay)	L-tyrosine amino acid	Cation exchange in aqueous solution	Mallakpour et al. (2013)
2014			
Bentonite	Ethanolamine (EA) Cetyltri-methyl-ammonium bromide Tetra-octadecyl-ammonium bromide (TO)	Cation exchange in aqueous solution	Zawrah et al. (2014)
Sodium montmorillonite	Cetyltrimethyl-ammonium bromide	Cation exchange in aqueous solution	Gürses et al. (2014)
Bentonite	Dodecyl-, tetradecyl-, hexadecyl-, octadecyl-trimethyl-ammonium bromide	Cation exchange in water/acid	Zaghouane-Boudiaf et al. (2014)
Sodium montmorillonite Cloisite (Southern Clay)	Quaternary dimethyl benzyl hydrogenated tallow ammonium	Cation exchange in aqueous solution	Papageo-rgiou et al. (2014)
Sodium montmorillonite	Cetyltrimethyl-ammonium bromide	Cation exchange in aqueous solution	Abd El-Ghaffar et al. (2014)
Calcium montmorillonite	Octadecyltri-methyl-ammonium chloride	Cation exchange in aqueous solution	Zhu et al. (2014)

Source: Reproduced by updating from de Paiva et al., Copyright 2008, with permission from Elsevier.

3.4 Applications of Organoclays as Sorbents

Organoclays have been used as sorbents in many applications. Some studies have showed that replacing the inorganic exchangeable cations of clay minerals with organic cations can result in greatly enhanced abilities to remove organic contaminants from water. The applications of organoclays as sorbents have been given in detail in the following section.

3.4.1 Aromatic Compounds

Aromatic wastes have become an ever-increasing concern for the environment. Organoclays have been utilized in a variety of studies aimed at environmental cleanup applications. Some researchers have used organoclays to remove aromatic pollutants (Sheng et al., 1996). HDTMA-exchanged clays were shown to be able to adsorb benzene, nitrobenzene, chlorobenzene, trichloroethylene, and carbon tetrachloride. It was found that organoclays are effective sorbents for neutral organic contaminants (NOCs). In some studies, the sorptive characteristics of organoclays were compared in terms of magnitude and mechanism to those of natural soil organic matter. The data presented supports multiple mechanisms controlling the sorption of aqueous-phase NOCs on organoclays, including solvation of the cationic ammonium centers, the alkyl chains of HDTMA, the mineral surfaces, and solute partitioning. In contrast, solute partitioning appears to be the singular mechanism for the sorption of aqueous-phase NOCs by natural soil organic matter. Organic contaminant sorption by the surfactant-derived organic matter of organoclays is substantially higher than that of natural soil organic matter due to the existence of multiple sorptive mechanisms for organoclays and the greater solvency of their organic phases.

Organoclays have also been used as a barrier to prevent the transport of hazardous contaminants in landfills. Generally speaking, clay minerals are known to be effective sorbents for mostly inorganic contaminants, while organoclays are mainly used for organic contaminants. Since organoclays are basically clay particles modified with cationic surfactants, there is an optimal coverage of cationic surfactant on the clay particles to adsorb both inorganic and organic contaminants. From an investigation of chlorobenzene and lead as contaminants on HDTMA/bentonites, it has been observed that when either chlorobenzene or lead ions exist as a single contaminant, chlorobenzene sorption increases with increasing HDTMA-to-bentonite ratios. Sorption of chlorobenzene was a function of HDTMA coverage on the bentonites, while lead sorption was much more influenced by the initial lead concentration rather than the mass of HDTMA added to the bentonites.

Organophilic clays have also been shown to play a significant role in removing contaminants from liquid wastes. Gitipour et al. (1997) used modified organophilic bentonites that are prepared by replacing the exchangeable inorganic cations with quaternary ammonium salts (HDTMA) to remove aromatic organics such as benzene, toluene, ethylbenzene, and *o*-xylene (BTE compounds) from contaminated soil. The results reveal that organophilic bentonite can act as a successful adsorbent for removing aromatic organics from contaminated soil.

3.4.1.1 Phenol and its derivatives

Besides removing aromatic contaminants, there is some research that concerns adsorption of phenol or naphthol. Lawrence et al. (1998) undertook studies on sorption of phenol and 2-, 3-, and 4-chlorophenol from water by tetramethylammonium (TMA)-smectite and tetramethylphosphonium (TMP)-smectite. Sorption of phenolic compounds appears to occur on the aluminosilicate surfaces between the neighboring organic cations (TMA or TMP). It was found again that TMP-smectite is a better sorbent than TMA-smectite, which did not measurably adsorb any of the phenolic compounds. This disparity in sorption efficiency was attributed to differences in hydration of the interlayer cations (Kukkadapu and Boyd, 1995). The authors also thought that the higher degree of hydration of the TMA cation in smectite renders the organoclay unable to accommodate space for the phenols in its interlayer region, causing the interlayer pore size to be smaller in TMA-smectite than in TMP-smectite. TMP-smectite showed selective sorption within the group of chlorinated phenols studied. Phenol and 4-chlorophenol were effectively sorbed by TMP-smectite, whereas 2- and 3-chlorophenol were not sorbed. The selectivity appeared to be size- and shape dependent and not strongly influenced by water solubility.

Dentel et al. (1995) and Jiang et al. (2002) have used HDTMA montmorillonite as an adsorbent to remove phenol and also studied dimethyldistearylammoniumchloride (DDAM) and DDAM partially re-exchanged with Ca^{2+} (Ca-DDAM) montmorillonite for sorption of phenol, 2,4,5-trichlorophenol, and tannic acid. This study also investigated the influence of pH on sorption of phenol and trichlorophenol. Adsorption of the organic molecules increased in the order phenol < tannic acid < trichlorophenol, which corresponds to the increase of hydrophobicity, indicating a partitioning mechanism. The authors also found that if a portion of the intercalated DDA is removed by exchanging with Ca^{2+}, the *d* (001) spacing will decrease but the sorption capacity of the organoclays will be improved.

3.4.1.2 Trichloroethylene

Trichloroethylene has been listed as apriority pollutant and is one of the pollutants present in water most frequently and at high concentrations. Organoclays were found to be promising adsorbends for trichloroethylene and other HOCs.

Zhao and Vance (1998) examined trichloroethylene sorption by organoclays HDTMA- and didodecyldimethylammonium (DODMA)-exchanged smectites (SWy-1 and Saz-1) in aqueous solutions, both in the presence and in the absence of humic substances. Their results indicated the uptake of humic substances by organoclays and the effects of humic substances on trichloroethylene sorption by organoclays were dependent on the type of organoclay and humic substance. The results of the study indicate that organoclays, such as HDTMA-Saz and DODMA-Saz, could be used to remove trichloroethylene from wastewaters that contain high concentrations of background organic solutes.

3.4.1.3 Pesticides and herbicides

The presence of pesticides in water and soil has raised concerns for the protection of the environment, in particular protection of drinking water quality. Celis et al. (2002) studied the adsorption of the herbicide hexazinone by montmorillonites modified with organic cations (hexadecyltrimethylammonium-saturated Arizona montmorillonite [HDTMA-SA]).

Carrizosa et al. (2001) studied the adsorption of the herbicide dicamba by organoclays Saz and SWy modified by octadecyl (C18), HDTMA, and DODMA from 0.5 to 1.0 CEC at different concentrations and pHs. Their results showed that the sorption capacity of organoclays is favored for high-layer charge and saturation with bulky organic cations close to the CEC. Hydrophobic interactions with polar contributions are responsible for adsorption of dicamba molecules, which need available polar interlayer room between alkylammonium in the organoclays. Large primary alkylammonium cations in low- and high-charge clay and very large quaternary alkylammonium ions in low-charge clay enhanced the irreversibility of the adsorption.

3.4.1.4 Anionic contaminants

Landfill leachate, often having high concentrations of metal cations, anions, and organic compounds, presents a great threat to nearby groundwater. Due to negative charges on soil particles, regular clay liners cannot effectively retard the movement of anionic contaminants such as chromate. Li et al. (2005) found that when illite is modified by cationic surfactants with a long hydrophobic tail group and in a low critical micelle concentration (CMC), for example, by HDTMA, the modified illite is an effective sorbent for anionic contaminants such as chromate. Modified illite surfactant loading was beyond its CEC, which greatly assists in preventing migration of contaminant anions in applications such as landfill liners (Xi, 2006).

References

Abd El-Ghaffarb, M. A., Youssef, A. M., Abd El-Hakim, A. A. (2014). Polyaniline nanocomposites via in situ emulsion polymerization based on montmorillonite: Preparation and characterization. *Arabian J. Chem.* (Article in press).

Akçay, G., Kilinç, E., Akcay, M. (2009). The equilibrium and kinetics studies of flurbiprofen adsorption onto tetrabutyl ammonium montmorillonite (TBAM). *Colloids Surf., A*, **335**, pp. 189–193.

Akçay, M. (2004). Characterization and determination of the thermodynamic and kinetic properties of p-CP adsorption onto organophilic bentonite from aqueous solution. *J. Colloid Interface Sci.*, **280**, pp. 299–304.

Akçay, M. (2006). Characterization and adsorption properties of tetrabutil ammonium montmorillonite (TBAM) clay thermodynamic and kinetic calculations. *J. Colloid Interface Sci.*, **296**, pp. 16–21.

Akyüz, S., Akyüz, T. (2003). FT-IR spectroscopic investigations of surface and intercalated 2-aminopyrimidine adsorbed on sepiolite and montmorillonite from anatolia. *J. Mol. Struct.*, **651–653**, pp. 205–210.

Alkaram, U. F., Mukhlis, A. A., Al-Dujaili, A. H. (2009). The removal of phenol from aqueous solutions by adsorption using surfactant-modified bentonite and kaolinite. *J. Hazard. Mater.*, **169**, pp. 324–332.

Araújo, E. M., Melo, T. J. A., Oliveira, A. D., Araújo, H. L. D., Araújo, K. D., Barbosa, R. (2006). Preparação de argilas organofílicas e desenvolvimento de nanocompósitos com matrizes poliméricas de polietileno e nylon6. Part e 1: comportamento mecânico. *Polímeros Ciência Tecnologia*, **16**, pp. 38–45.

Arroyo, M., López-Manchado, M. A., Herrero, B. (2003). Organo-montmorillonite as substitute of carbon black in natural rubber compounds. *Polymer*, **44**, pp. 2447–2453.

Auerbach, S. M., Carrado, K. A., Dutta, P. K. (2004). *Handbook of Layered Materials*. USA: Marcel Dekker. ISBN: 0-8247-5349-6.

Awad, W. H., Gilman, J. W., Nyden, M., Harris Jr., R. H., Sutto, T. E., Callahan, J., Truvole, P. C., DeLong, H. C., Fox, D. M. (2004). Thermal degradation studies of alkyl-imidazolium salts and their application in nanocomposites. *Thermochim. Acta*, **409**, pp. 3–11.

Baldassari, S., Komarneni, S., Mariani, E., Villa, C. (2006). Microwave versus conventional preparation of organo clays from natural and synthetic clays. *Appl. Clay Sci.*, **31**, pp. 134–141.

Beall, G. W., Goss, M. (2004). Self-assembly of organic molecules on montmorillonite. *Appl. Clay Sci.*, **27**, pp. 179–186.

Beltrán, M. I., Benavente, V., Marchante, V., Marcilla, A. (2013). The influence of surfactant loading level in a montmorillonite on the thermal, mechanical and rheological properties of EVA nanocomposites. *Appl. Clay Sci.*, **83–84**, pp. 153–161.

Belver, C., Aranda, P., Martín-Luengo, M. A., Ruiz-Hitzky, E. (2012). New silica/alumina–clay heterostructures. Properties as acid catalysts. *Microporous Mesoporous Mater.*, **147**, pp. 157–166.

Bergaya, F., Lagaly, G. (2001). Surface modifications of clay minerals. *Appl. Clay Sci.*, **19**, pp. 1–3.

Bergaya, F., Theng, B. K. G., Lagaly, G. (2006). *Handbook of Clay Science*. 1st Ed. Amsterdam, The Netherlands: Elsevier.

Bongiovanni, R., Monchietto, R., Sangermano, M., Malucelli, G., Priola, A. (2003). Epoxy coatings, via cationic UV-curing, containing nanodispersed phyllosilicates. *Annais of XVI. Pisa, Italy. Convegno Italiano di Scienza e Tecnologia delle Macromolecole*, pp. 245–246.

Borisover, M., Bukhanovsky, N., Lapides, I., Yariv, S. (2010). Thermal treatment of organo clays: Effect on the aqueous sorption of nitrobenzene on n-hexadecyltrimethyl ammonium montmorillonite. *Appl. Surf. Sci.*, **256**, pp. 5539–5544.

Bottino, F., Fragalà, I., Malandrino, G., Orestano, A., Pollicino, A. (2003). Polystyrene clay nanocomposites prepared with polymerizable imidazolium surfactants. *Annais of XVI. Pisa, Italy.Convegno Italiano di Scienza e Tecnologia delle Macromolecole*, pp. 229–230.

Bouberka, Z., Khenifi, A., Mahamed, H. A., Haddou, B., Belkaid, N., Bettahar, N., Derriche, Z. (2009). Adsorption of supranol yellow 4 GL from aqueous solution by surfactant-treated aluminum/chromium-intercalated bentonite. *J. Hazard. Mater.*, **162**, pp. 378–385.

Breakwell, K. I., Homer, J., Lawrence, M. A. M., McWhinnie, W. R. (1995). Studies of organophilic clays: The distribution of quaternary ammonium compounds on clay surfaces and the role of impurities. *Polyhedron*, **14**, pp. 2511–2518.

Burmistr, M. V., Sukhy, K. M., Shilov, V. V., Pissis, P., Spanoudaki, A., Sukha, I. V., Tomilo, V. I., Gonza, Y. P. (2005). Synthesis, structure, thermal and mechanical properties of nanocomposites based on a linear polymers and layered silicates modified by polymeric quaternary ammonium salts (ionenes). *Polymer*, **46**, pp. 12226–12232.

Carrizosa, M. J., Koskinen, W. C., Hermosin, M. C., Cornejo, J. (2001). Dicamba adsorption–desorption on organo clays. *Appl. Clay Sci.*, **18**, pp. 223–231.

Celis, R., Hermosian, M. C., Carrizosa, M. J., Cornejo, J. (2002). Inorganic and organic clays as carriers for controlled release of the herbicide hexazinone. *J. Agric. Food Chem.*, **50**, pp. 2324–2330.

Chaiko, D. J. (2003). New poly(ethylene oxide)-clay composites. *Chem. Mater.*, **15**, pp. 1105–1110.

Chang, J. H., An, Y. V., Cho, D., Giannelis, E. P. (2003). Poly(lactic acid) nanocomposites: Comparison of their properties with montmorillonite and synthetic mica (II). *Polymer*, **44**, pp. 3715–3720.

Chigwada, G., Wang, D. D., Wilkie, C. A. (2006a). Polystyrene nanocomposites based on quinolinium and pyridinium surfactants. *Polym. Degrad. Stab.*, **91**, pp. 848–855.

Chigwada, G., Wang, D. D., Jiang, D. D., Wilkie, C. A. (2006b). Styrenic nanocomposites prepared using a novel biphenyl-containing modified clay. *Polym. Degrad. Stab.*, **91**, pp. 755–762.

Contreras, V., Cafiero, M., Da Silva, S., Rosales, C., Perera, R., Matos, M. (2006). Characterization and tensile properties of ternary blends with PA-6 nanocomposites. *Polym. Eng. Sci.*, **46**, pp. 1111–1120.

Costa Filho, A. P., Gomes, A. S., Lucas, E. S. (2005). Preparação e caracterização de organobentonita modificada com ionenos alifáticos. *Polímeros Ciência Tecnologia*, **15**, pp. 212–217.

Dammaka, N., Fakhfakh, N., Fourmentin, S., Benzina, M. (2013). Natural clay as raw and modified material for efficient o-xylene abatement. *J. Environ. Chem. Eng.*, **1**, pp. 667–675.

Dentel, S. K., Bottero, J. Y, Khatib, K., Demougeot, H., Duguet, J. P, Anselme, C. (1995). Sorption of tannic acid, phenol and 2,4,5 trichloro phenol on organoclays. *Water Res.*, **29**, pp. 1273–1280.

de Paiva, L. B., Morales, A. R., Francisco, R., Díaz, V. (2008). Organoclays: Properties, preparation and applications. *Appl. Clay Sci.*, **42**, pp. 8–24.

Di Gianni, A., Amerio, E., Monticelli, O., Bongiovanni, R. (2008). Preparation of polymer/clay mineral nanocomposites via dispersion of silylated montmorillonite in a UV curable epoxy matrix. *Appl. Clay Sci.*, **42**, pp. 116–124.

Elban, W. L., Howarter, J. A., Richardson, M. C., Stutzman, P. E., Forster, A. M., Nolte, A. J., Holmes, G. A. (2012). Influence of solvent washing on interlayer structure of alkylammonium montmorillonites. *Appl. Clay Sci.*, **61**, pp. 29–36.

Fatimah, I., Huda, T. (2013). Preparation of cetyltrimethylammonium intercalated Indonesian montmorillonite for adsorption of toluene. *Appl. Clay Sci.*, **74**, pp. 115–120.

Filho, N. L. D., Carmo, D. R. (2006). Study of an organically modified clay: Selective adsorption of heavy metal ions and voltammetric determination of mercury (II). *Talanta*, **68**, pp. 919–926.

Filippi, S., Mameli, E., Marazzato, C., Magagnini, P. (2007). Comparison of solution-blending and melt-intercalation for the preparation of poly(ethylene-co-acrylicacid)/organo clay nanocomposites. *Eur. Polym. J.*, **43**, pp. 1645–1659.

Gammoudi, S., Frini-Srasra, N., Srasra, E. (2012). Influence of exchangeable cation of smectite on HDTMA adsorption: Equilibrium, kinetic and thermodynamic studies. *Appl. Clay Sci.*, **69**, pp. 99–107.

Gieseking, J. E. (1939). The mechanism of cation exchange in the montmorillonite–beidellite–nontronite type of clay minerals. *Soil Sci.*, **47**, pp. 1–14.

Gitipour, S., Bowers, M. T., Bodocsi, A. (1997). The use of modified bentonite for removal of aromatic organics from contaminated soil. *J. Colloid İnterface Sci.*, **196**(2), pp. 191–198.

Gök, O., Ozcan, A. S., Ozcan, A. (2008). Adsorption kinetics of naphthalene onto organo-sepiolite from aqueous solutions. *Desalination*, **220**, pp. 96–107.

Gorassi, G., Tortora, M., Vittoria, V., Kaempfer, D., Mülhaupt, R. (2003). Transport properties of organic vapor in nanocomposites of organophilic layered silicate and syndiotactic polypropylene. *Polymer*, **44**, pp. 3679–3685.

Gürses, A., Günes, K., Mindivan, F., Ejder Korucu, M., Açıkyıldız, M., Doğar, C. (2014). The investigation of electrokinetic behaviour of micro-particlesproduced by CTA^+ ions and Na-montmorillonite. *Appl. Surf. Sci.*, **318**, pp. 79–84.

Hasegawa, N., Okamoto, H., Kato, M., Usuki, M. (2000). Preparation and mechanical properties of polypropylene-clay hybrids based on modified polypropylene and organophilic clay. *J. Appl. Polym. Sci.*, **78**, pp. 1918–1922.

He, H., Duchet, J., Galy, J., Gerard, J. F. (2005). Grafting of swelling clay materials with 3-aminopropyltriethoxysilane. *J. Colloid Interface Sci.*, **288**, pp. 171–176.

He, H., Zhou, Q., Frost, R. L., Wood, B. J., Duong, L. V., Kloprogge, J. T. (2007). A X-ray photoelectron spectroscopy study of HDTMAB distribution within organoclays. *Spectrochim. Acta A*, **66**, pp. 1180–1188.

Herrera, N. N., Letoffe, J. M., Reymond, J. P., Bourgeat-Lami, E. (2005). Silylation of laponite clay particles with monofunctional and trifunctional vinyl alkoxysilanes. *J. Mater. Chem.*, **15**, pp. 863–871.

Hongping, H., Ray, F., Jianxi, Z. (2004). Infrared study of $HDTMA^+$ intercalated montmorillonite. *Spectrochim. Acta A*, **60**, pp. 2853–2859.

Hu, Z., He, G., Liu, Y., Dong, C., Wu, X., Zhao, W. (2013). Effects of surfactant concentration on alkyl chain arrangements in dry and swollen organic montmorillonite. *Appl. Clay Sci.*, **75–76**, pp. 134–140.

Huang, J., Wang, X., Jin, Q., Liu, Y., Wang, Y. (2007). Removal of phenol from aqueous solution by adsorption onto OTMAC-modified attapulgite. *J. Environ. Manage.*, **84**, pp. 229–236.

Jiang, J. Q., Cooper, C., Ouki, S. (2002). Comparison of modified montmorillonite adsorbents, part I: Preparation, characterization and phenol adsorption. *Chemosphere*, **47**, pp. 711–716.

Jović-Jovičić, N., Milutinović-Nikolić, A., Banković, P., Mojović, Z., Žunić, M., Gržetić, I., Jovanović, D. (2010). Organo-inorganic bentonite for simultaneous adsorption of acid orange 10 and lead ions. *Appl. Clay Sci.*, **47**, pp. 452–456.

Kang, Q., Zhou, W., Li, Q., Gao, B., Fan, J., Shen, D. (2009). Adsorption of anionic dyes on poly(epicholorohydrin dimethylamine) modified bentonite in single and mixed dye solutions. *Appl. Clay Sci.*, **45**, pp. 280–287.

Kim, J. H., Koo, C. M., Choi, Y. S., Wang, K. H., Chung, I. J. (2004). Preparation and characterization of polypropylene/layered silicate nanocomposites using an antioxidant, *Polymer*, **45**, pp. 7719–7727.

Kim, J. Y., Komarneni, S., Parette, R., Cannon, F., Katsuki, H. (2011). Perchlorate uptake by synthetic layered double hydroxides and organo-clay minerals. *Appl. Clay Sci.*, **51**, pp. 158–164.

Klapyta, Z., Fujita, T., Iyi, N. (2001). Adsorption of dodecyl and octadecyltrimethyl ammonium ions on a smectite and synthetic micas. *Appl. Clay Sci.*, **19**, pp. 5–10.

Koswojo, R., Utomo, R. P., Ju, Y. H., Ayucitra, A., Soetaredjo, F. E., Sunarso, J., Ismadji, S. (2010). Acid Green 25 removal from wastewater by organo-bentonite from Pacitan. *Appl. Clay Sci.*, **48**, pp. 81–86.

Kozak, M., Domba, L. (2004). Adsorption of the quaternary ammonium salts on montmorillonite. *J. Phys. Chem. Solids*, **65**, pp. 441–445.

Kukkadapu, R. K., Boyd, S. (1995). Tetramethylphosphonium- and tetramethyl ammonium smectites as adsorbents of aromatic and chlorinated hydrocarbons: Effect of water on adsorption efficiency. *Clays Clay Miner.*, **43**(3), pp. 318–323.

Kwolek, T., Hodorowicz, M., Standnick, K., Czapkiewicz, J. (2003). Adsorption isotherms of homologous alkyldimethyl-benzylammonium bromides on sodium montmorillonite. *J. Colloid Interface Sci.*, **264**, pp. 14–19.

Lawrence, M. A. M., Kukkadapu, R. K., Boyd, S. A. (1998). Adsorption of pjenol and chlorinated phenols from aqueous solution by tetramethylammonium- and tetramethylphosphonium-exchanged montmorillonite. *Appl. Clay Sci.*, **13**, pp. 13–20.

Lee, J. Y., Lee, H. K. (2004). Characterization of organobentonite used for polymer nanocomposites. *Mater. Chem. Phys.*, **85**, pp. 410–415.

Lee, S., Tiwari, M. D. (2012). Organo and inorgano-organo-modified clays in the remediation of aqueous solutions: An overview. *Appl. Clay Sci.*, **59-60**, pp. 84–102.

Lee, S. Y., Echo, W. J., Kim, K. J., Ahn, J. H., Lee, M. (2005). Interaction between cationic surfactants and montmorillonites under nonequilibrium condition. *J. Colloid Interface Sci.*, **284**, pp. 667–673.

Lee, S. Y., Kim, S. J. (2002). Expansion reserved. Characteristics of organoclay as a precursor to nanocomposites. *Colloids Surf., A*, **211**, pp. 19–26.

Li, H., Yu, Y., Yang, Y. (2005). Synthesis of exfoliated polystyrene/montmorillonite nanocomposite by emulsion polymerization using a zwitterion as the clay modifier. *Eur. Polym. J.*, **41**, pp. 2016–2022.

Li, J., Ton-That, M. T., Tsai, S. J. (2006). PP-based nanocomposites with various intercalant types and intercalant coverages. *Polym. Eng. Sci.*, **46**, pp. 1060–1068.

Litina, K., Miriouni, A., Gournis, D., Karakassides, M. A., Georgiou, N., Klontzas, E., Ntoukas, E., Avgeropoulos, A. (2006). Nanocomposites of polystyrene-b-polyisoprene copolymer with layered silicates and carbon nanotubes. *Eu. Polym. J.*, **42**(9), pp. 2098–2107.

Liu, X., Wu, Q. (2001). PP/clay nanocomposites prepared by grafting–melt intercalation. *Polymer*, **42**, pp. 10013–10019.

Majdan, M., Maryuk, O., Pikus, S., Olszewska, E., Kwiatkowski, R., Skrzypek, H. (2005). Equilibrium, FTIR, scanning electron microscopy and small wide angle X-ray scattering studies of chromates adsorption on modified bentonite. *J. Mol. Struct.*, **70**, pp. 203–211.

Mallakpour, S., Mohammad, D. (2013). Novel nanocomposites based on reactive organoclay of L-tyrosine and amine end-capped poly(amide-imide): Synthesis and characterization. *Appl. Clay Sci.*, **75–76**, pp. 67–73.

Mandalia, T., Bergaya, F. (2006). Organo clay mineral-melted polyolefin nanocomposites effect of surfactant/CEC ratio. *J. Phys. Chem. Solids*, **67**, pp. 836–845.

Marsal, A., Bautista, E., Ribosa, I., Pons, R., García, M. T. (2009). Adsorption of polyphenols in waste water by organo-bentonites. *Appl. Clay Sci.*, **44**, pp. 151–155.

McLauchlin, A., Bao, X., Zhao, F. (2011). Organoclay polybutyleneterephthalate nanocomposites using dual surfactant modified montmorillonite prepared by the master batch method. *Appl. Clay Sci.*, **53,** pp.749–753.

Merinska, D., Malac, Z., Pospisil, M., Weiss, Z., Chmielova, M., Capkova, P., Simonik, J. (2002). Polymer/clay nanocomposites based on MMT/ODA intercalates. *Compos. Interfaces*, **9**, pp. 529–540.

Mirmohamadsadeghi, S., Kaghazchi, T., Soleimani, M., Asasian, N. (2012). An efficient method for clay modification and its application for phenol removal from wastewater. *Appl. Clay Sci.*, **59–60**, pp. 8–12.

Moraru, V. N. M. (2001). Structure formation of alkylammonium montmorillonites in organic media. *Appl. Clay Sci.*, **19**, pp. 11–26.

Moronta, A., Ferrer, V., Quero, J., Arteaga, G., Chore, E. (2002). Influence of preparation method on the catalytic properties of acid-activated tetramethyl ammonium exchanged clays. *Appl. Catal. A Gen.*, **230**, pp. 127–135.

Nafees, M., Waseem, A., Khan, A. (2013). Comparative study of laterite and bentonite based organoclays: Implications of hydrophobic compounds remediation from aqueous solutions. *Sci. World J.*, **2013**, p. 6.

Nematollahi, M., Heidarian, M., Peikari, M., Kassiriha, S. M., Arianpouya, N., Esmaeilpour, M. (2010). Comparison between the effect of nanoglass flake and montmorillonite organo clay on corrosion performance of epoxy coating. *Corros. Sci.*, **52**, pp. 1809–1817.

Ogawa, M., Handa, T., Kuroda, K., Kato, C. (1990). Formation of organo ammonium montmorillonites by solid–solid reactions. *Chem. Lett.*, **19**, pp. 71–74.

Papageorgiou, G. Z., Karandrea, E., Giliopoulos, D., Papageorgiou, D. G., Ladavos, A., Katerinopoulou, A., Achilias, D. S., Triantafyllidis, K. S., Bikiaris, D. N. (2014). Effect of clay structure and type of organomodifier on the thermalproperties of poly(ethylene terephthalate) based nanocomposites. *Thermochim. Acta*, **576**, pp. 84–96.

Park, S. J., Seo, D. I., Lee, J. R. (2002). Surface modification on montmorillonite on surface acid–base characteristics of clay and thermal stability of epoxy/clay nanocomposites. *J. Colloid Interface Sci.*, **251**, pp. 160–165.

Park, Y., Ayoko, G. A., Frost, R. L. (2011). Characterisation of organoclays and adsorption of p-nitrophenol: Environmental application. *J. Colloid Interface Sci.*, **360**, pp. 440–456.

Patel, H. A., Somani, R. S., Bajah, H. C., Jasra, R. V. (2007). Preparation and characterization of phosphonium montmorillonite with enhanced thermal stability. *Appl. Clay Sci.*, **35**, pp. 194–200.

Phothitontimongkol, T., Siebers, N., Sukpiroma, N., Unob, F. (2009). Preparation and characterization of novel organo-clayminerals for Hg(II) ions adsorption from aqueous solution. *Appl. Clay Sci.*, **43**, pp. 343–349.

Pinto, M. R. O., Jr., M. M. A., Silva, S. M., Carvalho, L. H. (2005). Influência da modificação química de uma carga mineral nanoparticulada nas propriedades mecânicas e no envelhecimento térmico de compósitos de poliuretano/sisal. *Polímeros: Ciência Tecnologia*, **15**, pp. 313–319.

Pozsgay, A., Fráter, T., Százdi, L., Müler, P., Sajó, I., Pukánszky, B. (2004). Gallery structure and exfoliation of organophilized montmorillonite: Effect on composite properties. *Eur. Polym. J.*, **40**, pp. 27–36.

Qin, H., Zhang, S., Zhao, C., Hu, G., Yang, M. (2005). Flame retardant mechanism of polymer/clay nanocomposites based on polypropylene. *Polymer*, **46**, pp. 8386–8395.

Ramos Filho, F. G., Mélo, T. J. A., Rabello, M. S., Silva, S. M. L. (2005). Thermal stability of nanocomposites based on polypropylene and bentonite. *Polym. Degrad. Stab.*, **89**, pp. 383–392.

Richards, S., Bouazza, A. (2007). Phenol adsorption inorgano-modified basaltic clay and bentonite. *Appl. Clay Sci.*, **37**, pp. 133–142.

Rodríguez-Sarmiento, D. C., Pinzón-Bello, J. A. (2001). Adsorption of sodium dodecylbenzene sulfonate on organophilic bentonites. *Appl. Clay Sci.*, **18**, pp. 173–181.

Ruiz-Hitzky, E., Meerbeek, A. V. (2006). Clay mineral and organoclay-polymer nanocomposite, in *Handbook of Clay Science. Developments in Clay Science*, Vol. 1. Amsterdam, The Netherlands: Elsevier.

Rytwo, G., Gonen, Y., Afut, S. (2008). Preparation of berberine-montmorillonite-metolachlor formulations from hydrophobic/ hydrophilicmixtures. *Appl. Clay Sci.*, **41**, pp. 47–60.

Sánchez-Solís, A., Ibarra-Romero, I., Estrada, M. R., Calderas, F., Manero, O. (2004). Mechanical and rheological studies on polyethylene terphtalate–montmorillonite nanocomposites. *Polym. Eng. Sci.*, **44**, pp. 1094–1102.

Seliem, M. K., Komarneni, S., Cho, Y., Lim, T., Shahien, M. G., Khalil, A. A., Abd El-Gaid, I. M. (2011). Organo silicas and organo-clay minerals assorbents for tolüene. *Appl. Clay Sci.*, **52**, pp. 184–189.

Shen, D., Fan, J., Zhou, W., Gao, B., Yue, Q., Kang, Q. (2009). Adsorption kinetics and isotherm of anionic dyes onto organo-bentonite from single and multi-solute systems. *J. Hazard. Mater.*, **172**, pp. 99–107.

Shen, Y. H. (2001). Preparations of organobentonite using nonionic surfactants. *Chemosphere*, **44**, pp. 989–995.

Sheng, G., Xu, S., Boyd, S. A. (1996). Mechanism(s) controlling sorption of neutral organic contaminants by surfactant derived and natural organic matter. *Environ. Sci. Technol.*, **30**, pp. 1553–1557.

Shishan, W., Dingjung, J., Xiaodong, O., Fen, W., Jian, S. (2004). The structure and properties of PA6/MMT nanocomposites prepared by melt compounding. *Polym. Eng. Sci.*, **44**, pp. 2070–2074.

Sinha Ray, S., Okamoto, M. (2003). Polymer/layered silicate nanocomposites: A review from preparation to processing. *Prog. Polym. Sci.*, **28**, pp. 1539–1641.

Someya, Y., Shibata, M. (2004). Morphology, thermal, and mechanical properties of vinylester resin nanocomposites with various organo-modified montmorillonites. *Polym. Eng. Sci.*, **44**, pp. 2041–2046.

Százdi, L., Pukánszky Jr., B., Földes, E., Pukánszky, B. (2005). Possible mechanism of interaction among the components in MAPP modified layered silicate PP nanocomposites. *Polymer*, **46**, pp. 801–810.

Tamura, K., Nakazawa, H. (1996). Intercalation of n-alkyl trimethyl ammonium into swelling fluoro-mica. *Clays Clay Miner.*, **44**, pp. 501–505.

Tang, Y., Hu, Y., Song, L., Gui, Z., Chen, Z., Fan, W. (2003). Preparation and thermal stability of polypropylene/montmorillonite nanocomposites. *Polym. Degrad. Stab.*, **82**, pp. 127–131.

Tonlé, I. K., Ngameni, E., Tcheumi, H., Tchié, V., Carteret, C., Walcarius, A. (2008). Sorption of methylene blue on an organoclay bearing thiol groups and application to electrochemical sensing of the dye. *Talanta*, **74**, pp. 489–497.

Undabeytia, T., Nir, S., Sánchez-Verdejoa, T., Villaverde, J., Maqueda, C., Morillo, E. (2008). A clay–vesicle system for water purification from organic pollutants. *Water Res.*, **42**, pp. 1211–1219.

Vaia, R. A., Teukolsky, R. K., Giannelis, E. P. (1994). Interlayer structure and molecular environment of alkyl ammonium layered silicates. *Chem. Mater.*, **6**, pp. 1017–1022.

Valenzuela Diaz, F. R. (1994). *Preparação a nível de laboratório de algumas argilas esmectíticas organofílicas*. Doctorate Thesis. Chemical Engineering. São Paulo University.

Vazquez, A., López, M., Kortaberria, G., Martín, L., Mondragon, I. (2008). Modification of montmorillonite with cationic surfactants, thermal and chemical analysis including CEC determination. *Appl. Clay Sci.*, **41**, pp. 24–36.

Vianna, M. M. G. R. (2005). *Desenvolvimento de argilas organofílicas visando a remediação de áreas contaminadas por componentes da gasolina*. Doctorate Thesis. Chemical Engineering. São Paulo University.

Vidal, N. C., Volzone, C. (2009). Analysis of tetramethylammonium–montmorillonite and retention of tolüene from aqueous solution. *Appl. Clay Sci.*, **45**, pp. 227–231.

Volzone, C., Rinaldi, J. O., Ortiga, J. (2006). Retention of gases by hexadecyltrimethyl ammoniom-montmorillonite clays. *J. Environ. Manage.*, **79**, pp. 247–252.

Vougaris, D., Petridis, D. (2002). Emulsifying effect of dimethyldioctadecylammonium-hectorita in polystyrene/poly(ethyl methacrylate) blends. *Polymer*, **43**, pp. 2213–2218.

Wang, K. H., Choi, M. H., Koo, C. M., Choi, Y. S., Chung, I. J. (2001). Synthesis and characterization of maleated polyethylene/clay nanocomposites. *Polymer*, **42**, pp. 9819–9826.

Wang, Li, Wang, A. (2008). Adsorption properties of congo red from aqueous solution onto surfactant-modified montmorillonite. *J. Hazard. Mater.*, **160**, pp. 173–180.

Wang, S., Hu, Y., Tang, Y., Chen, Z., Fan, W. (2004). Preparation and characterization of flame retardant ABS/montmorillonite nanocomposite. *Appl. Clay Sci.*, **25**, pp. 49–55.

Wang, W. S., Chen, H. S., Wu, Y. W., Tsai, T. Y., Chen-Yang, Y. W. (2008). Properties of novel epoxy/clay nanocomposites prepared with a reactive phosphorus-containing organoclay. *Polymer*, **49**, pp. 4826–4836.

Xi, Y. (2006). *Synthesis, Characterisation and Application of Organoclays.* Doctor of Philosophy. Queensland University of Technology, China.

Xi, Y., Ding, Z., He, H., Frost, R. L. (2004). Structure of organoclays: An X-ray diffraction and thermogravimetric analysis study. *J. Colloid Interface Sci.*, **277**, pp. 116–120.

Xi, Y., Ding, Z., Hongping, H., Frost, R. L. (2005). Infrared spectroscopy of organoclays synthesized with the surfactant octadecyl trimethyl ammonium bromide. *Spectrochim. Acta, A*, **61**, pp. 515–525.

Xi, Y., Mallavarapu, M., Naidu, R. (2010). Preparation, characterization of surfactants modified clay minerals and nitrate adsorption. *Appl. Clay Sci.*, **48**, pp. 92–96.

Yan, L. G., Shan, X. Q., Wen, B., Zhang, S. (2007). Effect of lead on the sorption of phenol onto montmorillonites and organo-montmorillonites. *J. Colloid. Interface Sci.*, **308**, pp. 11–19.

Yao, H., Zhu, J., Morgan, A. B., Wilkie, C. A. (2002). Crown ether-modified clays and their polystyrene nanocomposites. *Polym. Eng. Sci.*, **42**, pp. 1808–1814.

Yapar, S., Özbudak, V., Lopes, A. (2005). Effect of adsorbent of concentration to the adsorption of phenol on hexadecyl trimethyl ammonium-bentonite. *J. Hazard. Mater.*, **121**, pp. 135–139.

Yariv, S., Cross, H. (2002). *Organo-Clay Complexes and Interactions*. New York: Marcel Dekker.

Yilmaz, N., Yapar, S. (2004). Adsorption properties of tetradecyl- and hexadecyl trimethylammonium bentonites. *Appl. Clay Sci.*, **27**, pp. 223–228.

Yoo, Y., Shah, R. K., Paul, D. R. (2007). Fracture behavior of nanocomposites based on poly(ethylene-co-methacrylicacid) ionomers. *Polymer*, **48**, pp. 4867–4873.

Yoshimoto, S., Osashi, F., Kameyama, T. (2005). X-ray diffraction studies of intercalation compounds prepared from aniline salts and montmorillonite by mechanical processing. *Solid State Commun.*, **136**, pp. 251–256.

Yu, Y. H., Yeh, J. M., Liou, S. J., Chang, Y. P. (2004). Organo-soluble polyimide (TBAPPODPA)/clay nanocomposite materials with advanced anticorrosive properties prepared from solution dispersion technique. *Acta Mater.*, **52**, pp. 475–486.

Yuehong, M., Jianxi, Z., Hongping, H., Peng, Y., Wei, S., Dong, L. (2010). Infrared investigation of organo-montmorillonites prepared from different surfactants. *Spectrochim. Acta A*, **76**, pp. 122–129.

Zaghouane-Boudiaf, H., Boutahala, M., Sahnoun, S., Tiar, C., Gomri, F. (2014). Adsorption characteristics, isotherm, kinetics, and diffusion of modified natural bentonite for removing the 2,4,5-trichlorophenol. *Appl. Clay Sci.*, **90**, pp. 81–87.

Zanetti, M., Lomakin, S., Camino, G. (2000). Polymer layered silicate nanocomposites. *Macromol. Mater. Eng.*, **279**, pp. 1–9.

Zawrah, M. F., Khattab, R. M., Saad, E. M., Gado, R. A. (2014). Effect of surfactant types and their concentration on the structural characteristics of nano clay. *Spectrochim. Acta A*, **122**, pp. 616–623.

Zhang, J., Jiang, D. D., Wilkie, C. A. (2005). Polyethylene and polypropylene nanocomposites based upon an oligomerically modified clay. *Thermochim. Acta*, **430**, pp. 107–113.

Zhang, J., Jiang, D. D., Wilkie, C. A. (2006a). Polyethylene and polypropylene nanocomposites based on a three component oligomerically-modified clay. *Polym. Degrad. Stab.*, **91**, pp. 641–648.

Zhang, J., Jiang, D. D., Wilkie, C. A. (2006b). Thermal and flame properties of polyethylene and polypropylene nanocomposites based on an oligomerically-modified clay. *Polym. Degrad. Stab.*, **91**, pp. 298–304.

Zhang, Y.-Q., Lee, J.-H., Jang, H.-J., Nah, C.-W. (2004a). Preparing PP/clay nanocomposites using a swelling agent. *Composites: Part B*, **35**, pp. 133–138.

Zhang, Y.-Q., Lee, J.-H., Rhee, J. M., Rhee, K. Y. (2004b). Polypropylene-clay nano composites prepared by in situ grafting–intercalating in melt. *Comp. Sci. Technol.*, **64**, pp. 1373–1389.

Zhang, Y., Zhao, Y., Zhu, Y., Wu, H., Wang, H., Lu, W. (2012). Adsorption of mixed cationic-nonionic surfactant and its effect on bentonite structure. *J. Environ. Sci.*, **24**(8), pp. 1525–1532.

Zhao, H. T., Vance, G. F. (1998). Sorption of trichloroethylene by organo-clays in the presence of humic substances. *Water Res.*, **32**, pp. 3710–3716.

Zheng, X., Jiang, D. D., Wang, D., Wilkie, C. A. (2006). Flammability of styrenic polymer clay nanocomposites based on a methyl methacrylate oligomerically-modified clay. *Polym. Degrad. Stab.*, **91**, pp. 289–297.

Zhou, Q., Deng, S., Yu, Q., Zhang, Q., Yu, G., Huang, J., He, H. (2010). Sorption of perfluorooctane sulfonate on organo-montmorillonites. *Chemosphere*, **78**, pp. 688–694.

Zhu, J., He, H., Zhu, L., Wen, X., Deng, F. (2005). Characterization of organic phases in the interlayer of montmorillonite using FTIR and ^{13}C NMR. *J. Colloid Interface Sci.*, **286**, pp. 239–244.

Zhu, J., Wang, T., Zhu, R., Ge, F., Wei, J., Yuan, P., He, H. (2011). Novel polymer/surfactant modified montmorillonite hybrids and the implications for the treatment of hydrophobic organic compounds in wastewaters. *Appl. Clay Sci.*, **51**, pp. 317–322.

Zhu, L., Tian, S., Zhu, J., Shi, Y. (2007). Silylated pillared clay (SPILC): A novel bentonite-based inorgano–organo composite sorbent synthesized by integration of pillaring and silylation. *J. Colloid Interface Sci.*, **315**, pp. 191–199.

Zhu, R., Zhu, L. (2008). Thermodynamics of naphthalene sorption to organoclays: Role of surfactants packing density. *J. Colloid Interface Sci.*, **322**, pp. 27–32.

Zhu, S., Chen, J., Li, H., Cao, Y., Yang, Y., Feng, Z. (2014). Preparation and properties of montmorillonite/poly(ethylene glycol) grafted polypropylene/polypropylene nanocomposites. *Appl. Clay Sci.*, **87**, pp. 303–310.

Chapter 4

Polymer–Clay Nanocomposite Synthesis Methods

4.1 Introduction

The composite in which one dimension of at least one component is in the nanometer-size scale (<100 nm) is defined as nanocomposite material. Although nanocomposite as a term represents a lot exciting field in terms of material science and technology, nanocomposite materials have actually been used for centuries in the nature. In nature, there are many nanocomposites such as bones, shells, and wood synthesized by using natural polymeric materials such as carbohydrates, lipids, and proteins (Hussain et al., 2006). Nowadays, nanocomposite technology has emerged as an efficient and powerful strategy, having a potential to upgrade the structural and functional properties of synthetic polymers. This new view that is referred as nanocomposite science to recently in nanotechnology was started with the superior improvement in the properties of nylon 6 by the preparation of nylon 6–clay nanocomposites. Then, the more advanced studies related to the production of the nanocomposites, such as carbon nanotubes, carbon nanofillers, the exfoliated graphite, nanocrystaline metals, and the fibers modified with inorganic fillers emerged (Olad, 2011; Paul and Robeson, 2008; Usuki et al., 1993; Vaia et al., 1993). Currently, worldwide in order to use in the

Introduction to Polymer–Clay Nanocomposites
Ahmet Gürses

ISBN 978-981-4613-02-6 (Hardcover), 978-981-4613-03-3 (eBook)
www.panstanford.com

production of composites of each type of polymers, intensive efforts are being expended.

Nowadays, the polymer-layered silicate (PLS) nanocomposites are of great interest, in terms of industrial applications and scientific researches, because they often exert remarkable improvement in material properties when compared to virgin polymer or conventional micro- and macrocomposites. These improvements can include high moduli, increased strength (Biswas and Sinha Ray, 2001; Giannelis 1996; LeBaron et al., 1999; Okada et al., 1990) and heat resistance (Giannelis, 1998), decreased gas permeability and flammability (Gilman, 1999; Gilman et al., 1997, 2000), and increased biodegradability of biodegradable polymers (Sinha Ray et al., 2002). On the other hand, there has been considerable interest in theory and simulations addressing the preparation and properties of these materials (Dabrowski et al., 2000; Ginsburg and Balazs, 2000; Vaia and Giannelis, 1997), and they are also considered to be unique model systems to study the structure and dynamics of polymers in confined environments (Hackett et al., 1998; Manias and Kuppa, 2001). Although the intercalation chemistry of polymers when mixed with appropriately modified layered silicate and synthetic layered silicates has long been known, the field of PLS nanocomposites has gained momentum recently (Blumstein, 1965; Lagaly and Beneke, 1991; Sinha Ray and Okamoto, 2003).

The dispersion of nanoscale filler within the polymer matrix leads to effective interfacial interactions between the polymer and the inorganic filler, which may result in more superior properties than those of the bulk polymer phase. If the dimensions of filler particles are decreased to the nanoscale dimension, their properties will change significantly and this change is well known as the nanoeffect, i.e., the effect of nanofiller on the polymer properties is different from that predicted with thermodynamic considerations for reduced-particle-size fillers (Gacitua et al., 2005; Kumar et al., 2009). The enhanced properties of nanocomposites are strongly dependent on the particular features of the nanofiller system, such as content, aspect ratio, and the ratio of filler mechanical properties to those of the matrix. A broad spectrum of polymer properties, such as diffusion barrier, durability, chemical stability, flame retardancy, scratch/wear resistance, and biodegradability, as well as optical, magnetic, and electrical properties can be improved by nanocomposite technology (Aranda et al., 2003; Armentano et al., 2010; Cosoli et al., 2008; Ma et al., 2006; Pandey et al., 2005; Wu and Lerner, 1993).

The eventual properties of nanocomposites depend on their composition, their microstructures, and the interfacial interactions between their components. However, the properties of nanocomposites are significantly influenced by the dimensions of filler as well as its microstructure; that is, the filler nature has a predominant effect on the eventual morphology and properties of the polymer nanocomposite. Clays are an important group of nanofillers that have been widely used for the preparation of polymer nanocomposites. Recently, there has been a growing interest for the development of polymer–clay nanocomposites due to their dramatic improved properties compared to the conventional filled polymers at the very low filler ratio (Le Baron et al., 1999; Thostenson et al., 2005). Polymer–clay nanocomposites have received an intense attention depending on the unique properties that can never be obtained by micro-size fillers or especially by other nanofillers. The value-added properties enhanced without sacrificing pure polymer processability, mechanical properties, and lightness make the clays more and more important in the modern polymer industry. Clay minerals constitute the main group of silicates having a layered structure, known as layered silicates (Olad, 2011).

4.2 Types of Polymer–Clay Nanocomposites

In general, layered silicates have a layer thickness on the order of 1 nm and a very high aspect ratio (i.e., 10–1000). A small percent fraction of layered silicates that are properly dispersed throughout the polymer matrix thus creates a much higher surface area for polymer/filler interaction as compared to conventional composites. Depending on the nature of the components (layered silicate, organic cation, and polymer matrix) used and the method of preparation, three main types of composites may be obtained when layered clay is associated with a polymer (see Fig. 4.1) (Alexandre and Dubois, 2000; Sinha Ray and Okamoto, 2003):

(a) **Phase-separated structure:** When the organic polymer is interacted with inorganic clay (unmodified clay), the polymer is unable to intercalate within the clay layers and the clay is dispersed as aggregates or regular stacks of layers formed by stacking together within the polymer matrix. The obtained composite structure is considered "phase separated." The properties of phase-separated polymer–clay composites are in the range of traditional microcomposites (Olad, 2011).

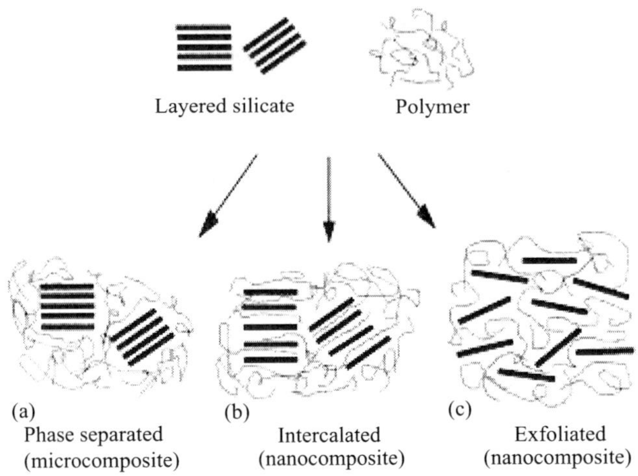

Figure 4.1 Schematic illustration of three different types of thermodynamically achievable PLS nanocomposites. Reprinted from Alexandre and Dubois, Copyright 2000, with permission from Elsevier.

(b) Intercalated nanocomposites: Intercalated structures are formed when a single or sometimes more extended polymer chain was intercalated to the basal spacing between the silicate layers. The result is a regular multilayer structure formed by alternating polymeric and inorganic layers with the increased interlayer distance. Intercalation causes less than 20–30 Å separation between the platelets (Pavlidou and Papaspyrides, 2008; Alexandre and Dubois, 2000; Chin et al., 2001; Beyer 2002; Kim et al., 2001; Dennis et al., 2001).

(c) Exfoliated nanocomposites: The exfoliated nanocomposites are obtained when the clay layers are well separated from one another and individually dispersed in the continuous polymer matrix. In this case, the polymer separates the clay platelets by 80–100 Å or more (Dennis et al., 2001). That is, the level of interlayer expansion is comparable of the radius of gyration of the polymer rather than that of an extended chain, as in the case of the intercalated hybrids (Pavlidou and Papaspyrides, 2008).

The formation and consequent morphology of the nanocomposites are related to enthalpic (e.g., molecular interactions) and entropic (changes in the configurations of the components) factors. Hence, efforts have been made to describe these systems. As an example, Vaia and Giannelis (1997) developed a model to predict the structure above according to the free-energy variation of the polymer–clay mixture in function of the clay layer separation. The free-energy variation, ΔG, associated with the clay layer separation and polymer incorporation is divided into two terms: the term related to the energy variation, ΔH, and the entropy variation, ΔS, associated with the configuration changes of various components.

$$\Delta G = G(h) - G(h_0) = \Delta H - T\Delta S \quad (4.1)$$

where h and h_0 are the initial and final separations of the clay layers. Then, when $\Delta G < 0$, the intercalation process is favorable. Such a model presents as a limitation the separation of the configuration term, the intermolecular interactions, and the separation of the entropy terms of various components. Other mathematical models were also developed for studies of simulation of the thermodynamics of polymer–clay nanocomposites. These models consider the nanocomposite thermodynamics and architecture and the interaction between clay and polymer to determine the free energy and the polymer and clay conformation. Specifically for polyamide 6– and 6,6–clay nanocomposites, the study of molecular dynamics was employed, which uses the bond energy between the various components that compose the nanocomposite (Zeng et al., 2008).

The kinetics of polymer–clay nanocomposite formation is also a very important issue to predict the resulting nanocomposite. Studies of molecular dynamics were also employed to understand the system kinetics. Other mathematical models were also used to describe the system kinetics, but kinetics is less understood than thermodynamics.

There is still the need for developing models that are explored in individual time and length scales, besides the integration of concepts that permeate from smaller to larger scales, that is, in the quantum, molecular, mesoscopic, and macroscopic dominium (Anadao, 2012).

4.3 Preparation of Polymer–Clay Nanocomposites

Many efforts have been made for the preparation of intercalated and exfoliated polymer–clay nanocomposites with improved properties. A variety of polymer characteristics, including polarity, molar mass, hydrophobicity, and reactive groups, as well as clay characteristics, such as charge density and modified structure and polarity, are effective on the intercalation of polymer chains within the clay galleries. Therefore, different synthetic approaches have been used for the preparation of polymer–clay nanocomposites. In general, there are four preparation methods: in situ template synthesis, solution intercalation, in situ intercalative polymerization, and melt intercalation (Olad, 2011; Tjong, 2006).

4.3.1 Template Synthesis (Sol–Gel Technology)

In this technique, clay minerals are synthesized within the polymer matrix using an aqueous solution (or gel) containing the polymer and silicate-building blocks (Kim et al., 2007). As precursors for the clay, silica sol, magnesium hydroxide sol, and lithium fluoride are used. During the process, the polymer aids the nucleation and growth of the inorganic host crystals and gets trapped within the layers as they grow. Although theoretically this method has the potential of promoting the dispersion of the silicate layers in a one-step process without needing the presence of the counter ions, it presents serious disadvantages. First of all, the synthesis of clay minerals generally requires high temperatures, which may decompose the polymers. An exception is the synthesis of hectorite-type clay minerals, which can be performed under relatively mild conditions. Another problem is the aggregation tendency of the growing silicate layers. Therefore, this technique, although widely used for the synthesis of double-layer hydroxide-based nanocomposites, is far less developed for layered silicates. However, it should be mentioned that this technique has been successfully applied for the preparation of nanocomposite materials (Pavlidou and Papaspyrides, 2008). The sol–gel process is well known to proceed by hydrolysis and condensation reactions, which are controlled by the value of pH, the solvent-water-to-alkoxide ratio, concentration of the reactants, catalyst, and temperature. An organic component, involved during the sol–gel process, influences the reaction kinetics, morphology, and physical properties of resultant hybrid materials. The structure–property relationship of organic–inorganic sol–gel composites with either hydrogen bonding or weaker interaction between organic and inorganic components has received considerable attention (Chiang and Ma, 2002; Joseph et al., 1996). Recent polymer-based nanocomposite systems prepared via the in situ intercalative polymerization technique are given in Table 4.1.

Table 4.1 Recent polymer-based nanocomposite systems prepared via the sol–gel technique

Sol–gel technique	Applications
Polycaprolactone (PCL)/tetraethyl orthosilicate (TEOS)	Bone-biodegradable polymer composites for skeletal tissue repair
Polyimide/tetraethyl orthosilicate (TEOS)	Microelectronics
Polymethylmethacrylate (PMMA)/silica	Dental application, optical devices
Poly(ethyl acrylate) (PEA)/silica	Catalysis support, stationary phase for chromatography
Poly(ethylene oxide) (PEO)/tetraethyl orthosilicate (TEOS)	Electrolyte and highly conductive polymer
Poly(*p*-phenylene vinylene) (PPV)/ tetramethyl orthosilicate (TMOS)	Nonlinear optical material for optical wave guides
Poly(amide–imide)/TiO_2	Composite membranes: gas separation applications
Polycarbonate/tetraethyl orthosilicate (TEOS)	Abrasion-resistant coating

Source: Reprinted from Avella et al., Copyright 2001, with permission from John Wiley and Sons.

Organic–inorganic hybrids were prepared using diglycidyl ether of bisphenol (DGEBA)-type epoxy and tetraethoxysilane via the sol–gel process. The sol–gel technique was used successfully to incorporate silicon and phosphorus into the network of hybrids, increasing flame retardance (Chiang and Ma, 2002).

Kim et al. (2007) investigated the dispersion behavior of nanosilica colloids in methacryl oligosiloxane resin in the fabrication of methacryl hybrid nanocomposites containing multiscale nanosized silica phases (~15 nm silica particles and ~2 nm oligosiloxanes). Colloidal silica sols were dispersed in methacryl oligosiloxane resin synthesized by the sol–gel reaction of methacryloxypropylmethoxysilane and diphenylsilanediol. Using transmission electron microscopy (TEM) and small-angle neutron scattering (SANS), it was found that the fabricated multiscale nanocomposites have better mechanical properties, higher thermal stability, and lower coefficients of thermal expansion (CTEs) than a base nanohybrid that does not incorporate silica colloids.

4.3.2 Solution Intercalation

This method is based on a solvent system in which the polymer or prepolymer is soluble and the silicate layers are swellable. The layered silicate is first swollen in a solvent, such as water, chloroform, or toluene. When the polymer and layered silicate in solution are mixed, the polymer chains intercalate and displace the solvent within the interlayer of the silicate. Upon solvent removal, the intercalated structure remains, resulting in a PLS nanocomposite (Ray and Bousima, 2005). This process can be seen schematically in Fig. 4.2 (Pavlidou and Papaspyrides, 2008).

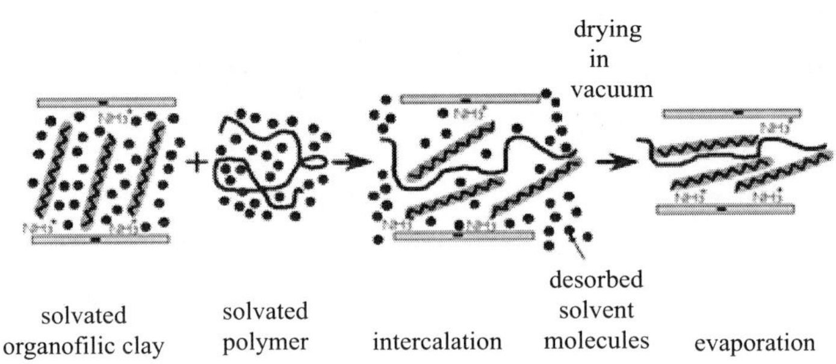

Figure 4.2 Schematic view of solution intercalation. Reprinted from Pavlidou and Papaspyrides, Copyright 2008, with permission from Elsevier.

Water-soluble polymers, such as poly(ethylene oxide) (PEO) (Aranda and Ruiz-Hitzky, 1992; 1994; 1999), poly(vinyl alcohol) (PVA) (Greenland, 1963; Ogata et al., 1997), poly(*N*-vinyl pyrrolidone) (PVP), poly(acryl amide) (PAA), and poly(ethylene vinylacetate) (PEVA) (Zhao et al., 1989), have been intercalated into the clay galleries using the solution intercalation method. Examples from nonaqueous solvents are nanocomposites of polycaprolactone (PCL)–clay (Jimenez et al., 1997) and poly(L-lactide) (PLA)–clay in chloroform as a cosolvent and high-density polyethylene (HDPE) with xylene and benzonitrile (Jeon et al., 1998). Nematic liquid crystal PLS nanocomposites have also been prepared using this method in various organic solvents, such as toluene and dimethylformamide (DMF) (Kawasumi et al., 1998; Sinha Ray and Okamoto, 2003).

Aranda and Ruiz-Hitzky (1992) reported the first preparation of PEO/montmorillonite (MMT) nanocomposites by this method. They conducted a series of experiments to intercalate PEO (M_w=10^5 g/mol) into Na-MMT using different polar solvents: water, methanol, acetonitrile, and mixtures (1:1) of water/methanol and methanol/acetonitrile. In this method, the nature of solvents is critical in facilitating the insertion of polymers between the silicate layers, polarity of the medium being a determining factor for intercalations. The high polarity of water causes swelling of Na-MMT, provoking cracking of the films. Methanol is not suitable as a solvent for high-molecular-weight PEO, whereas water/methanol mixtures appear to be useful for intercalations, although cracking of the resulting materials is frequently observed.

The poly(propylene carbonate) (PPC)/organomontmorillonite (OMMT) nanocomposites were prepared with solution intercalation by benzene. An intercalated and flocculated structure of PPC/OMMT nanocomposites was confirmed by X-ray diffraction (XRD) and TEM. The thermal and mechanical properties of PPC/OMMT nanocomposites were investigated by thermogravimetric analysis (TGA), differential scanning calorimetry (DSC), and electronic tensile testers. Due to the nanometer-sized dispersion of layered silicate in a polymer matrix, PPC/OMMT nanocomposites exhibit improved thermal and mechanical properties than pure PPC (Shi and Gan, 2007). Silicone/expanded graphite (EG) composites were prepared by melt mixing and solution (toluene) intercalation. The effect of EG content and preparation methods of composites on thermal conductivity of composites was investigated, and it was found that with increasing EG, the thermal conductivity increased, and the solution intercalation method showed a better effect than the melt-mixing method in improving the thermal conductivity of composites (Mu and Feng, 2007).

The Toyota Research Group was the first to use the solution intercalation technique to produce polyimide (PI) nanocomposites (Lan et al., 1994). The PI–MMT nanocomposite has been synthesized by mixing in dimethylacetamide, a modified MMT, with the PI precursor, that is, a polyamic acid obtained from the step polymerization of 4,4′-diaminodiphenyl ether with pyromellitic dianhydride. The organomodified MMT was prepared by previous intercalation with dodecylammonium hydrochloride. After elimination of the solvent, an organoclay-filled polyamic acid film was recovered, which was thermally treated up to 300°C to trigger the imidization reaction and to produce the PI nanocomposite. The wide-angle X-ray diffraction (WAXD) patterns of these filled PI films do not show any diffraction-peak-typical structure of an intercalated morphology. This led the authors to conclude that an exfoliated structure was formed and explained the excellent gas barrier properties of the resultant films. Enhanced properties and excellent gas barrier performance are attributable to the exfoliated nanocomposite structure.

XRD data of the resulting PI-based nanocomposites show no noticeable peak, indicating a possible exfoliated structure for the MMT and the synthetic mica. For both hectorite and saponite, a board peak, centered on a value of 15 Å, is observed, indicating that for these layered silicates, polymer intercalation occurs, probably mixed with some exfoliation. For saponite, the measured interlayer spacing is even smaller than the value measured for the starting organically modified clay (18 Å), suggesting that the organic cation may have been expelled from the clay interlayer during the imidization reaction (Alexandre and Dubois, 2000).

4.3.3 In situ Intercalative Polymerization

In situ polymerization was the first method used to synthesize polymer–clay nanocomposites based on nylon 6. In this method, the layered silicate is swollen within the liquid monomer or a monomer solution so that the polymer formation can occur between the intercalated sheets. Polymerization can be initiated by heat or radiation, by the diffusion of a suitable initiator, or by an organic initiator or catalyst fixed through cation exchange inside the interlayer before the swelling step. This process is schematically illustrated in Fig. 4.3. Recent polymer-based nanocomposite systems prepared via the in situ intercalative polymerization technique are given in Table 4.2 (Avella et al., 2001; Pavlidou and Papaspyrides, 2008; Solomon et al., 2001).

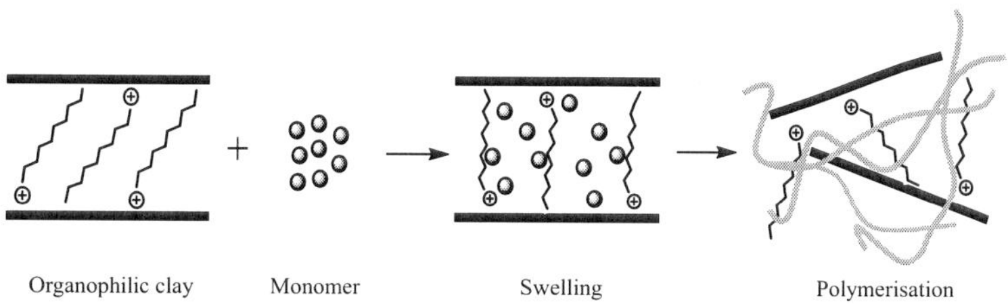

Figure 4.3 Schematic representation of PLS obtained by an in situ process.

Table 4.2 Recent polymer-based nanocomposite systems prepared via the in situ intercalative polymerization technique

In situ intercalative polymerization	Applications
ε-PCL/Cr^{3+} fluorohectorite and montmorillonite	Biodegradable/biocompatible materials, packaging (enhanced barrier properties)
Epoxy/organomodified montmorillonite	Improved properties (e.g., aeronautics)
Polyimide/organomodified montmorillonite	Materials for microelectronics with reduced thermal expansion coefficient and moisture absorption
Polystyrene/organomodified montmorillonite	Improved properties
Copolymer butadiene/acrylonitrile/organomodified montmorillonite	Rubber with enhanced barrier properties (H_2, H_2O)
iPP/organoclay	Improved properties
Starch/organomodified montmorillonite	Enhanced barrier properties
Nylon/organomodified montmorillonite	Improvement of structural, mechanical, thermal, and barrier characteristics without significant loss in clarity or strength

Source: Reprinted from Avella et al., Copyright 2001, with permission from John Wiley and Sons.

4.3.3.1 Thermoplastic nanocomposites

Many "interlamellar" polymerization reactions were studied by various research groups in the 1960s and 1970s using layered silicate (Eastman et al., 1999), but it is the work initiated by the Toyota research team about 15 years ago that the study of PLS nanocomposites became recognized. The Toyota team studied the Na-MMT organically modified by protonated α,ω-amino acid ($^{+}H_3N\text{-}(CH_2)_{n\text{-}1}\text{-}COOH$, with $n = 2, 3, 4, 5, 6, 8, 11, 12, 18$) and swollen by the ε-caprolactam (ε-CLa) monomer at 100°C. Subsequently, the nanomodified ε-CLa was transformed by ring-opening polymerization to nylon-6-based nanocomposites (Fukushima et al., 1999; Kornmann, 2000; Lagaly, 1999; Usuki et al., 1993). A clear difference occurs in the swelling behavior between the MMT with relatively short ($n < 11$) and longer alkyl chains, as shown in Table 4.3, indicating that a larger amount of monomer can be intercalated when longer alkyl chains are used (Alexandre and Dubois, 2000). An MMT modified by 12-aminolauric acid ($n = 12$) was selected during the intercalative ring-opening polymerization of ε-CLa (Kojima et al., 1993; Okada and Usuki, 1995). The modified MMT (12-Mont) was mixed with the monomer containing a small amount of 6-aminocaproic acid as a polymerization accelerator. The mixture was heated first at 100°C for 30 min and then at 250°C for 6 h. The cooled and solidified product was crushed, washed with water at 80°C, and then dried. Depending on the amount of 12-Mont introduced, either an exfoliated (for less than 15 wt.%) or a delaminated structure (from 15 wt.% to 70 wt.%) was obtained, as evidenced by WAXD and TEM measurements. Comparison of the titrated amounts of COOH and NH_2 end groups present in the synthesized nanocomposites with given values, such as the cation exchange capacity (CEC) of the MMT used (119 meq/100 g), led to the conclusion that the COOH end groups present along the 12-Mont surface are responsible for the initiation of polymerization. Nearly all of the $-NH_3^+$ end groups present in the matrix should be interacting with the MMT anions. Finally, the ratio of bonded to nonbonded polymer chains increased with the amount of incorporated MMT (from 32.3% of bonded chains for 1.5 wt.% MMT to 92.3% of bonded chains for 59.6 wt.% clay) (Alexandre and Dubois, 2000; Shelley et al., 2002).

Table 4.3 Basal spacing on organomodified montmorillonite in the presence of ε-CLa at 100°C

$^{+}H_3N\text{-}(CH_2)_{n-1}\text{-}COOH$ (n)	Interlayer spacing of the modified clay (Å)	Interlayer spacing when swollen by ε-CLa at 100°C (Å)
2	12.7	14.1
3	13.1	19.7
4	13.2	19.9
5	13.2	20.4
6	13.2	23.4
8	13.4	26.4
11	17.4	35.7
12	17.2	38.7
18	28.2	71.2

Source: Reprinted from Alexandre and Dubois, Copyright 2000, with permission from Elsevier.

4.3.3.2 Thermoset nanocomposites

In situ polymerization has also been explored to create thermoset-based nanocomposites. This method has been extensively used for the production of both intercalated and exfoliated epoxy-based nanocomposties (Alexandre and Dubois, 2000). Messersmith and Giannelis (1994) have analyzed the effects of different curing agents and conditions in the formation of epoxy nanocomposites based on diglycidyl ether of bisphenol A (DGEBA) and an MMT modified by the bis(2-hydroxyethyl)methyl hydrogenated tallow alkyl ammonium cation. The observed increase in viscosity at relatively low shear rates and the transition of opaque to semitransparent suspension indicated that the modified MMT clay dispersed quite readily in DGEBA when sonicated for a short period of time. The increase in viscosity was attributed to the formation of a "house of cards" structure, in which edge-to-edge and edge-to-face interactions between dispersed layers form percolation structures. WAXD patterns of uncured clay/DGEBA also indicate that intercalation occurred during mixing and that this intercalation improves as the temperature is increased from room temperature to 90°C (Alexandre and Dubois, 2000).

4.3.3.3 Rubber-modified epoxy nanocomposites

Lan et al. (1994) examined the formation of nanocomposites with a rubber-epoxy matrix obtained from a DGEBA derivative (Epon 828) and cured with a polyether diamine (Jeffamine D2000) to obtain low-glass-transition-temperature materials. Two MMTs modified by the protonated *n*-octylamine and the protonated *n*-octadecylamine, respectively, were used in this study. It was shown that depending on the alkyl chain length of the modified clays, an intercalated and partially exfoliated (*n*-octyl) or a totally exfoliated *n*-octadecyl nanocomposite can be obtained. These authors have also studied other parameters such as the nature of alkylammonium cations present in the gallery and the effect of the CEC of the clay when Epon 828 was cured with *m*-phenylene diamine. It was demonstrated that when mixed with the epoxide and variable length of the protonated primary alkylamine (with 8, 10, 12, 16, or 18 carbons), the modified clays adopt a structure where the carbon chains are fully extended and oriented perpendicularly to the silicate plane incorporating a maximum of one monolayer of epoxide molecules.

4.3.4 Melt Intercalation

Melt intercalation involves annealing, statically or under shear, a mixture of a polymer and an organically modified layered silicate (OMLS) above the softening point of the polymer. This method has great advantages over either in situ intercalative polymerization or polymer solution intercalation. First, this method is environmentally benign due to the absence of organic solvents. Second, it is compatible with current industrial process, such as extrusion and injection molding. The melt intercalation method allows the use of polymers that were previously not suitable for in situ polymerization or solution intercalation (Sinha Ray and Okamoto, 2003). This process is schematically illustrated in Fig. 4.4. In this technique, no solvent is required. Due to its great potential in industrial applications, the melt intercalation process has become increasingly popular.

The early literature seemed to suggest that intercalation would be useful and perhaps a precursor to exfoliation. Subsequent research has suggested alternative ideas about how the exfoliation process may occur in melt processing and how the details of the mixing equipment and conditions alter the state of dispersion achieved (Fornes et al., 2001). These ideas are summarized in a full-size sketch shown in Fig. 4.5. As made commercially, the particles of an organoclay powder are about 8 μm in size and consist of aggregates of tactoids, or stacks of platelets; the stresses imposed during melt mixing break up aggregates and can shear the stack into smaller ones, as suggested in Fig. 4.5. However, there evidently is a limit to how finely the clay can be dispersed just by mechanical forces. If the polymers and organoclay have an affinity for one another, the contact between polymers and organoclay can be increased by peeling the platelets from these stacks one by one until, given enough time in the mixing device, all the platelets are individually dispersed, as suggested in Fig. 4.5.

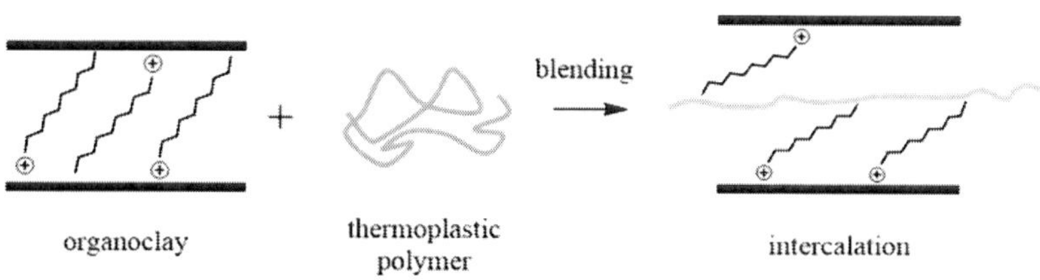

Figure 4.4 Schematic representation of the melt intercalation process.

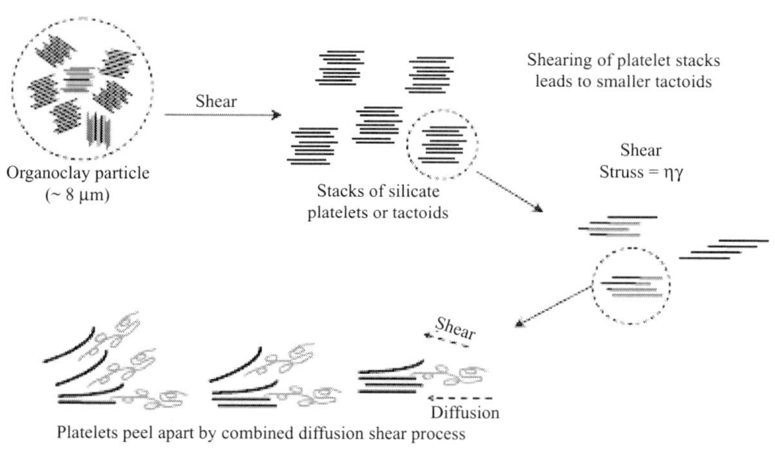

Figure 4.5 Mechanism of organoclay dispersion and exfoliation during melt processing. Reprinted from Fornes et al., Copyright 2001, with permission from Elsevier.

4.3.4.1 Thermoplastic nanocomposites

Vaia and Giannelis (1997) have studied polystyrene (PS) as the matrix for dispersing different types of clays. Li-fluorohectorite (CEC = 150 meq/100 g), saponite (CEC = 100 meq/100 g), and sodium MMT (CEC = 80 meq/100 g) were modified accordingly using various ammonium cations: dioctadecyldimethylammonium, octadecyltrimethylammonium, and a series of primary alkylammonium cations with carbon chains of 6, 9–16, and 18 carbon atoms. The nanocomposites were synthesized by statically annealing (without any mixing or shearing) a pelletized intimate mix of the modified silicate in PS (M_w 30,000, M_w/M_n 1.06) under vacuum at 170°C, a temperature well above the PS glass transition, as shown in Table 4.4. Comparison of the first three entries indicates that, for a given alkyl surfactant, increasing the cation exchange from 80 meq/100 g to 120 meq/100 g allows PS intercalation to occur. At a low CEC and for single-alkyl-chain buildup cations (entries 1 and 2), no intercalation is observed. Under these conditions, the aliphatic alkyl chain of the organic cation adopts a pseudomonolayer arrangement characterized by low gallery height. Intercalation can take place at lower CECs when the clay surface possesses an ammonium cation bearing two long alkyl chains (entries 5 and 6). However, excessive packing of the chains all along the layer surface (high CEC and two long alkyl chains per cationic head, entry 7) leads to the formation of a nonintercalated structure, as predicted by the theory introduced by Vaia and Giannelis (1997). The authors also found that polymer intercalation depends on the length of the exchange ammonium cation as well as on the annealing temperature (Koo, 2006; Vaia et al., 1993).

4.3.4.2 Elastomer nanocomposites

Burnside and Giannelis (1995) have described the two-step preparation of silicon-rubber-based nanocomposites. First, silanol-terminated polydimethylsiloxane (PDMS, M_w 18,000) was melt-blended at room temperature with dimethylditallowammonium-exchanged MMT. Then the silanol end groups were crosslinked with tetrathylorthosilicate (TEOS) in the presence of tin bis(2-ethylhexanoate) as a catalyst at room temperature. To obtain exfoliated nanocomposites (as determined by the absence of peak WAXD patterns), several conditions were required, such as mixing the modified clay and PDMS under sonication (for better mixing) and adding a small quantity of water extending/crosslinking TEOS end groups. The nature of both the silicone matrix and the clay modifier plays an important role. Neither exfoliation nor intercalation can occur if MMT is modified with a different surfacing agent, such as benzyl dimethyloctadecylammonium cations, or if too much water is added. On the other hand, only intercalation was observed when a PDMS–polydiphenylsiloxane random copolymer containing 14–18 mol.% diphenylsiloxane units was used. The results stress the key importance of selecting the right match between the matrix and the organoclay to optimize the layered silicate exfoliation (Burnside and Giannelis, 1995; Gupta and Bhattacharya, 2008).

Table 4.4 Characteristics of polystyrene melt intercalation within octadecylammonium-modified clays

Entry	Clay (CEC in meq/ 100 g)[a]	Ammonium cation	Initial gallery height (nm)	Final gallery height (nm)	Net change (nm)
1	M (80)	PODA[b]	0.75	0.75	0
2	S (100)	PODA[b]	0.83	0.83	0
3	F (120)	PODA[b]	1.33	2.16	0.83
4	F (120)	QODA[c]	1.57	2.69	1.12
5	M (80)	DODMDA[d]	1.43	2.25	0.82
6	S (100)	DODMDA[d]	1.50	2.35	0.85
7	F (120)	DODMDA[d]	2.85	2.85	0

Source: Reprinted from Vaia and Giannelis, Copyright 1997, with permission from American Chemical Society.

Note: [a]M = montmorillonite, S = saponite, and F = fluorohectorite

[b]PODA: primary octadecylammonium

[c]QODA: quaternized octadecylammonium

[d]DODMA: dioctadecyldimethylammonium

4.4 Characterization of Polymer–Clay Nanocomposites

The improvement in the properties of nanocomposites is related to the dispersion and nanostructure of the layered silicate in the polymer. Therefore, it is of paramount importance to determine whether the exfoliation and intercalation of pristine layered silicates can be achieved. There are several techniques that are used to elucidate the nanostructure of PLS nanocomposites, including (Davis et al., 2000; Gunter et al., 2000; Jacob, 2008; Morgan and Gilman, 2003; Powell and Beall 2006):

- WAXD
- TEM and spectroscopy
- Scanning electron microscopy (SEM)
- Small-angle X-ray scattering (SAXS)
- TGA
- DSC
- Dynamic mechanical thermal analyses (DMTA)
- Mechanical properties (tensile, flexural, and compressive strength and modulus, and Izod impact)
- Flammability (cone calorimeter [CC] and UL 94)

Other techniques include:

- Nuclear magnetic resonance (NMR)
- Scanning probe microscopy (SPM): atomic force microscopy (AFM) and scanning tunneling microscopy (STM)
- Neutron scattering
- Optical microscopy

4.4.1 X-Ray Diffraction

WAXD measures the spacing between the ordered crystalline layers of an organoclay by using Bragg's law

$$n\lambda = 2d\sin\theta \qquad (4.2)$$

where d is the spacing between atomic planes in the crystalline phase and λ is the X-ray wavelength. The intensity of the diffracted

X-ray is measured as a function of the diffraction angle 2θ and the specimen's orientation. This diffraction pattern is used to identify the specimen's crystalline phases and to measure its structural properties.

The diffractograms produced by different types of hybrid are illustrated in Fig. 4.6. The influence of intercalation of the polymer on the order of the silicate is indicated by changes in the intensity and shape of the basal reflections (Vaia and Giannelis, 1997). Spacing change (increase or decrease) information can be used to determine the type of polymer nanocomposite formed.

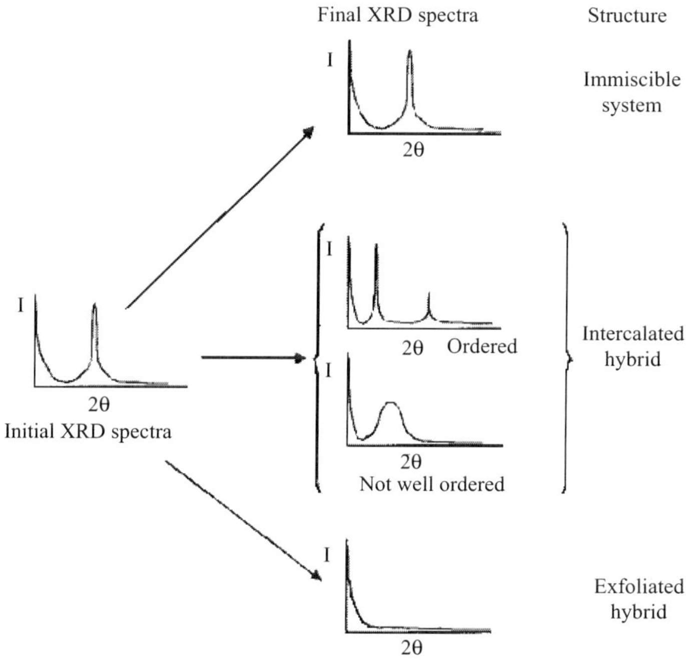

Figure 4.6 Schematic illustration depicting the expected XRD patterns for immiscible, intercalated, and exfoliated hybrid structures. Reprinted from Vaia and Giannelis, Copyright 1997, with permission from American Chemical Society.

For immiscible polymer/OMLS mixtures, the structure of the silicate is not affected, and thus, the characteristics of the OMLS basal reflections do not change (microcomposite). On the other hand, in comparison with the spacing of the organoclay used, the intercalation of the polymer chains increases the interlayer spacing, leading to a shift in the diffraction peak toward a lower angle, according to Bragg's law. In such intercalated nanocomposites, the repetitive multilayer structure is well preserved, allowing the interlayer spacing to be determined. In contrast, the extensive layer separation associated with exfoliated structures disrupts the coherent layer stacking and results in a featureless diffraction pattern. Thus, for exfoliated structures no more diffraction peaks are visible in the XRD diffractograms either because of a much too large spacing between the layers (i.e., exceeding 8 nm in the case of an ordered exfoliated structure) or because the nanocomposite does not present ordering (Beyer, 2002; Pavlidou and Papaspyrides, 2008; Vaia and Giannelis, 1997).

4.4.2 Transmission Electron Microscopy and Scanning Electron Microscopy

TEM is a very effective technique for the study of polymer morphology. Both XRD and TEM are essential tools for evaluating the nanocomposite structure. However, TEM provides direct visual information of the morphology, atom arrangement, and spatial distribution of the phases and structural defects of a selected area of the sample (Zanetti et al., 2000).

TEM only gives a qualitative understanding of the internal structure, spatial distribution, and dispersion of the nanoparticles within a polymer matrix and views of the defect structure through direct visualization. Special care must be exercised to guarantee a representative cross section of the sample (Journet et al., 1997; Kumar et al., 2009).

TEM is often used with thin samples of materials in which different phases within the samples absorb electron differently. It includes an illumination system, a specimen stage, an objective lens system, a magnification system, a data recording system, and a chemical analysis system. The illumination system in which the electron gun is used is composed of condenser lenses, which are very important for forming a fine electron probe. The objective lens determines the limit of image resolution. There are intermediate lenses and projection lenses in the magnification system, which give magnification up to 1.5 million. The data recording system allows quantitative data processing and quantification.

In modern TEM equipment, software is attached to make statistical calculations of the particles, which is helpful for obtaining the average size and distribution of particles. Usually, 100 particles are measured for statistical calculations. In modern TEM equipment, the particle crystallite diameter–particle number curve is usually obtained in the final results. This curve is called the particle distribution pattern, in which the particle size corresponding to the peak of the curve is used as the mean particle size (Ke and Stroeve, 2005).

TEM allows for the observation of the overall organoclay dispersion in a polymer nanocomposite sample. The clay dispersion and structure observed under the microscope can determine the nature of a clay nanocomposite as follows:

- **Immiscible:** Usually large clay tactoids and undispersed clay particles
- **Intercalated:** Clay layers in ordered stacks
- **Exfoliated:** Single clay layers

Global microscale dispersion, as well as nanoscale dispersion and structure, can be determined.

TEM has several limitations as follows:

- In sampling, heterogeneous samples will give false results.
- Labor-intensive analysis is required, involving expensive analytical instruments.
- TEM cannot measure the *d*-spacing of clay; therefore, it cannot easily determine the difference between an intercalated clay nanocomposite and a well-dispersed, immiscible nanocomposite (Koo, 2006).

TEM measures the overall characteristics of clay dispersion, such as uniformity, tactoids, or nonuniformity of the clay within the material. It is recommended that TEM measurements be combined with WAXD data for thorough analyses of the nanomaterial.

Figure 4.7 shows the TEM images of the composites that have been prepared using the urea formaldehyde (UF) resin reinforced by organoclay with *Rosa Canina* sp. seeds (Karagöz, 2013).

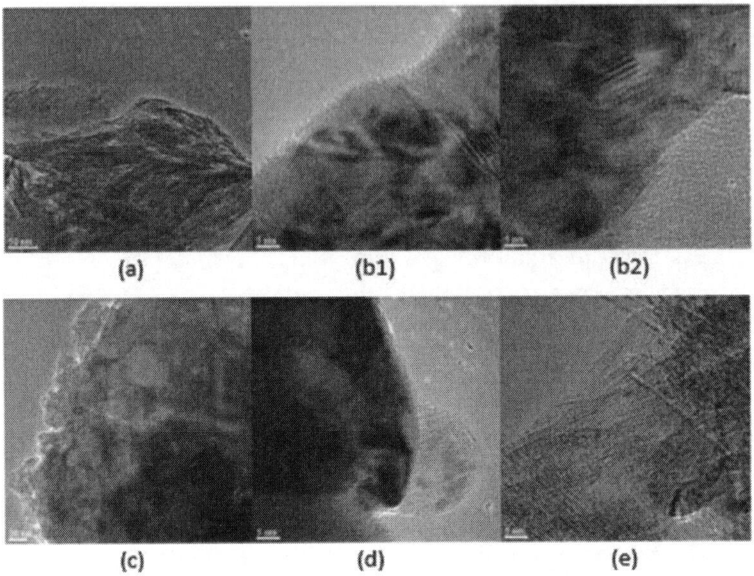

Figure 4.7 TEM images of the organoclay (a), and the composites consisting of *Rosa Canina* sp. seeds and the UF reinforced with the different organoclay contents: 0.2 wt.% (b1 and b2), 0.5 wt.% (c), 1.6 wt.% (d) and 2.5 wt.% (e).

SEM for studying the morphology of nanocomposites is quite different from TEM. In SEM, electrons are emitted in a direction parallel to the sample's surface. SEM constitutes one of the older, and one of the most widely used, tools for surface analysis. Because SEM provides a three-dimensional visual image, the qualitative analysis is relatively straightforward. SEM forms its image by scanning a focused electron beam probe across the specimen of interest. In this technique, a fine electron beam (5–10 nm in diameter) is scanned across the sample surface in synchronization with a beam from a cathode ray tube. The scattered electrons produced can then result in a signal that modulates this beam (Ke and Stroeve, 2005; Stuart, 2002). As polymers tend not to be good conductors, they need to be coated with a thin layer of a conducting material such as gold or silver. SEM has been used for a broad range of polymer studies and applications, including surface roughness, adhesive failures, fractured surfaces, networks, and phase boundaries in blends. SEM is limited to a surface view only. It does not give any information about the interior of the specimen (Sperling, 2006).

4.4.3 Atomic Force Microscopy

AFM was found to be a powerful tool to investigate the nanocomposite surface on a micrometer and in some cases even on a nanometer scale (Feng et al., 2001). AFM involves moving a finely tipped stylus across a surface and monitoring the tip movements as it traces the surface topography. In studying adsorbed dendrimers, samples can be scanned repeatedly and in a variety of directions. When this is done, it is found that all the images are the same. True dendrimers form objects of only one size. AFM has also been used to study adsorbed dendrimers of increasing generation number. When this is done, it is found that there is a steady and regular rise in the diameters of the adsorbed dendrimers with each additional generation number but that each generation of dendrimers remains highly homogeneous (Nicholson, 2005).

4.4.4 Differential Scanning Calorimetry and Thermogravimetric Analysis

DSC is the most useful indirect method for analysis of polymer blends. DSC is used to quantitatively measure phenomena associated with both thermodynamic and kinetic processes such as crystallization, melting, glass transition, curing, and aging of polymers and their blends. Interactions between polymers in a blend will change some of these thermal characteristics due to the morphological changes in the blend (Simon, 2003). Several important blend attributes need to be understood; in each case, DSC can provide a measure by means of the heat flow of the process. DSC allows determination of heat capacity versus temperature or time, measurement of heats of fusion, identification of crystalline and liquid crystalline phases, degrees of crystallinity, etc. Glass transition measurement allows characterization of ageing and blend compatibilities. Heats of reaction allow cure and degradation studies

The thermal stability of polymeric materials is usually studied by TGA. The weight loss due to the formation of volatile products after degradation at high temperature is monitored as a function of temperature (and/or time). When heating occurs under an inert gas flow, non-oxidative degradation occurs, while the use of air or oxygen allows oxidative degradation of the samples (Bower, 2002; Hunt and James, 1993; Pavlidou and Papaspyrides, 2008).

Zanetti et al. (2004) reported TGA curves of a nanocomposite polyethylene (PE)/ethylvinyl acetate (EVA) copolymer/OMMT and the corresponding PE/EVA matrix. Under nitrogen, these samples do not show great differences of stability. However, in air, the PE/EVA blend is subject to a marked weight loss above 350°C, to form a 5 wt.% residue at 450°C, which is completely oxidized to volatile products between 470°C and 550°C. The nanocomposite, on the other hand, displays a different pattern. The presence of 5 wt.% OMMT is enough to change the polymer's thermo-oxidative behavior, and between 350°C and 480°C, the amount of residue is higher than that observed in a nitrogen flow. According to the authors, the organoclay shields the polymer from the action of oxygen, dramatically increasing the thermal stability under oxidative conditions.

4.4.5 Fourier Transform Infrared Spectroscopy

Fourier transform infrared spectroscopy (FTIR) is one of the most commonly used analysis methods for determining the composition of nanocomposites. FTIR may be able to identify differences between the bonding in a mixture and the bonding in a related nanocomposite, but as these variations are extremely small, even when intercalation has taken place, at present, FTIR is an unreliable method of characterization in most cases. The FTIR technique corresponds to the vibrational energies of atoms or specific groups of atoms within a molecule, as well as rotational energies. It identifies components by comparing the spectrum of a sample to reference spectra. An IR spectrum is commonly obtained by passing IR radiation through a sample and determining what fraction of the incident radiation is absorbed at a particular energy. The energy of any peak appearing in the absorption spectrum corresponds to the frequency of any vibration mode exhibited by the sample molecules. The absorption of IR radiation by a molecule can occur if the spectral interaction leads to a specific change in the electrical dipole moment. The interactions of IR radiation with matter may be taken into consideration in terms of changes in molecular dipoles associated with vibration modes. The stretching and bending vibrations can be described as the movements in which the atoms forming molecules can move relative to one another, that is, bond lengths can vary or one atom can move out of its present plane. Vibrations can involve either a change in bond length (stretching) or bond angle (bending). Some bonds can stretch in phase (symmetrical stretching) or out of phase (asymmetric stretching). If a molecule has different terminal atoms, the amount of coupling will vary. Varying proportions of the stretching motion of each group is observed (Stuart, 2004).

4.4.6 Nuclear Magnetic Resonance

The characterization of PLS nanocomposites by solid-state nuclear magnetic resonance (NMR) (^{1}H and ^{13}C) technique was first used by VanderHart et al. (2001). This technique can be considered a tool for gaining greater insight into the morphology, surface chemistry, and, to a very limited extent, the dynamics of exfoliated polymer–clay nanocomposites. They were especially interested in developing NMR methods to quantify the level of clay exfoliation, a very important aspect of the characterization of nanocomposites. The main objective in solid-state NMR measurement is to connect the measured longitudinal relaxations, T1s, of ^{1}H and ^{13}C nuclei, with the quality of clay dispersion because the extent and homogeneity of the dispersion of the silicate layers within the polymer matrix are very important for determining physical properties.

Zhang et al. (2000) prepared polypropylene/MMT nanocomposites via melt intercalation using OMMT and twin-screw extrusion and also investigated the dispersibility of layers of MMT in the composites. It was determined that the silicate layers dispersed at the nanometer scale according to the XRD and TEM results, and also the impact strength of nanocomposites significantly increased compared to that of polypropylene and conventional filled composites, but their tensile strengths did not increase.

4.4.7 Energy-Dispersive X-Ray Spectroscopy

X-rays emitted from atoms are characteristic of the elements, and the intensity distribution of the X-rays represents the thickness-projected atom densities in the specimen. This is known as energy-dispersive X-ray spectroscopy (EDS), which is important for microanalysis, particularly for heavier elements. EDS is a key tool for identifying the chemical composition of specimen from a nanocomposite. Also, high-resolution electron microscopes are capable of producing a fine electron probe, smaller than 2 nm, allowing direct identification of the local composition of an individual element (Wang, 2001).

4.4.8 Small-Angle X-Ray Scattering

Small-angle X-ray scattering (SAXS) is an analytical X-ray technique using for the structural characterization of solid and fluid materials in the nanometer scale. When a nonhomogeneous dispersion, such as proteins, in water is irradiated, structural information about the scattering particles can be derived from the intensity distribution of the scattered beam at very low scattering angles.

It is possible to study the monodisperse in which the size, shape, and internal structure of the particle can be determined, and the polydisperse systems in which a size distribution could be calculated, assuming that all particles have the same shape, by using SAXS. SAXS is used to investigate structural details in the 0.5–50 nm size range in materials such as (Koo, 2006):

- Suspended nanopowders
- Materials in life sciences and biotechnology (proteins, viruses, DNA complexes)
- Polymer films and fibers
- Catalyst surface per volume
- Microemulsions
- Liquid crystals

SAXS can also provide the following information:

- Lamella repeat distance
- Radius of gyration
- Particle size and shape
- Large-scale structure and long-range order

4.4.9 Cone Calorimeter

The potential fire hazard for any material depends on many factors, which are the ignitability, the rate of surface flame spread, the heat-release-rate (HRR) (peak, average, and total), mass loss rate, smoke emission and the emission of toxic gas (e.g., carbon monoxide). The cone calorimeter (CC) is regarded as the most significant bench-scale instrument for evaluating the reaction associated with fire properties of materials. The data obtained from the CC test is available for engineering units that are conducive to extrapolation to other fire sizes and scenarios.

The CC is a fire test instrument based on the principle of oxygen consumption, as a result of the observation that the net heat of combustion of any organic material is directly related to the amount of oxygen required for combustion. Approximately 13.1 MJ of heat is released per kilogram of oxygen consumed. At the core of the instrument is a radiant electrical heater in the shape of a truncated cone (hence the name). This heating element irradiates a flat horizontal sample (100 × 100 mm and up to 50 mm thick) placed beneath it at a preset heat flux of up to 100 kW/m^2. The sample is placed on a load cell for continuous monitoring of its mass as it burns. Ignition is provided by an intermittent spark igniter located 13 mm above the sample.

The gas stream containing the combined combustion products is captured through an exhaust duct system consisting of a high-temperature centrifugal fan, a hood, and an orifice-plate flow meter. The typical air flow rate is 0.024 m^3/s. Oxygen concentration in the exhaust stream is measured with an oxygen analyzer capable of an accuracy of 50 ppm, and the HRR is determined by comparing oxygen concentration with the value obtained when no sample is burning. Smoke obscuration measurements are made in the exhaust duct using a helium–neon laser, with silicon photodiodes as the main beam and reference detectors, and appropriate electronics to derive the extinction coefficient and set the zero reading. Locations in the exhaust duct also provide additional sampling probes to determine concentrations of other combustion products such as carbon dioxide and carbon monoxide. All data are, continuously recorded at fixed intervals of a few seconds during the test.

The CC is used to determine the principal flammability properties such as HRR per unit area (kW/m^2), cumulative heat released (kW), effective heat of combustion (MJ/kg), time to ignition (s), mass loss rate (kg/s), total mass loss (kg), and smoke density (m^2/kg).

Ignitability is defined as the propensity to ignition, as measured by the time in seconds to sustain flaming (the existence of flame on or over most of the specimen surface for periods of at least 4 s) at a specified heat flux. Ignitability is thus a measure of how quickly a material will ignite in a fire situation. Heat release is defined as the heat generated in a fire due to various chemical reactions occurring within a given weight or volume of material. The major contributors are those reactions where carbon monoxide and carbon dioxide are generated and oxygen is consumed. Heat release data provide a relative fire hazard assessment for materials. Materials with low heat release per unit weight or volume will do less damage to the surroundings than materials with a high HRR. The HRR, especially the peak amount, is the primary characteristic determining the size, growth, and suppression requirements of a fire environment. The rate of heat release is determined by the measurement of oxygen consumption as determined by the oxygen concentration and the flow rate in the exhaust product stream.

The effective heat of combustion, mass remaining, smoke ratio, carbon dioxide, and carbon monoxide constitute the additional CC data obtained during the experiment (Koo, 2006).

4.4.10 Mass Loss Calorimeter

The mass loss calorimeter or cone (MLC) is used to determine the mass loss rate and ignitability at a standard heat flux condition.

The MLC, which is constructed as a two-part component system, the cone heater assembly (with a conical heater, shutter mechanism, spark ignition) and the load cell, may also serve as a quality control tool. It can also be constructed as an enclosed system to measure combustion products and toxicity yields or to use in low-oxygen conditions. An HRR instrument can be used to determine the effect of combustion as a ratio of the HRR and the mass loss rate. Thus, if the effective heat of combustion is known, mass loss rate data can be used to calculate approximate values of the HRR. The MLC is an inexpensive way of assessing fire properties of materials. Heat release is calculated using thermopile in the MLC (Koo, 2006).

4.5 Properties of Polymer–Clay Nanocomposites

The addition of clay minerals to polymers improves polymer properties and makes possible the production of polymer–

clay nanocomposites with desired characteristics for various applications. Commonly, researchers want to obtain a nanocomposite with desired characteristics as well as to eliminate the disadvantages of polymeric materials, while maintaining the intrinsic advantages of the primary polymer matrix. Because of the low price, availability, high aspect ratio, and desirable nanostructure and interfacial interactions, clays can provide dramatic and adjustable improved properties at very low loadings and also can behave as reinforcement filler for retaining the polymer's original useful properties. The structures and characteristics of components used, as well as preparation methodology and conditions, affect the eventual properties of the polymer–clay nanocomposites (Ray, 2010).

4.5.1 Mechanical Properties

The first and important goal in the application of fillers to a polymer is improvement in the mechanical properties; therefore, fillers are commonly called reinforcement agents (Jiang et al., 2005). The reinforcement mechanism is based on the fact that the filled materials have a higher resistance against stretching due to their higher elastic modulus values. When any rigid filler is added to the elastic polymer matrix, it may carry the major portion of applied load to the polymer matrix under stress conditions, if the interfacial interactions between filler and matrix are adequate (Gorrasi et al., 2003; Tortora et al., 2002).

Therefore, the larger the interface between the filler and the matrix, and also the stronger the interfacial interactions, the greater the reinforcement effect. Moreover, the modulus increases with an increase in the filler ratio in the composite or nanocomposite composition. The rigid structure of clay layers with high aspect ratio plays a very effective role on the increased modulus and stiffness values of polymer matrix in well-dispersed conditions. In fact, with the nanoclay fillers with very higher aspect ratio compared to conventional fillers such as glass fiber, the dramatic improvements in the mechanical properties of nanocomposites may be achieved, even if very lower amounts of nanoclay have been added (Le Baron et al., 1999).

Another approach suggests that the polymer chains adsorbed with strong physical interactions onto rigid clay platelets can behave as an integral portion of the rigid material, and thus their higher modulus values were emerged. Considering the very large interfacial area in well-dispersed nanocomposite structures, a significant improvement in the modulus can be expected. However, the significant improvement in the polymer–clay interfacial contact may result in a better stress transfer in the nanocomposites. It has been demonstrated that the enhancement of interfacial adhesion properties between clay and polymer using a suitable polar compatibilizer can also improve the mechanical properties of the nanocomposites. It should be noted that the high loadings of compatibilzers with the low molar mass may cause to some degree the plasticization of the nanocomposites and thus lead to their decreased values of modulus (Mittal, 2008; Osman et al., 2005). Also the modulus and stiffness of the nanocomposites can be increased by using the organically modified clay which would provide the higher intercalation/exfoliation level. Any parameter capable of facilitating the intercalation of polymer chains within the clay galleries can cause to emerge the more high exfoliation and interfacial interactions, providing the greater improvement in the modulus values. However, it is difficult to achieve complete exfoliation of clay layers, and the different clay platelet stacks with varied thicknesses may occur, depending on the number of layers stacked in a polymer matrix. Fornes and Paul (2003), who investigated the effect of incomplete exfoliation on nanocomposite properties, reported that the tensile modulus increased with an increase in the filler volume fraction of a nanocomposite. However, more clay loading than the threshold limit may affect Young's modulus, which is due to the formation of a partially exfoliated/intercalated structure after that fully exfoliated structure formation. The tensile strength, which may decrease with the addition of clay minerals, also depends on the eventual morphology of the nanocomposites.

As in the modulus, any factor affecting the degree of intercalation/exfoliation can affect the tensile strength of a nanocomposite (Alexandre and Dubois, 2000; Finnigan et al., 2004; Mittal, 2007; Ray et al., 2003). The effect of clay fillers on the stress at break values for polymer nanocomposites depends on the interfacial interactions between polymer and clay layers.

The stronger interfacial interaction increases stress at break values, and the weak interfacial forces may lead to some decrease in stress at break values for the nanocomposite. The more polar polymers, such as nylon and rubber-based polymers, have more strong interfacial interactions with the charged clay layers, and also their nanocomposites exhibit the more improved values of the stress at break compared to the virgin polymer (Wang et al., 1998). Polyamide–clay nanocomposites have shown highly improved values of stress at break, which is associated with the higher degree exfoliated structures, emerged as a result of strong ionic interactions between polymer chains and clay layers. The nanocomposites of any nonpolar polymer with clay have weaker interfacial interactions as well as a lower degree of exfoliation in which decreased or slightly improved values of stress at break may be obtained. However, the modification of polymer chain or clay layers may provide the higher values of stress at break for characteristic nonpolar polymers. The values of elongation at break for polymer–clay nanocomposites also similarly depend on the interfacial interactions of the polymer–clay system (Finnigan et al., 2004; Yao et al., 2002). The impact strength of polymer–clay nanocomposites was also studied and compared with that of virgin polymer systems. It is usually reported that the impact strength is increased by the addition of a very low amount of clay (0.1 wt.%) (Oya et al., 2000; Zhang et al., 2000). However, it is noted that the addition of excess clay also reduces the impact strength of polymer–clay nanocomposites (Phang et al., 2005). Dynamic mechanical analysis studies on polymer–clay nanocomposites have shown that usually the storage modulus and glass transition temperature (T_g) are improved by the addition of clay nanofillers (Krikorian and Pochan, 2003; Laus et al., 1997; Nam et al., 2001).

4.5.2 Thermal Properties

Thermal gravimetric analysis (TGA) is used commonly for the thermal characterization of modified fillers and polymer nanocomposites (Koh et al., 2008). It has been reported that the incorporation of the layered silicate into polymeric matrix provided the enhancement of thermal stability at polymer material. Also it has been postulated that the layered silicates would behave as the superior insulator and mass transporting barriers against the volatile products generated during decomposition. The silicates may also assist in the formation of carbonaceous char after thermal decomposition (Becker et al., 2002; Mittal, 2009; Ray and Bousima, 2005; Sinha Ray and Okamoto, 2003; Tortora et al., 2002; Vyazovkin et al., 2004; Zanetti et al., 2004; Zhu et al., 2001).

Inorganic clay minerals are almost stable in the temperature ranges in which the organic polymers are degraded to volatile compounds. Therefore, during the thermal progress, the clay content of nanocomposites remains as a residue after the heating program of TGA (Gilman, 1999).

Blumstein (1965) first reported that the thermal stability of PMMA/MMT nanocomposites is higher compared to that of the virgin polymer, and that the polymer intercalated into Na-MMT has 40–50°C higher decomposition temperature. Vyazovkin et al. (2004) also showed that the polystyrene (PS)–clay nanocomposites have 30–40°C higher degradation temperature compared to virgin PS.

In some other studies, comparative examinations of the thermal decomposition behavior of nanocomposite and virgin polymers under oxidative and non-oxidative decomposition conditions have been carried out (Berta et al., 2006; Beyer, 2002; Zhao et al., 2005). The effect of polymer compatibilizer on the thermal degradation behavior of the polymer–clay nanocomposites has also been investigated, and the results showed that the incorporation of compatibilizer with low molar mass does not create a negative effect on the thermal degradation of nanocomposites (Mittal, 2008).

4.5.3 Flame Retardancy

Precise control over the ignition (burn) characteristics of polymers that are widely used in all areas of daily life is required to make them safer. Conventional chemical additives known as flame retardants are used to control the ignition (burn) characteristics. Also it has been determined that in the polymer–

clay nanocomposites, clay minerals play a role in flame retardancy, beyond an improvement in the physical and mechanical properties of polymers (Giannelis, 1998; Kiliaris and Papaspyrides, 2010). However, detailed studies indicate that clays are effective for controlling the flames from spreading in developing fires but not significantly effective for controlling fully developed fires (Levchik and Weil, 2000; Zanetti, 2006). Therefore, the use of clay minerals together with a low fraction of conventional flame retardants is more suitable (Wang et al., 2004). However, it is known that clay minerals and conventional flame retardants have a considerable synergistic effect on the reduction of ignitability of polymers, and that the organically modified polymer–clay nanocomposites combined with halogenated flame retardants and phosphorous flame retardants also exhibit a similar effect. On the other hand, the synergic effect of clays on the flame retardancy of halogenated and phosphorous compounds may cause a reduction in the volume fraction of the polymer–clay nanocomposite systems (Hao et al., 2006; Wang et al., 2005; Weil et al., 1992; Zhang and Horrocks, 2003). There are many studies in which various phosphorous flame retardants in clay galleries have been made to intercalate. The clays modified with phosphorous compounds have been used in the preparation of nanocomposites with various polymers. It has been considered that the phosphorous flame retardants contained in the composites are shielded with the clay layers and thus it prevented their evaporation during preparation of nanocomposites, especially by the melt intercalation method, which may provide more efficient flame retardancy in the eventual product (Chigwada and Wilkie, 2003). Schartel et al. (2006) have used tetraphenyl phosphonium-modified clay for the preparation of nanocomposites based on bisphenol A diglycidyl ether (DGEBA) epoxy resin. Toldy et al. (2006) have also used phosphorous compounds both as curing agents and flame retardants in the preparation of epoxy–clay nanocomposites.

4.5.4 Barrier Properties

One of the most considerable effects of clays on the matrixial properties of polymer is the extraordinary improvement emerging in the barrier properties of polymers (Krishnamoorti et al., 1996). The clay layers that are naturally impermeable can improve the barrier properties of polymers by creating a maze or tortuous path that retards the diffusion of gas molecules throughout the polymer matrix (Paul and Robeson, 2008; Yano et al., 1993) (Fig. 4.8).

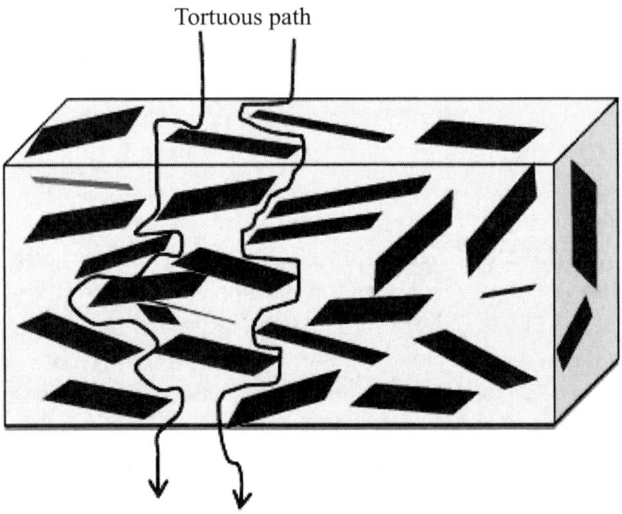

Figure 4.8 Schematic presentation of the barrier improvement mechanism by the addition of clay platelets. Reprinted from Paul and Robeson, Copyright 2008, with permission from Elsevier.

The degree of enhancement in the barrier properties depends on the degree of tortuosity formed by clay sheets across the diffusion path of molecules through the polymer film. The tortuous factor is determined by the ratio of the actual distance to the shortest distance in which the diffusive molecule is able to be diffused (polymer film thickness). This factor is affected by the aspect ratio of clay dispersed in the matrix. Increasing the lateral length of the clay sheet as well as increasing the exfoliation or dispersion degree causes greater barrier enhancement in the polymer matrix. Many studies on the barrier properties of polymer–clay nanocomposites against the diffusion of gases and vapors are reported (Giannelis, 1996; Lange and Wyser, 2003; Tortora et al., 2002).

4.5.5 Anticorrosive Properties

Polymers are widely used as anticorrosive coatings on metals to prevent corrosion. Primarily polymeric coatings act as a physical barrier against the diffusion of aggressive species to the metal surface. However, most polymer coatings show some degree of permeations, especially in long periods of contact with aggressive species. Various techniques have been applied to improve the barrier effect of polymeric coatings. One of the effective methods is the preparation of the coatings of polymer based composites or nanocomposites by the addition of fillers suitable to the polymer coating matrix. It has been reported that the addition of platelet fillers such as layered silicates improved effectively the anticorrosive barrier effect of polymer coatings by extending the diffusion pathways of aggressive species. Clay minerals having a layered structure and high aspect ratio, provided that they are well dispersed, decrease the permeability of polymer coating films parallel to the increase in the diffusion pathways. The nanocomposites prepared by using the unmodified and organically modified clays with various polymeric materials, such as polystyrene (PS), poly(styrene-co-acrylonitrile) (SAN), polyaniline (PANI), polypyrrole (PPy), polysulfone (PSU), polyacrylates, and epoxy resins, have been investigated as anticorrosive coatings on metals (Olad and Rashidzadeh, 2008; Yeh and Chang, 2008).

4.5.6 Chemical Properties of Polymer–Silicate Nanocomposites

In most cases, polymeric nanomaterials exhibit a behavior with multifunctionality. The superior resistance against solvent and moisture is known as one of the functions of these materials. In 1963, an organoclay was incorporated into a thermoplastic polyolefin matrix by Nahin et al., and the organoclay-modified composite exhibited improved solvent resistance. Ten years later, polymer–clay nanocomposites that exhibited increased solvent resistance and decreased thermal expansion behavior were produced on a larger scale using nylon 6, by in situ intercalative polymerization technique, similar to the Unitika process (Nyugen and Baird, 2006).

In a study conducted on natural rubber–clay nanocomposites, it was found that the virgin rubber exhibited much higher toluene uptake than that of the nanocomposites at 25°C, which implies a higher swelling resistance. Also a similar swelling resistance was observed with silicon–clay nanocomposites, showing a significant reduction in toluene uptake, even at very low loadings (1.0 wt.%). This improvement in solvent uptake resistance of rubber nanocomposites has been explained with the intensive interactions between the rubber matrix and organoclay platelets and also with the increased average diffusion pathways for the solvents, depending on the large surface area of impermeable clay layers in the nanocomposites (Mohammad and Simon, 2006).

Poly(etherimide) (PEI) resin, which is amorphous thermoplastic polymer, has interesting physical and chemical properties as well as gained chain flexibility and good melt flow characteristics with ether bridges, which makes it an attractive candidate for many applications. Although PEI has good melt processability and solubility in some solvents, its applications in aerospace and automotive fields are limited due to its relatively poor solvent resistance and thermal stability. The solvent resistance of virgin PEI and PEI/MMT nanocomposites has been studied by using two different series of solvents. One series was conventional solvents, such as water, ethanol, and xylene, while the other was nonprotonic polar solvents, such as *N*-methyl-2-pyrrolidone (NMP), *N,N*-dimethylacetamide (DMA), and *N,N*-dimethylsulfoxide (DMSO). In addition, virgin PEI could dissolve in nonprotonic polar solvents, while the PEI/MMT nanocomposites could not be dissolved in these solvents at 25°C. The dramatic decrease in solvent uptake of the PEI/MMT nanocomposites compared to virgin PEI is due to the strong interactions between PEI and MMT, the excellent barrier properties of the large-aspect-ratio silicate layers, and the exfoliated structure of MMT layers that maximizes the available surface area of the reinforcing phase. The interaction between PEI and MMT leads to the formation of a "bound polymer," a polymer in close proximity to the reinforcing filler (MMT) that is either physisorbed or chemisorbed and, therefore, restricts the solvent uptake. Conventional nonreinforcing fillers prevent the formation of a bound polymer, allowing large vesicles to form during solvent exposure, resulting in increased solvent uptake. Solvent uptake is inversely proportional to the filler-reinforcing strength, which strongly depends on filler surface area and agglomerate structure.

Therefore, the exfoliated structure and large surface of the layered silicate (750 m^2/g) are favorable to the solvent resistance properties of PEI/MMT nanocomposites (Huang et al., 2001).

One of the main factors that determine the durability of a material is its chemical stability. The material should not dissolve, erode, or corrode, nor should it leach important constituents into oral fluids. Erosion in terms of dental applications is a process that combines the chemical process of dissolution with a mild mechanical action. These properties are particularly important for all restorative materials since high solubility or poor resistance to erosion will severely limit the effective lifetime of the restoration. Some soft acrylic polymers used for cushioning the fitting surfaces of dentures rely on the absence of relatively large quantities of a plasticizer in the acrylic resin for their softness. The slow leaching of the plasticizer causes the resin to become hard and, therefore, ineffective as a cushion (McCabe and Walls, 2008).

Kato et al. (2003) has successfully prepared PE–clay nanocomposites by melt intercalation with maleic anhydride-grafted polyethylene (MA-g-PE) organoclay, and polyethylene. It was found that in these PE–clay hybrids, the silicate layers were exfoliated and dispersed well in the polymer matrix. The hybrids exhibited higher tensile yield strengths and tensile moduli than those of PE matrices and those of PE–inorganic clay composites. When 5.0 wt.% clay was loaded, the tensile yield strength and tensile modulus of the hybrid were, respectively, 1.4 and 1.8 times higher than those of the PE/MA-g-PE mixture. The gas permeability of the clay hybrid decreased 30% compared to the PE/MA-g-PE mixture.

Kato et al. (2011) has developed successfully PP–clay nanocomposite using organoclay and modified PP. In this work, organoclay, PP oligomer carrying polar groups and PP were melt-blended. This is a direct polymer intercalation process for preparing polymer nanocomposites by melt compounding, a useful process from an industrial standpoint. Other polyolefin polymers (polyethylene), polyolefin rubbers (ethylene–propylene elastomer, ethylene–propylene diene monomer), and nanocomposites have also been developed successfully by this direct polymer intercalation process or by techniques based on direct polymer intercalation process. These prepared polyolefin nanocomposites exhibit superior mechanical, thermal, and gas barrier properties.

Hotta and Paul (2004) prepared PE–clay nanocomposites by melt compounding with various combinations of a maleic anhydride-grafted linear low-density polyethylene (LLDPE-g-MA), linear low-density polyethylene (LLDPE), and two organoclays. They selected two types of organoclays to show the effect of the number of alkyl groups attached to the nitrogen of the organic modifier on exfoliation and improvement of mechanical properties. Nanocomposites derived from the organoclay having two alkyl tails exhibited better dispersion and improvement of mechanical properties than nanocomposites based on the organoclay having one alkyl tail. In addition, the rheological properties and gas permeability of the nanocomposites derived from the organoclay having two alkyl tails were investigated. Both melt viscosity and melt tension (melt strength) increased with increased content of clay (MMT) and LLDPE-g-MA. Gas permeability decreased by the addition of MMT.

Liang et al. (2004) prepared PE/PE-g-MA/OMMT nanocomposites by direct melt blending and solution blending. Three cationic surfactants, hexadecyltrimethylammoniumchloride, hexadecyldimethylbenzylammonium chloride, and octadecyl-trimethylammonium chloride, were used to modify MMT. It was found that the intercalation effect of PE/PE-g-MA/OMMT could be enhanced by increasing the content of PE-g-MMT using the silicate modified by a cationic surfactant with a benzyl group or a long alkyl chain, adopting the solution-blending method or using high-density PE as a matrix. The maximum value in tensile strength was achieved when the concentration of PE-g-MA was 6 wt.%. The impact strength increased concomitantly with the content of PE-g-MA.

Liu et al. (1999) processed nylon 6–clay nanocomposites by melt intercalation. XRD and DSC results showed that the crystal structure and crystallization behavior of the nanocomposites were different from those of nylon 6. Mechanical and thermal tests showed that the properties of the nanocomposites were superior to nylon 6 in terms of the heat distortion temperature, strength, and modulus without sacrificing their impact strength.

The mechanical behavior of compression-molded polyamide 6 (PA6) reinforced with 2 wt.% of organonanoclay was studied and compared to that of PA6 by Bureau et al. (2002). The strength and Young's modulus of the PA6–clay were 15% higher than those of PA6. DSC, FTIR, and XRD analysis showed that the crystalline structures of PA6 and PA6/clay differed considerably. A crystallinity of 25% with a dual structure composed of the γ and α forms was obtained in PA6/clay, while a crystallinity of 31% with the γ form as the dominant crystalline structure was obtained in PA6.

Khatua et al. (2004) studied the effect of organoclay platelets on the morphology of polyamide 6 and ethylene propylene rubber (EPR) blends by SEM and TEM. They observed that the dispersed domain size of the EPR phase in 20 wt.% EPR-containing blends decreased significantly even if a small amount of organoclay was added. The extent of the decrease in domain size was similar to polyamide 6/EPR-g-MA blends. Their results indicated that if the clay becomes exfoliated in the polymer matrix, the exfoliated clay platelets prevent the coalescence of the dispersed domains.

Tjong and Bao (2005) investigated the impact fracture toughness of polyamide-6/MMT nanocomposites toughened with a maleic anhydride-grafted styrene-ethylene-butylene-styrene (SEBS-g-MA) elastomer. Their results showed that SEBS-g-MA addition improved the tensile ductility and impact strength of polyamide 6/MMT nanocomposites at the expense of its tensile strength and ductility.

SEBS-MA (up to 40 wt.%) toughened polyamide 6 nanocomposites with 3 wt.% organoclay were studied by González et al. (2006). The morphology of the polyamide 6 matrix did not change on blending with an elastomer. However, rubber particle sizes in nanocomposites were larger than their corresponding blends. Supertough nanocomposites were obtained with 30 wt.% SEBS-MA. The interactions between the organoclay surface and elastomer bearing maleic anhydride groups led to a decrease in compatibility, thus decreased the particle size.

Polymer–clay composites have been prepared by melt blending an organobentonite with linear polymers (polyamide, polystyrene, and polypropylene) in a disk-screw extruder by Burmistr et al. (2005). The first time the organoclay was prepared by surface treatment of Na-bentonite with polymeric quaternary ammonium salts (PQAS). The XRD analysis indicated that organobentonite layers were exfoliated and dispersed into polyamide and PS. With the incorporation of 2 wt.% organobentonites (optimal concentration) into PS, the tensile strength increased to 28% and the Charpy impact value, which determines the amount of energy absorbed by a material during fracture, increased to 25%. In polypropylene–organobentonite composites, the delamination and hence any reinforcement effect has not been observed. TGA showed that the polyamide and PS nanocomposites have a higher decomposition temperature in comparison to the original polymers.

Somwangthanaroj et al. (2012) investigated nylon 6–clay nanocomposite films prepared by melt mixing of nylon 6 with an organoclay using a twin-screw extruder attached to a blown film die. It has been expected that the type of surfactant used in the pretreatment of the organoclay would affect the degree of clay dispersion, depending on the changing of the degree of crystallinity, crystalline phase, and bulk properties of the polymer composite. Hence, for the modification of the surface of MMT clay, two different surfactants, trimethyl tallow quaternary ammonium chloride (M_3T, a surfactant with single-chain) and dimethyl bis (hydrogenated tallow) ammonium chloride [$M_2(HT)_2$, a surfactant with double-chain], were used. The M_3T organoclay modified with the single-chain surfactant shows a more disordered structure than the $M_2(HT)_2$ organoclay modified with the double-chain surfactant. XRD analysis confirmed that the M_3T nanocomposite films exhibited a higher degree of clay dispersion than the corresponding $M_2(HT)_2$ nanocomposite films. In addition, both DSC and XRD results revealed that the addition of layered silicates induced the γ-crystalline form of nylon 6. The degree of crystallinity of the M_3T nanocomposite films was shown to be higher than that of the corresponding $M_2(HT)_2$ nanocomposite films at the same organoclay loading. Since good dispersion of the clay platelets and a high degree of crystallinity of the nylon 6 matrix led to the enhancement of the mechanical and barrier properties of the nanocomposite films, the tensile modulus, yield strength, and gas barrier properties of the M_3T films were much higher than those of the corresponding $M_2(HT)_2$ films. Furthermore, all nylon 6–clay nanocomposite films exhibited higher mechanical properties in the machine direction than the transverse direction.

Lan et al. (1994) observed in polyimide (PI) nanocomposites that the relative permeability for O_2, CO_2, and H_2O fit the tortuous path model reasonably well, but measurements on ethyl acetate exhibited a very large dependence on relative humidity (RH). The relative permeability of ethyl acetate at 0% RH was 0.19, and at 50% RH, it was 0.09. In contrast, pure PI was more than double its permeability going from 0% to 50% RH.

Chaiko and Leyva (2005) reported that the oxygen permeability of a polypropylene wax–clay composite decreased 62-fold with the clay loading of 5.0 wt.%.

Uhl et al. (2004) has investigated the effect of clay on cross-link density on a polyurethane nanocomposite. The control polyurethane had a crosslink density of 6.4 (10^{-3} mol/cm^3), while the composites containing 5% clay gave an average of 8.1. This is a 26% increase, which is very close to the average increase in the modulus of 28%. Accordingly, it can be claimed that the clay either acted as a catalyst for the crosslinking reaction, or acted as a crosslinker.

The nylon 6/ MMT composites were produced by using in situ intercalative polymerization technique, with the clay modified with 12-aminolauric acid by Okada et al. (1990). The tensile strength, tensile modulus, and heat distortion temperature of the composites exhibited increase by 55%, 91%, and 134%, respectively. The impact strength does not show a statistically significant change. The TEM micrographs of nanocomposites show that the clay is almost completely exfoliated. A work reported by Fornes et al. (2001) on the melt compounding of organoclays and nylon 6 polymers having various molar masses makes the comparison of the present data. The interpolation of the data to the clay content of the in situ polymerized nylon 6 predicts that the tensile strength and tensile modulus of the composites increased by 26% and 59%, respectively. These numbers are lower than the in situ polymerized composite. A detailed TEM analysis of the compounded samples shows that a large portion of the plates was exfoliated but a significant number of aggregates containing two and three plates was still present. The impact strength is not as that of the in situ material. The values of elongation at break decreased significantly in the compounded material.

Fornes and Paul (2004) have reported the data on the nanocomposites of nylon 11 and 12. It was found that at a clay content of 4.0 wt.%, increases in the tensile strength and modulus of nylon 11 and 12 were 18% and 17%, and 56% and 51%, respectively. The comparisons with nylon 6 nanocomposites indicate that the level of exfoliation in nylon 11 and 12 is much lower than that of nylon 6.

References

Alexandre, M., Dubois, P. (2000). Polymer-layered silicate nanocomposites: Preparation, properties and uses of a new class of materials. *Mater. Sci. Eng.*, **28**, pp. 1–63.

Anadao, P. (2012). Polymer–clay nanocomposites, concepts, researches, applications and trends for the future. In *Nanocomposites: New Trends and Developments*, Ebrahimi, F. (Ed.). Croatia: InTech, pp. 1–16.

Aranda, P., Mosqueda, Y., Perez-Cappe, E., Ruiz-Hitzky, E. (2003). Electrical characterization of PEO-clay nanocomposites prepared by microwave irradiation. *J. Polym. Sci. B*, **41**, pp. 3249–3263.

Aranda, P., Ruiz-Hitzky, E. (1992). Poly(ethylene oxide)-silicate intercalation materials. *Chem. Mater.*, **4**, pp. 1395–1403.

Aranda, P., Ruiz-Hitzky, E. (1994). New polyelectrolyte materials based on smectite polyoxyethylene intercalation compounds. *Acta Polym.*, **45**, pp. 59–67.

Aranda, P., Ruiz-Hitzky, E. (1999). Poly(ethylene oxide)/$NH4^+$ nanocomposites. *Appl. Clay Sci.*, **15**, pp. 119–135.

Armentano, I., Dottori, M., Fortunati, E., Mattioli, S., Kenny, J. M. (2010). Biodegradable polymer matrix nanocomposites for tissue engineering: A review. *Polym. Degrad. Stab.*, **95**, pp. 2126–2146.

Avella, M., Errico, M., Martelli, S., Martuscelli, E. (2001). Preparation methodologies of polymer matrix nanocomposites. *Appl. Organomet. Chem.*, **15**, pp. 435–439.

Becker, O., Varley, R., Simon, G. (2002). Morphology, thermal relaxations and mechanical properties of layered silicate nanocomposites based upon high-functionality epoxy resins. *Polym*er, **43**, pp. 4365–4373.

Becker, O., Varley, R. J., Simon, G. P. (2004). Thermal stability and water uptake of high performance epoxy layered silicate nanocomposites. *Eur. Polym. J.*, **40**, pp. 187– 195.

Berta, M., Lindsay, C., Pans, G., Camino, G. (2006). Effect of chemical structure on combustion and thermal behavior of polyurethane elastomer layered silicate nanocomposites. *Polym. Degrad. Stab.*, **91**, pp. 1179–1191.

Beyer, G. (2002). Nanocomposites: A new class of flame retardants for polymers. *Plast. Addit. Compd.*, **4**(10), pp. 22–27.

Biswas, M., Sinha Ray, S. (2001). Recent progress in synthesis and evaluation of polymer-montmorillonite nanocomposites. *Adv. Polym. Sci.*, **155**, pp. 167–221.

Blumstein, A. (1965). Polymerization of adsorbed monolayers. II. Thermal degradation of the inserted polymers. *J. Polym. Sci. A*, **3**, pp. 2665–2670.

Bower, D. J. (2002). *An Introduction to Polymer Physics.* USA: Cambridge University Press.

Bureau, M. N., Denault, J., Cole, K. C., Enright, G. D. (2002). The role of crystallinity and reinforcement in the mechanical behavior of polyamide-6/clay nanocomposites. *Polym. Eng. Sci.*, **42**, pp. 1897–1906.

Burmistr, M. V., Sukhyy, K. M., Shilov, V. V., Pissis, P., Spanoudaki, A., Sukha, I. V., Tomilo, V. I., Gomza, Y. P. (2005). Synthesis, structure, thermal and mechanical properties of nanocomposites based on linear polymers and layered silicates modified by polymeric quaternary ammonium salts (ionenes). *Polym*er, **46**, pp. 12226–12232.

Burnside, S. D., Giannelis, E. P. (1995). Synthesis and properties of new poly(dimethylsiloxane) nanocomposites. *Chem. Mater.*, **7**, pp. 1597–1600.

Chaiko, D. J., Leyva, A. A. (2005). Thermal transitions and barrier properties of olefinic nanocomposites. *Chem. Mater.*, **17**, pp. 13–19.

Chiang, C. L., Ma, C. C. M. (2002). Synthesis, characterization and thermal properties of novel epoxy containing silicon and phosphorus nanocomposites by sol-gel method. *Eur. Polym. J.*, **38**, pp. 2219–2224.

Chigwada, C., Wilkie, C. A. (2003). Synergy between conventional phosphorus fire retardants and organically-modified clays can lead to fire retardancy of styrenics. *Polym. Degrad. Stab.*, **80**, pp. 551–557.

Chin, I. J., Thurn-Albrecht, T., Kim, H. C., Russell, T. P., Wang, J. (2001). On exfoliation of montmorillonite in epoxy. *Polymer*, **42**, pp. 5947–5952.

Cho, J. W., Paul, D. R. (2001). Nylon 6 nanocomposites by melt compounding. *Polymer*, **42**, pp. 1083–1094.

Cosoli, P., Scocchi, G., Pricl, S., Fermaglia, M. (2008). Many-scale molecular simulation for ABS-MMT nanocomposites: Upgrading of industrial scraps. *Microporous Mesoporous Mater.*, **107**, pp. 169–179.

Dabrowski, F., Bourbigot, S., Delbel, R., Bras, M. L. (2000). Kinetic molding of the thermal degradation of polyamide-6 nanocomposite. *Eur. Polym. J.*,

36, pp. 273–284.

Davis, R. D., Jarrett, W. L., Mathias, L. J. (2000). Observation of α and γ crystal forms and amorphous regions of nylon 6-clay nanocomposites using solid state ^{15}N nuclear magnetic resonance. *Polym. Mater. Sci. Eng.*, **82**, p. 272–273.

Dennis, H. R., Hunter, D. L., Chang, D., Kim, S., White, J. L., Cho, J. W, Paul, D. R. (2001). Effect of melt processing conditions on the extent of exfoliation in organoclay-based nanocomposites. *Polymer*, **42**, pp. 9513–9522.

Eastman, M. P., Bain, E., Porter, T. L., Manygoats, K., Whitehorse, R., Parnell, R. A., Hagerman, M. E. (1999). The formation of poly(methyl-methacylate) on transition metal-exchanged hectorite. *Appl. Clay Sci.*, **15**, pp. 173–185.

Finnigan, B., Martin, D., Halley, P., Truss, R., Campell, K. (2004). Morphology and properties of thermoplastic polyurethane nanocomposites incorporating hydrophilic layered silicates. *Polymer*, **45**, pp. 2249–2260.

Fornes, T. D., Paul, D. R. (2003). Modeling properties of nylon 6/clay nanocomposites using composite theories. *Polymer*, **4**, pp. 4993–5013.

Fornes, T. D., Paul, D. R. (2004). Structure and properties of nanocomposites based on Nylon-11 and -12 compared with those based on Nylon-6. *Macromolecules*, **37**, pp. 7698–7709.

Fornes, T. D., Yoon, P. J., Keskkula, H., Paul, D. R. (2001). Nylon 6 nanocomposites: The effect of matrix molecularweight. *Polymer*, **42**, pp. 9929–9940.

Feng, J., Weng, L.-T., Chan, C.-M., Xhie, J., Li, L. (2001). Imaging of sub-surface nanoparticles by tapping-mode atomic force microscopy. *Polymer*, **42**, pp. 2259–2262.

Fukushima, Y., Okada, A., Kawasumi, M., Kurauchi, T., Kamigaito, O. (1988). Swelling behavior of montmorillonite by poly-6-amide. *Clay Miner.*, **23**, pp. 27–34.

Gacitua, W. E., Ballerini, A. A., Zhang, J. (2005). Polymer nanocomposites: Synthetic and natural fillers, a review. *Ciencia Technologia*, **7**, pp. 159–178.

Giannelis, E. P. (1996). Polymer layered silicate nanocomposites. *Adv. Mater.*, **8**, pp. 29–35.

Giannelis, E. P. (1998). Polymer-layered silicate nanocomposites: Synthesis, properties and applications. *Appl. Organomet. Chem.*, **12**, pp. 675–680.

Gilman, J. W. (1999). Flammability and thermal stability studies of polymer layered-silicate (clay) nanocomposites. *Appl. Clay Sci.*, **15**, pp. 31–49.

Gilman, J. W., Jackson, C. L., Morgan, A. B., Harris, Jr. R., Manias, E., Giannelis, E. P., Wuthenow, M., Hilton, D., Phillips, S. H. (2000). Flammability properties of polymer-layered silicate nanocomposites. Propylene and polystyrene nanocomposites. *Chem. Mater.,* **12**, pp. 1866–1873.

Gilman, J. W., Kashiwagi, T., Lichtenhan, J. D. (1997). Flammability studies of polymer-layered silicate nanocomposites. *SAMPE J.*, **33**, pp. 40–45.

Ginsburg, V. V., Balazs, A. C. (2000). Calculating phase diagrams for nanocomposites: The effect of adding end-functionalized chains to polymer–clay mixture. *Adv. Mater.*, **12**, pp. 1805–1809.

González, I., Eguiazábal, J. I., Nazábal, J. (2006). Rubber-toughened polyamide 6/clay nanocomposites. *Comp. Sci. Technol.*, **66**, pp. 1833–1843.

Gorrasi, G., Tortora, M., Vittoria, V., Pollet, E., Lepoittevin, B., Alexandre, M. (2003). Vapor barrier properties of polycaprolactone montmorillonite nanocomposites: Effect of clay dispersion. *Polymer*, **44**, pp. 2271–2279.

Greenland, D. J. (1963). Adsorption of poly(vinyl alcohols) by montmorillonite. *J. Colloid Sci.*, **18**, pp. 647–664.

Gunter, M., Reichert, P., Mulhaupt, R., Gronski, W. (2000). Morphology and mechanical properties, and mechanism of reinforcement of rubber nanocomposites. *Polym. Mater. Sci. Eng.*, **82**, pp. 228–229.

Hackett, E., Manias, E., Giannelis, E. P. (1998). Molecular dynamics simulations of organically modified layered silicates. *J. Chem. Phys.*, **108**, pp. 7410–7415.

Hao, X., Gai, G., Liu, J., Yang, Y., Zhang, Y., Nan, C. (2006). Flame retardancy and antidripping effect of OMT/PA nanocomposites. *Mater. Chem. Phys.*, **96**, pp. 34–41.

Hotta, S., Paul, D. R. (2004). Nanocomposites formed from linear low density polyethylene and organoclays. *Polymer,* **45**, pp. 7639–7654.

Huang, J.-C., Zhu, Z.-K., Yin, J., Qian, X.-F., Sun, Y.-Y. (2001). Poly(etherimide)/montmorillonite nanocomposites prepared by melt intercalation: Morphology, solvent resistance properties and thermal properties. *Polymer*, **42**(3), pp. 873–877.

Hunt, B. J., James, M. I. (1993). *Polymer Characterisation*. Glasgow, UK: Blackie Academic Professional.

Hussain, F., Hojjati, M., Okamoto, M., Gorga, R. E. (2006). Review article: Polymer-matrix nanocomposites, processing, manufacturing, and application; an overview. *J. Compos. Mater.*, **40**, pp. 1511–1565.

Jacob, A. (2008). Rubber-layered clay nanocomposites. PhD thesis. Department of Polymer Science and Rubber Technology, Cochin

University of Science and Technology, Kerala, India.

Jeon, H. G., Jung, H. T., Lee, S. W., Hudson, S. D. (1998). Morphology of polymer silicate nanocomposites. High density polyethylene and a nitrile. *Polym. Bull.*, **41**, pp. 107–113.

Jiang, L., Lam, Y. C., Tam, K. C., Chua, T. H., Sim, G. W., Ang, L. S. (2005). Strengthening acrylonitrile-butadiene-styrene (ABS) with nano-sized and micron-sized calcium carbonate. *Polymer*, **46**, pp. 243–252.

Jimenez, G., Ogata, N., Kawai, H., Ogihara, T. (1997). Structure and thermal/mechanical properties of poly(1-caprolactone)–clay blend. *J. Appl. Polym. Sci.*, **64**, pp. 2211–2220.

Joseph, R., Zhang, S., Ford, W. (1996). Structure and dynamics of a colloidal silica–poly(methyl methacrylate) composite by ^{13}C and ^{29}Si MAS NMR spectroscopy. *Macromolecules*, **29**, pp. 1305–1312.

Journet, C., Maser, W. K., Bernier, P., Loiseau, A., de la Chapell, M. L., Lefrant, S. (1997). Large-scale production of single-walled carbon nanotubes by the electric-arc technique. *Nature*, **388**, pp. 756–758.

Karagöz, S. (2013). *Preparation and Characterization of Composites Material with Urea/Formaldehyde Resins*. Master Thesis. Atatürk University, Erzurum, Turkey.

Kato, M., Okamoto, H., Hasegawa, N., Tsukigase, A., Usuki A. (2003). Preparation and properties of polyethylene-clay hybrids. *Polym. Eng. Sci.*, **43**(6), pp. 1312–1316.

Kato, M., Usuki, A., Hasegawa, N., Okamoto, H., Kawasumi, M. (2011). Development and applications of polyolefin– and rubber–clay nanocomposites. *Polym. J.*, **43**, pp. 583–593.

Kawasumi, M., Hasegawa, N., Usuki, A., Okada, A. (1998). Nematic liquid crystal/clay mineral composites. *Mater. Sci. Eng. C*, **6**, pp. 135–143.

Ke, Y. C., Stroeve, P. (2005). *Polymer-Layered Silicate and Silica Nanocomposites*. Amsterdam, The Netherlands: Elsevier.

Khatua, B. B., Lee, D. J., Kim, H. Y., Kim, J. K. (2004). Effect of organoclay platelets on morphology of nylon-6 and poly(ethylene-*ran*-propylene) rubber blends. *Macromolecules*, **37**, pp. 2454–2459.

Kiliaris, P., Papaspyrides, C. D. (2010). Polymer/layered silicate (clay) nanocomposites, an overview of flame retardancy. *Prog. Polym. Sci.*, **35**, pp. 902–958.

Kim, G. M., Lee, D. H., Hoffmann, B., Kressler, J., Stoppelmann, G. (2001). Influence of nanofillers on the deformation process in layered silicate/polyamide 12 nanocomposites. *Polymer*, **42**, pp. 1095–1100.

Kim, J. H., Ko, J. H., Bae, B. S. (2007). Dispersion of silica nano-particles in sol-gel hybrid resins for fabrication of multi-scale hybrid nanocomposite. *J. Sol-Gel. Sci. Technol.*, **41**, pp. 249–255.

Koh, H. C., Park, J. S., Jeong, M. A., Hwang, H. Y., Hong, Y. T., Ha, S. Y., Nam, S. Y. (2008). Preparation and gas permeation properties of biodegradable polymer/layered silicate nanocomposite membranes. *Desalination*, **233**, pp. 201–209.

Kojima, Y., Usuki, A., Kawasumi, M., Fukushima, Y., Okada, A., Kurauchi, T., Kamigaito, O. (1993). Mechanical properties of nylon-6/clay hybrid. *J. Mater. Res.*, **8**, pp. 1179–1184.

Kojima, Y., Usuki, A., Kawasumi, M., Okada, A., Kurauchi, T., Kamigaito, O. (1993). Synthesis of nylon-6-clay hybrid by montmorillonite intercalated with ε-caprolatam. *J. Polym. Sci. A*, **31**, pp. 983–986.

Koo, H. J. (2006). *Polymer Nancomposites. Processing, Characterization, and Application*. USA: McGraw-Hill.

Koo, J. H., Stretz, H., Bray, A., Weispfenning, J., Luo, Z. P., Wootan, W. (2004). Nanocomposite rocket ablative materials: Processing, microstructure, and performance. AIAA-2004-1996 paper, 44th AIAA/ASME/ASCE/AHS Structures. *Structural Dynamics, and Materials Conference, Palm Springs*, CA, April. pp. 19–22.

Kornmann, X. (2000). *Synthesis and Characterisation of Thermoset-Clay Nanocomposites*. Sweden: Luleå University of Technology.

Krikorian, V., Pochan, D. (2003). Poly(l-lactide acid)/layered silicate nanocomposite, fabrication, characterization, and properties. *Chem. Mater.*, **15**, pp. 4317–4324.

Krishnamoorti, R., Vaia, R. A., Giannelis, E. P. (1996). Structure and dynamics of polymer-layered silicate nanocomposites. *Chem. Mater.*, **8**, 1728–1734.

Kumar, A. P., Depan, D., Tomer, N. S., Singh, R. P. (2009). Nanoscale particles for polymer degradation and stabilization: Trends and future perspectives. *Prog. Polym. Sci.*, **34**, pp. 479–515.

Lagaly, G. (1999). Introduction: From clay mineral-polymer interactions to clay mineral-polymer nanocomposites. *Appl. Clay Sci.*, **15**, pp. 1–9.

Lagaly, G., Beneke, K. (1991). Intercalation and exchange reactions of clay minerals and non-clay layer compounds. *Colloid Polym. Sci.*, **269**, pp. 1198–1211.

Lan, T., Kaviratna, P. D., Pinnavaia, T. J. (1994). On the nature of polyimide-clay hybrid composites. *Chem. Mater.*, **6**, pp. 573–575.

Lan, T., Kaviratna, P. D., Pinnavaia, T. J. (1995). Mechanism of clay tactoid exfoliation in epoxy-clay nanocomposites. *Chem. Mater.*, **7**, pp. 2144–2150.

Lan, T., Pinnavaia, T. J. (1994). Clay-reinforced epoxy nanocomposites. *Chem. Mater.*, **6**, pp. 2216–2219.

Lange, J., Wyser, Y. (2003). Recent innovations in barrier technologies for plastic packaging: A review. *Packag. Technol. Sci.*, **16**, pp. 149–158.

Laus, M., Francesangeli, O., Sandrolini, F. (1997). New hybrid nanocomposites based on an organophilic clay and poly(styrene-b-butadiene) copolymers. *J. Mater. Res.*, **12**, pp. 3134–3139.

LeBaron, P. C., Wang, Z., Pinnavaia, T. J. (1999). Polymer-layered silicate nanocomposites: An overview. *Appl. Clay Sci.*, **15**, pp. 11–29.

Levchik, S. V., Weil, E. D. (2000). Combustion and fire retardancy of aliphatic nylons. *Polym. Int.*, **49**, pp. 1033–1073.

Liang, G., Xu, J., Bao, S., Xu, W. (2004). Polyethylene/maleic anhydride grafted polyethylene/organic-montmorillonite nanocomposites. I. Preparation, microstructure, and mechanical properties. *J. Appl. Polym. Sci.*, **91**, pp. 3974–3980.

Liu, L., Qi, Z., Zhu, X. (1999). Studies on nylon 6/clay nanocomposites by melt-intercalation process. *J. Appl. Polym. Sci.*, **71**, pp. 1133–1138.

Ma, H., Xu, Z., Tong, L., Gu, A., Fang, Z. (2006). Studies of ABS-graft-maleic anhydride/clay nanocomposites: Morphologies, thermal stability and flammability properties. *Polym. Degrad. Stab.*, **91**, pp. 2951–2959.

Manias, E., Kuppa, V. (2001). Relaxation of polymers in 2-nm slitpores: Confinement induced segmental dynamics and suppression of the glass transition. *Colloids Surf., A*, **187–188**, pp. 509–521.

McCabe, J. F., Walls, A. W. G. (eds.) (2008). *Applied Dental Materials*. Oxford, UK: Blackwell Publishing.

Messersmith, P. B., Giannelis, E. P. (1994). Synthesis and characterization of layered silicate-epoxy nanocomposites. *Chem. Mater.*, **6**, pp. 1719–1725.

Mittal, V. (2007). Gas permeation and mechanical properties of polypropylene nanocomposites with thermally-stable imidazolium modified clay. *Eur. Polym. J.*, **43**, pp. 3727–3736.

Mittal, V. (2007). Polypropylene-layered silicate nanocomposites, filler matrix interactions and mechanical properties. *J. Thermoplast. Compos. Mater.*, **20**, pp. 575–599.

Mittal, V. (2008). Mechanical and gas permeation properties of compatibilized polypropylene–layered silicate nanocomposites. *J. Appl. Polym. Sci.*, **107**, pp. 1350–1361.

Mittal, V. (2009). Polymer layered silicate nanocomposites: A review. *Materials*, **2**, pp. 992–1057.

Morgan, A. B., Gilman, J. W. (2003). Characterization of poly-layered silicate (clay) nanocomposites by transmission electron microscopy and X-ray diffraction: A comparative study. *J. Appl. Polym. Sci.*, **87**, pp. 1329–1338.

Mu, Q., Feng, S. (2007). Thermal conductivity of graphite/silicone rubber prepared by solution intercalation. *Thermochim. Acta*, **462**, pp. 70–75.

Mohammad, A., Sımon, G. P. (2006). Rubber-clay nanocomposites. In: Mai, Y. -W., Yu, Z. -Z. (Eds.), *Polymer Nanocomposites*, pp. 297–323. Cambridge, UK: CRC Press.

Nam, P. H., Maiti, P., Okamoto, M., Kotaka, T., Hasegawa, N., Usuki, A. (2001). A hierarchical structure and properties of intercalated polypropylene/ clay nanocomposites. *Polymer*, **42**, pp. 9633–9640.

Nicholson, J. W. (2005). *The Chemistry of Polymers.* Dorchester, Dorset, UK: RSC.

Nyugen, Q. T., Baird, D. G. (2006). Preparation of polymer-clay nanocomposites and their properties. *Adv. Polym. Technol.*, **25**, pp. 270–285.

Ogata, N., Kawakage, S., Ogihara, T. (1997). Poly(vinyl alcohol)/clay and poly(ethylene oxide)/clay blend prepared using water as solvent. *J. Appl. Polym. Sci.*, **66**, pp. 573–581.

Okada, A., Usuki, A. (1995). The chemistry of polymer-clay hybrids. *Mater. Sci. Eng.*, **C3**, pp. 109–115.

Okada, A., Kawasumi, M., Usuki, A., Kojima, Y., Kurauchi, T., Kamigaito, O. (1990). Synthesis and properties of nylon-6/clay hybrids. In *MRS Symposium Proceedings* (Vol. 171): *Polymer Based Molecular Composites*, Schaefer, D. W. and Mark, J. E. (Eds.), Pittsburgh: Materials Research Society, pp. 45–50.

Olad, A. (2011). Polymer–clay nanocomposites. In *Advances in Diverse Industrial Applications of Nanocomposites*, Reddy, B. (Ed.). Croatia: InTech, pp. 113–138.

Olad, A., Rashidzadeh, A. (2008). Preparation and anticorrosive properties of PANI/Na-MMT and PANI/O-MMT nanocomposites. *Prog. Org. Coat*, **62**, pp. 293–298.

Osman, M. A., Rupp, J. E. P., Suter, U. W. (2005). Effect of non-ionic surfactants on the exfoliation and properties of polyethylene-layered silicate nanocomposites. *Polymer*, **46**, pp. 8202–8209.

Osman, M. A., Rupp, J. E. P., Suter, U. W. (2005). Tensile properties of polyethylene-layered silicate nanocomposites. *Polymer*, **46**, pp. 1653–1660.

Oya, A., Kurokawa, Y., Yasuda, H. (2000). Factors controlling mechanical properties of clay mineral/polypropylene nanocomposites. *J. Mater. Sci.*, **35**, 1045–1050.

Pandey, J. K., Reddy, K. R., Kumar, A. P., Singh, R. P. (2005). An overview on the degradability of polymer nanocomposites. *Polym. Degrad. Stab.*, **88**, pp. 234–250.

Paul, D. R., Robeson, L. M. (2008). Polymer nanotechnology, nanocomposites. *Polymer*, **49**, pp. 3187–3204.

Pavlidou, S., Papaspyrides, C. D. (2008). A review on polymer-layered silicate nanocomposites. *Prog. Polym. Sci.*, **33**, pp. 1119–1198.

Phang, I. Y., Liu, T., Mohamed, A., Pramoda, K. P., Chen, L., Shen, L., Chow, S. Y., He, C., Lu, X., Hu, X. (2005). Morphology, thermal and mechanical properties of nylon 12/organoclay nanocomposites prepared by melt compounding. *Polym. Int.*, **54**, pp. 456–464.

Powell, C. E., Beall, G. W. (2006). Physical properties of polymer–clay nanocomposites. *Curr. Opin. Solid State Mater. Sci.*, **10**, pp. 73–80.

Ray, S. S. (2010). A new possibility for microstructural investigation of clay-based polymer nanocomposite by focused ion beam tomography. *Polymer*, **51**, pp. 3966–3970.

Ray, S. S., Bousima, M. (2005). Biodegradable polymers and their layered silicate nanocomposites in greening the 21st century materials world. *Prog. Mater. Sci.*, **50**, pp. 962–1079.

Ray, S. S., Yamada, K., Okamoto, M., Ogami, A., Ueda, K. (2003). New polylactide/layered silicate nanocomposites. 3. High performance biodegradable materials. *Chem. Mater.*, **15**, pp. 1456–1465.

Schartel, B., Knoll, U., Hartwig, A., Putz, D. (2006). Phosphonium-modified layered silicate epoxy resins nanocomposites and their combinations with ATH and organophosphorus fire retardants. *Polym. Adv. Technol.*, **17**, pp. 281–293.

Shelley, J. S., Mather, P. T., DeVries, K. L. (2002). Reinforcement and environmental degradation of nylon 6/clay nanocomposites. *Polymer*, **42**, pp. 5849–5858.

Shi, X., Gan, Z. (2007). Preparation and characterization of poly(propylene carbonate)/montmorillonite nanocomposites by solution intercalation. *Eur. Polym. J.*, **43**, pp. 4852–4858.

Simon, G. P. (2003). *Polymer Characterization Techniques and Their Application to Blends.* New York, USA: Oxford University Press.

Sinha Ray, S., Okamoto, M. (2003). Polymer/layered silicate nanocomposites: A review from preparation to processing. *Prog. Polym. Sci.*, **28**, pp. 1539–1641.

Sinha Ray, S., Yamada, K., Okamoto, M., Ueda, K. (2002). New polylactide/layered silicate nanocomposite: A novel biodegradable material. *Nano. Lett.*, **2**, pp. 1093–1096.

Solomon, M. J., Almusallam, A. S., Seefeldt, K. F., Somwangthanaroj, A., Varadan, P. (2001). Rheology of polypropylene/clay hybrid materials. *Macromolecules*, **34**, pp. 1864–1872.

Somwangthanaroj, A., Tantiviwattanawongsa, M., Tanthapanichakoon, W. (2012). Mechanical and gas barrier properties of nylon 6/clay nanocomposite blown films. *Eng. J.*, **16**, pp. 93–105.

Sperling, L. H. (2006). *Introduction to Physical Polymer Science*. 4th Ed. New Jersey: John Wiley and Sons.

Stuart, B. H. (2002). *Polymer Analysis*. UK: John Wiley and Sons.

Thostenson, E. T., Li, C., Chou, T. W. (2005). Nanocomposites in context. *Compos. Sci. Technol.*, **65**, pp. 491–516.

Tjong, S. C. (2006). Structural and mechanical properties of polymer nanocomposites. *Mater. Sci. Eng. R*, **53**, pp. 73–197.

Tjong, S. C., Bao, S. P. (2005). Impact fracture toughness of polyamide-6/montmorillonite nanocomposites toughened with a maleated styrene/ethylene butylene/styrene elastomer. *J. Polym. Sci. B*, **43**, pp. 585–595.

Toldy, A., Toth, N., Anna, P., Keglevich, G., Kiss, K., Marosi, G. (2006). Flame retardancy of epoxy resin with phosphorus-containing reactive amine and clay minerals. *Polym. Adv. Technol.*, **17**, pp. 778–781.

Tortora, M., Gorrasi, G., Vittoria, V., Galli, G., Ritrovati, S., Chiellini, E. (2002). Structural characterization and transport properties of organically modified montmorillonite/polyurethane nanocomposites. *Polymer*, **43**, pp. 6147–6157.

Tortora, M., Vittoria, V., Galli, G., Ritrovati, S., Chiellini, E. (2002). Transport properties of modified montmorillonite/poly(caprolactone) nanocomposites. *Macromol. Mater. Eng.*, **287**, pp. 243–249.

Uhl, F. M., Davuluri, S. P., Wong, S. C., Webster, D. C. (2004). Organically modified montmorillonites in UV curable urethane acrylate films *Polymer*, **45**, pp. 6175–6187.

Usuki, A., Kawasumi, M., Kojima, Y., Okada, A., Kurauchi, T., Kamigaito, O. J. (1993). Swelling behavior of montmorillonite cation exchanged for V-amino acids by Ecaprolactam. *Mater. Res.*, **8**(5), p. 1174.

Vaia, R. A., Giannelis, E. P. (1997). Polymer melt intercalation in organically modified layered silicates, model predictions and experiment. *Macromolecules*, **30**(25), pp. 8000–8009.

Vaia, R. A., Ishii, H., Giannelis, E. P. (1993). Synthesis and properties of two-dimensional nanostructures by direct intercalation of polymer melts in layered silicates. *Chem. Mater.*, **5**, pp. 1694–1696.

Vaia, R. A., Vasudevan, S., Krawiec, W., Scanlon, L. G., Giannelis, E. P. (1995). New polymer electrolyte nanocomposites: Melt intercalation of poly(ethylene oxide) in mica-type silicates. *Adv. Mater.*, **7**, pp. 154–156.

VanderHart, D. L., Asano, A., Gilman, J. W. (2001). NMR measurements related to clay dispersion quality and organic-modifier stability in nylon 6/ clay nanocomposites. *Macromolecules*, **34**, pp. 3819–3822.

VanderHart, D. L., Asano, A., Gilman, J. W. (2001). Solid-state NMR investigation of paramagnetic nylon-6 clay nanocomposites. 1. Crystallinity, morphology, and the direct influence of Fe^{3+} on nuclear spins. *Chem. Mater.*, **13**, pp. 3781–3796.

Vyazovkin, S., Dranka, I., Fan, X., Advincula, R. (2004). Kinetics of the thermal and thermo-oxidative degradation of a polystyrene–clay nanocomposite. *Macromol. Rapid Commun.*, **25**, pp. 498–503.

Wang, D., Echols, K., Wilkie, C. A. (2005). Cone calorimetric and thermogravimetric analysis evaluation of halogen-containing polymer nanocomposites. *Fire Mater.*, **29**, pp. 283–294.

Wang, L. Z. (2001). *Characterization of Nanophase Materials*. Germany: Wiley-VCH.

Wang, S., Hu, Y., Song, L., Wang, Z., Chen, Z., Fan, W. (2002). Preparation and thermal properties of ABS/montmorillonite nanocomposite. *Polym. Degrad. Stab.*, **77**, pp. 423–426.

Wang, S., Hu, Y., Zong, R., Tang, Y., Chen, Z., Fan, W. (2004). Preparation and characterization of flame retardant ABS/montmorillonite nanocomposite. *Appl. Clay Sci.*, **25**, pp. 49–55.

Wang, S. J., Long, C. F., Wang, X. Y., Li, Q., Qi, Z. N. (1998). Synthesis and properties of silicon rubber organo-montmorillonite hybrid nanocomposites. *J. Appl. Polym. Sci.*, **69**, pp. 1557–1561.

Weil, E. D., Hirschler, M. M., Patel, N. G., Shaki, S. (1992). Oxygen index, correlations to other fire tests. *Fire Mater.*, **16**, pp. 159–167.

Wu, J., Lerner, M. M. (1993). Structural, thermal, and electrical characterization of layered nanocomposites derived from Na-montmorillonite and polyethers. *Chem. Mater.*, **5**, pp. 835–838.

Xu, R., Manias, E., Synder, A. J., Runt, J. (2001). New biomedical poly(urethane uera)-layered silicate nanocomposites. *Macromolecules*, **34**, pp. 337–339.

Yano, K., Usuki, A., Okada, A., Kurauchi, T., Kamigaito, O. (1993). Synthesis and properties of polyimide-clay hybrid. *J. Polym. Sci.*, **31**, pp. 2493–2498.

Yao, K. J., Song, M., Hourston, D. J., Luo, D. Z. (2002). Polymer/layered clay nanocomposites, 2 polyurethane nanocomposites. *Polymer*, **43**, pp. 1017–1020.

Yeh, J. M., Chang, K. C. (2008). Polymer/layered silicate nanocomposite anticorrosive coatings. *J. Ind. Eng. Chem.*, **14**, pp. 275–291.

Zanetti, M. (2006). Flammability and thermal stability of polymer/layered silicate nanocomposites. In *Polymer Nanocomposites*, Mai, Y. W. and Yu, Z. Z. (Eds.). Cambridge: Woodhead, pp. 256–272.

Zanetti, M., Bracco, P., Costa, L. (2004). Thermal degradation behavior of PE/clay nanocomposites. *Polym. Degrad. Stabil.*, **85**, pp. 657–665.

Zanetti, M., Camino, G., Canavese, D., Morgan, A. B., Lamelas, F. J., Wilkie, C. A. (2002). Fire retardant halogen–antimony–clay synergism in polypropylene layered silicate nanocomposites. *Chem. Mater.*, **14**, pp. 189–193.

Zanetti, M., Lomakin, S., Camino, G. (2000). Polymer layered silicate nanocomposites. *Macromol. Mater. Eng.*, **279**, pp. 1–9.

Zeng, Q. H., Yu, A. B., Lu, G. Q. (2008). Multiscale modeling and simulation of polymer nanocomposites. *Prog. Polym. Sci.*, **33**, pp. 191–269.

Zhang Q., Fu Q., Jiang L., Lei, Y. (2000). Preparation and properties of polypropylene/montmorillonite layered nanocomposites. *Polymer Int.*, **49**, pp. 1561–1564.

Zhang, S., Horrocks, A. R. (2003). A review of flame retardant polypropylene fibres. *Progr. Polym. Sci.*, **28**, pp. 1517–1538.

Zhao, C., Qin, H., Gong, F., Feng, M., Zhang, S., Yang, M. (2005). Mechanical, thermal and flammability properties of polyethylene/clay nanocomposites. *Polym. Degrad. Stab.*, **87**, pp. 183–189.

Zhao, X., Urano, K., Ogasawara, S. (1989). Adsorption of poly(ethylene vinyl alcohol) from aqueous solution on montmorillonite clays. *Colloid Polym. Sci.*, **267**, pp. 899–906.

Zhu, J., Uhl, F. M., Morgan, A. B., Wilkie, C. A. (2001). Studies on the mechanism by which the formation of nanocomposites enhances thermal stability. *Chem. Mater.*, **13**, pp. 4649–4654.

Chapter 5

Applications of Polymer–Clay Nanocomposites

5.1 Introduction

Since 1960, a widespread potential has emerged for the use of advanced composite materials in aerospace, marine, and automobile industries, due to superior engineering properties such as high specific strength and stiffness, lower density, high fatigue endurance, high damping, and improved thermal behavior. Recently, civil engineers and the construction industry have begun to realize the potential of composites as strengthening material for many problems associated with the deterioration of infrastructures (Sandeep et al., 2008). Polymer composites are also, widely used in applications such as transportation, construction and, the production of electronic and biomedical products.

Nanocomposites have several advantages compared to conventional composites. Loading small amounts of inorganic nanoparticles usually produces superior properties in nanocomposites. From a theoretical point of view, nanoeffects of the nanoparticles in nanocomposites provide an important window to modify traditional petrochemicals and related materials and thus play an important role in reforming traditional industry. Also polymer-layered silicate and

Introduction to Polymer–Clay Nanocomposites
Ahmet Gürses

ISBN 978-981-4613-02-6 (Hardcover), 978-981-4613-03-3 (eBook)
www.panstanford.com

silica nanocomposite materials can be utilized in those fields where virgin polymers are being used (Kee and Stroeve, 2005).

5.2 Food Packing

To meet the increasing expectations of consumers, foods must be safe, cheap, of consistently good quality, healthy, and also have a long shelf life. These considerations have led to extensive investigations into suitable packaging materials for food products (Kilcast and Subramaniam, 2000; Ray et al., 2006).

Polymer nanocomposites usually exhibit much better polymer/filler interactions than those in conventional composites. The uniform dispersion of nanofiller into a polymer matrix can result in very large matrix/filler interfacial interactions and also lead to restricted mechanical mobility of polymer matrix; hence, improved properties such as higher mechanical strength, lower weight, higher heat resistance and flame retardancy; improved barrier properties against oxygen, carbon dioxide, ultraviolet, moisture, and volatiles; and more effective flavor conservation of drinks and beverages emerge (de Azeredo et al., 2011; Lange and Wyser, 2003; Ludueña et al., 2007; Moore et al., 1999; Moraru et al., 2003).

The ratio of the largest to the smallest dimension of any filler is an important property known as the aspect ratio. Fillers with higher aspect ratios have higher specific surface areas, providing better reinforcing effects (Azizi Samir et al., 2005; Dalmas et al., 2007). In addition to the effects of the nanoreinforcements themselves, an interphase region of decreased mobility surrounding the nanofiller results in a percolating interphase network in the composite, which plays an important role in improving the nanocomposite properties (Qiao and Brinson, 2009). For constant filler content, a reduction in particle size increases the number of filler particles, bringing them closer to one another; thus, the interface layers from adjacent particles overlap, altering the bulk properties more significantly (Jordan et al., 2005; de Azeredo et al., 2011).

Since clay layers constitute a barrier to gases and water, forcing them to follow a tortuous path, the introduction of nanoclays into polymer biostructures has been shown to greatly improve barrier properties, minimizing one of the main limitations of biopolymer films (Adame and Beall, 2009; Meneghetti and Qutubuddin, 2006).

Indeed, many studies have reported the effectiveness of nanoclays in decreasing oxygen and water vapor permeability (Brody, 2003, 2006; Cabedo et al., 2004; Cava et al., 2006; de Azeredo et al., 2011; Koh et al., 2011; Lotti et al., 2008; Mangiacapra et al., 2006).

The food package should hinder gain or loss of moisture, prevent microbial contamination, and act as a barrier against permeation of watervapor, oxygen, carbon dioxide, and other volatile compounds such as flavors and taints, in addition to the basic properties of packaging materials such as mechanical, optical, and thermal properties (Brody, 2007; Marsh and Bugusu, 2007; Singh and Anderson, 2004) (Fig. 5.1). Food packaging is not only used as a container but also acts as a protective barrier with some innovative functions. In that sense, food packaging is quite different from other durable goods such as electronics, home appliances, and furniture because of safety aspects and relatively very short shelf life. Basic packaging materials such as paper and paperboard, plastic, glass, metal, and a combination of materials having various chemical and physical structures are used to fulfill the needed functions and main requirements for packaging of foods, depending on their types. However, there has been ever-increasing effort in the development of different kinds of packaging materials in order to enhance their effectiveness in keeping the food quality with improved convenience for processing and final use (Rhim et al., 2013).

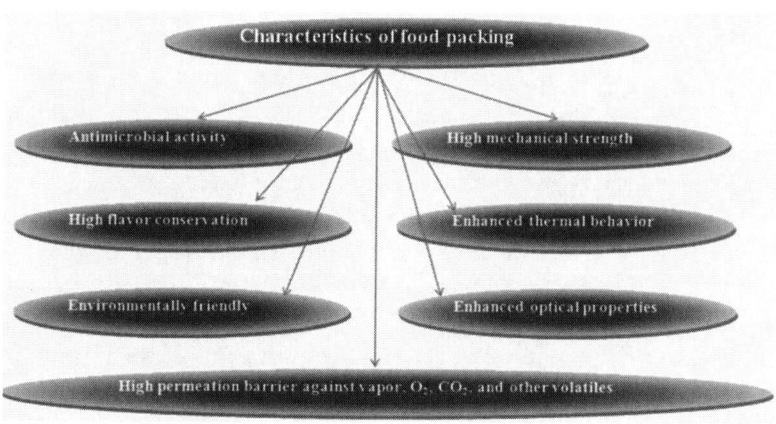

Figure 5.1 Main characteristics and limitations required for food-packaging materials.

The most widely used theories to explain the improved barrier properties of PCNs are based on a theory developed by Nielsen (1967), which focuses on a tortuous path around the clay plates, forcing the gas permeant to travel a longer path to diffuse through the film. The increase in path length is a function of the high aspect ratio of the clay filler and the volume fraction of the filler in the composite. Nielsen's model predicts permeability of systems at clay-loading rates of less than 1%, but experimental data deviate significantly from predicted values at higher loading rates and more extensively in certain polymers (Adame and Beall, 2009; de Azeredo et al., 2011).

Many deviations can be explained by factors such as poor clay orientation or less-than-complete exfoliation. However, even those factors cannot explain the many present experimental results that reported much lower permeability than predicted. Beall (2000) proposed a new model to predict permeability of polymer nanocomposites, focused on the polymer–clay interface as the governing factor in addition to the tortuous path. This model provides a correction factor applicable to Nielsen's model. It defines three regions around clay plates: the surface modifier region, the constrained polymer region, and the unconstrained polymer region. The surface modifier region (1–2 nm) is where the clay binds with the polymer; it is assumed to be small enough to have little effect on the permeability of the composite. The unconstrained polymer region is not affected significantly by the clay, its properties being that of the bulk polymer. The constrained polymer region, less well defined and indirectly confirmed, is in direct contact with the surface modifier and may extend 50–100 nm from the clay surface as a function of polymer interaction parameters; it is assumed to have a lower free volume and, therefore, a lower diffusion coefficient than that of the bulk polymer. Since the main effect of the constrained region is a lower free volume, and this effect is not significant in crystalline regions, the constrained region of semicrystalline polymers does not significantly affect permeability unless the crystallinity is decreased. Indeed, most of the large deviations from the simple tortuous path model involve amorphous polymers (Adame and Beall, 2009; de Azeredo et al., 2009).

Kim et al. (2005) found that the moisture permeability of octadecylamine-modified montmorillonite (MMT) was lower than that of quaternary alkylamine-modified MMT due to the larger interlayer distance in the first organoclay that extends the water molecules' diffusion path in the nanocomposites. The interlayer distances of both organoclays have also been measured as 2.38 and 1.96 nm, using X-ray diffraction (XRD) and transmission electron microscopy (TEM) techniques, respectively. Mixing with an epoxy resin followed by curing resulted in intercalation of the second organoclay with an interlayer distance of 3.3 nm, whereas a mixture of exfoliation and intercalation was manifested by the first organoclay with an interlayer distance larger than about 8 nm, as measured from the TEM micrograph.

The enhanced gas/vapor barrier properties of nanocomposites make them attractive and useful in food-packaging applications, both flexible and rigid. Specific examples include packaging for processed meats, cheese, confectionery, cereals, and boil-in-the-bag foods, in addition to extrusion-coating applications in association with paperboard for fruit juice and dairy products, together with co-extrusion processes for the manufacture of beer and carbonated drinks bottles. The use of nanocomposite formulations may result in a longer shelf life for many food types, because it provides gas barriers against carbon dioxide and oxygen; for example, three to six months for beers and fruit juices and one year for carbonated soft drinks (Ray et al., 2006).

Clays have been also reported to improve the mechanical strength of biopolymers, making their use feasible (Weiss et al., 2006). Park et al. (2003) generated thermoplastic starch (TPS)–clay nanocomposites with improved mechanical properties and decreased water vapor permeability by using only 5% (w/w) of clays. Findings obtained by Dean et al. (2007) indicate that optimum levels of both plasticizer and nanoclay exist for each clay to produce a gelatinized starch film with the highest exfoliation and best improvement in mechanical properties, and those optimum levels have some dependency on the cation exchange capacity (CEC) of the clay. Petersson and Oksman (2006) reported that bentonite was able to improve strength and modulus of a polylactic acid (PLA) matrix, but it drastically decreased elongation of the material. Similar results were reported by Lotti et al. (2008) for low-density polyethylene (LDPE) treated with an organoclay and also by Xu et al. (2006) for chitosan films with MMT. On the other hand, Marras et al. (2008) reported that the elongation of poly(ε-caprolactone) (PCL) was not impaired by MMT. Other authors observed improved mechanical properties of several polymers by addition of nanoclays (Cyras et al., 2008; de Azeredo et al., 2009; Mangiacapra et al., 2006).

Also, modified organoclays can have the potential to provide substantial improvements in the mechanical properties of polymeric matrix, even with a level of filler loading as low as 1–5 wt.%. With such a low ratio of filler loading, polymers such as nylon 6 have shown serious improvement in their mechanical properties, for example an increase of 103% in Young's modulus and 49% in tensile strength (Kojima et al., 1993; Ray et al., 2006).

There has been reporting on the other benefits of the obtained performances of different polymers, using clay nanoparticles, together with the increased glass transition and thermal degradation temperatures (Cabedo et al., 2004; de Azeredo et al., 2009; Paul et al., 2003). As a minor disadvantage of nanoclays on polymers, Yu et al. (2003) have described decreased transparency.

Yoo et al. (2004) observed that the thermal decomposition behavior and linear dimensional changes of recycled poly(vinyl chloride) (RPVC) could be improved by the incorporation of organically modified clays. In the case of RPVC–clay nanocomposites, the degree of improvement of the above properties was proportional to the clay content. They show that the coefficients of thermal expansion (CTEs) of RPVC–clay nanocomposites decreased with increasing clay content; for example, the coefficient of thermal expansion of RPVC prepared by using only 5% organoclay was decreased by about 3.5 times.

In addition, it has been reported that this novel class fillers could be useful to reduce flammability. Under fire-like conditions, the polymer–clay nanocomposites can form char having a multilayered carbonaceous silicate structure, which may improve flammability properties, depending on exfoliation or intercalation of clay lamellae. This multilayered carbonaceous silicate-char formed on the surface during burning can insulate the underlying material and may reduce the escape velocity of volatile products during decomposition (Gilman et al., 2000; Morgan et al., 2005). This property can be estimated by measuring the heat release rate (HRR) of nanocomposites. The improved flammability reductions in polystyrene, nylon 6, and maleic anhydride-grafted polypropylene (MA-g-PP) nanocomposites are indicators of lower HRR of 50 to 75% during combustion (Ray et al., 2006).

Also clay-based polymer nanocomposites have been used in plastic bottle manufacturing industries for improving the barrier, mechanical properties, and shelf life of the product. To improve the bottles' properties and seek an alternative solution, plastic bottles have been tried in the beer package industry. Nanocomposites have been used for beer bottle manufacture to solve many problems, such as the beer colloidal instability, including biological and nonbiological aspects such as oxygen permeation and bad taste due to light exposure. It was determined that the shelf life of a clay-based nanocomposite plastic beer bottle was more than six months (Ke and Stroeve, 2005). A study on nitrile rubber–organoclay nanocomposites (in 10 wt.% organoclay) shows that their relative vapor permeability against water and methanol is reduced up to 85% and 42%, respectively, compared to the virgin polymer (Anandhan and Bandyopadhyay, 2011).

One of the recently commercialized sports goods is the double-core Wilson tennis ball. The clay nanocomposite coating of Wilson tennis balls maintains the internal pressure for an extended period of time. The core is coated by a butyl rubber–clay (vermiculite) nanocomposite that acts as a gas barrier, doubling its shelf life. By using polymer nanocomposites, more flexible coatings with gas permeability 30–300 times lower than butyl rubber have been produced. These coatings have been shown to be undamaged by strains up to 20%. These new double-core tennis balls using this coating retain air longer and are able to bounce twice as long as ordinary balls (improvement in air retention). It is anticipated that this technology could be incorporated to the rubber industry as well as soccer balls or automobile and bicycle tires (Koo, 2006). It can also be said that another important application associated with sports goods is the basketball shoe pouch produced by using ethylene-vinyl acetate EVA–clay nanocomposites. A unique property of the clay nanocomposite pouch filled with helium inserted into a basketball shoe is that it may provide a good resilience to basketball players while jumping; in the meantime, it exhibits excellent gas barrier properties (Anandhan and Bandyopadhyay, 2011).

5.3 Biomedical Applications

Biomaterials are materials of natural or man-made origin, which are used directly, or as supplements, to perform the functions of living tissues in the human body (Black, 1992). The date on biomaterial usage goes far back to the ancient civilizations; for example, artificial eyes, ears, teeth, and noses on Egyptian mummies have been found, and also Chinese and Indians have used waxes, glues, and tissues in reconstructing missing or defective parts of the body (Williams, 1979). Over the centuries, advancements in synthetic materials, surgical techniques, and sterilization methods have permitted the use of biomaterials in many ways (Park, 1984; Ramakrishna et al., 2001). Therefore, the applicability of polymer nanotechnology and nanocomposites to biomedical/biotechnological applications is a rapidly emerging area (Berry, 2005; Ray et al., 2002). One field of intense research involves electrospinning for producing bioresorbable nanofiber scaffolds to use in tissue engineering applications. The scaffold that is available for cell growth might be a constructed nanocomposite. Another field involving nanofibers is also the utilization of electrically conducting nanofibers containing conjugated chains for regeneration of nerve growth within the biological living system (Paul and Robeson, 2008).

Polymers, the largest and most promising class of biomaterials, are widely used in various medical applications such as tissue engineering, implantation of artificial organ as well as prostheses, ophthalmology, dentistry, and bone repair (Kulshrestha and Mahapatro, 2008). The eventual goal of biomaterials research is to create the implant surfaces that can interact actively with biological systems and also fulfill the same functions as corporeal tissues do. Improvement in the interface compatibility of polymers selected for implantation by designed surface modification is extremely important in terms of biomaterial development. Also biomedical polymer–silicate nanocomposites have a great potential in the development of biomedical applications such as diagnostic and therapeutic devices, tissue regeneration, and drug delivery matrixes (Gürses et al., 2015). A fundamental obligation in terms of polymer–nanoparticle interactions is the control of the structure–property relationships of polymer nanocomposites that need to be handled within the chemical, physical, and biological constraints required by a biomedical application. Consequently, because of the interdisciplinary nature of the polymer nanocomposite biomaterials, researchers in the fields of polymer science, biology, materials and biomedical engineering, chemistry, and physics need to work together (Wu et al., 2010).

A variety of biopolymer nanocomposite materials are generated by the combination of inorganic nanoparticles with polymers of synthetic or natural origin. Applications of polymer–silicate nanocomposites used for this purpose are shown schematically in Fig. 5.2. Many nanocomposites made from biomedical polymers and silicate nanoparticles are reviewed by taking into account their superior aspects and shortcomings in the biomedical and biotechnological fields (Kevadiya et al., 2010; Wu et al., 2010).

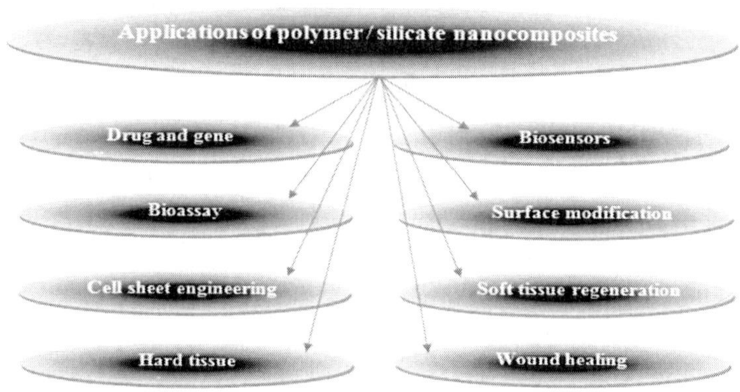

Figure 5.2 Schematic representation of biomedical applications of polymer–silicate nanocomposites.

Polymer/hydroxyapatite ($Ca_{10}(PO_4)_6(OH)_2$) nanocomposites have been investigated for bone repair and implantation. Besides, hydroxyapatite, which is a major component of the hard tissue, exhibits undesirable mechanical properties, if directly employed. The biodegradation of matrix is also desired to allow the infiltration of new bone growth at the repair site. Often natural polymers (polysaccharides, polypeptides, collagen, and chitosan) or synthetic biodegradable polymers are employed as the matrix. Collagen-derived gelatin and poly-2-hydroxyethyl methacrylate/poly(3-caprolactone) nanocomposites based on hydroxyapatite are studied as bone repair systems (Huang et al., 2007; Kim et al., 2005). The electrospinning of the solution of biodegradable polymer is a popular method used to produce nanofiber scaffolds for tissue engineering applications. In this method, for example, poly(L-lactic acid) (PLLA)/the exfoliated silicate (clay)/salt mixtures are electrospun followed by salt leaching gas foaming. The resultant scaffold structure contains nano- and micropores, offering a combination of cell growth and blood vessel invasion, microdimensions along with nanodimensions for nutrient and metabolic waste transport (Lee et al., 2005).

Polymer matrix nanocomposites have also been proposed for drug delivery/release applications. The addition of nanoparticles can provide an impediment to drug release allowing slower and more controlled release, and the reduced swelling and the improved mechanical integrity of hydrogel-based nanocomposites. In addition, ironoxide nanoparticles have been investigated for various applications, including drug delivery, magnetic resonance imaging contrast enhancement, immunoassays, and cellular therapy (Gupta and Gupta, 2005; Haraguchi et al., 2006; Zhang et al., 2007). In the investigations, often magnetite (Fe_3O_4) is dispersed in a polymeric microsphere or microcapsule involving biodegradable and/or natural polymers (Paul and Robeson, 2008). The poly(L-lysine) microspheres containing the magnetic nanoparticles prepared with co-acervation have been characterized for potential use in targeted drug delivery applications (Toprak et al., 2007). Iron and cobalt nanoparticles encapsulated in polydimethylsiloxane were indicated to be applicable for treating retinal detachment disorders (Rutnakornpituk et al., 2002; Vadala et al., 2004). A novel imprint lithography method based on a crosslinked perfluoroether templating mold was employed to fabricate uniform micro- and nanoparticles with biodegradable polymers such as poly(lactic acid) (Rolland et al., 2005).

Composite materials have been used extensively in dentistry and prosthesis design as well as in other fields. Typically, the matrix of polyethylene (UHMWPE) with ultrahigh molar mass is reinforced with carbon fibers. The carbon fibers that disperse randomly in the matrix, and with 15 micron in diameter, are produced by the pyrolysis of acrylic fibers to obtain an oriented graphite structure having high tensile strength and modulus of elasticity. In order to provide high modulus property to reinforcing fibers to strengthen the matrix, sufficient interfacial interaction between the fiber and the matrix must be achieved, during the manufacturing process. This fiber-reinforced composite can also be used to make various implants such as intramedullary rods and artificial joints. Since the mechanical properties of these composites are depended on the proportion of carbon fiber in the composite, it is possible to modify the design flexibility of material to suit the ultimate design of prostheses (Ramakrishna et al., 2001).

Aliphatic polyesters are materials chosen predominantly for degradable drug delivery systems (Kulshrestha and Mahapatro, 2008). Among the polyesters, poly(L-lactic acid) (PLLA) is the most promising polymer in biomedical fields (Oksman et al., 2006). Although some works associated with the use of PLLA were done in the 1960s, more intensive studies have been performed recently. PLLA was widely used for applications in sutures, drug delivery devices, prosthetics, scaffolds, vascular grafts, and bone screws, pins, and plates for temporary internal fixation (Kronenthal 1975; Tsivintzelis et al., 2007). Also, PLLA exhibits superior properties such as improved mechanical strength and degradation to nontoxic products (Hule and Pochan, 2007; Tsuji and Ikada, 2000).

Poly(glycolic acid) (PGA), which is an aliphatic polyester available for biomedical applications, is readily soluble in water, unlike PLLA. However, its higher solubility in water and its high melting point limit the use of PGA in the synthesis of bionanocomposites. The biodegradable and nontoxic aliphatic polyester used in the synthesis of biomedical nanocomposite is polycaprolactone (PCL), which was synthesized by the ring-opening polymerization of ε-caprolactone and has a melting temperature of 61°C and a T_g of 60°C. The rubbery behavior permits the diffusion of species with low molar mass at body temperature, and therefore PCL is accepted as a promising candidate for controlled release and soft-tissue engineering. The range of properties is enriched by copolymerization with other lactones such as glycolide, lactide, and poly(ethylene oxide) (PEO) or by using nanofiller and this has created a great potential for biomedical applications (Edlund and Albertsson, 2002; Hule and Pochan, 2007).

Poly(urethane urea) (PUU) segmented block copolymers are commonly used in ventricular assist devices and total artificial hearts as blood sacs. One of the main disadvantages of PUU in these devices is its relatively high permeability against air and watervapor, depending on the diffusion through the poly(tetramethylene oxide) soft segments, which are present as a major component of the copolymer. The use of organically modified layered silicates seems particularly attractive among the variety of approaches oriented to the reduction of permeability, by maintaining the desired biocompatibility and mechanical properties. This can be ascribed to highly improved barrier properties with the presence of intercalated or exfoliated high-aspect-ratio (~1000) clay platelets in the polymer matrix. The studies made on PUU–clay systems showed that the mechanical strength and modulus of the nanocomposites increased without the loss of ductility (even with increased ductility), unlike conventional composites (Xu et al., 2001). The full explanation for such a remarkable behavior is not present yet. Nonetheless, more reduced gas permeability with an increase in mechanical properties of the composites is also reported (Hule and Pochan, 2007).

The nanocomposites used for dental applications require a lot of basic qualifications, as the nanocomposites used for various biomedical applications have different requirements. Thermoset methacrylate-based composites are used commonly as dental restorative materials because of relatively high cure efficiency by free-radical polymerization and excellent esthetic qualities. However, the demands of the oral environment and the masticatory loads encountered by dental restoratives require further property improvements in these materials. Specifically, nanocomposites having properties such as improved modulus, better efficiency by free-radical polymerization, lower water sorption, improved processability, and low shrinkage are needed. Selective functionalization of the filler can lead to better interactions at the filler–matrix interphase and has been used in studies in which silica nanoparticles were silanized to varying extents using two different modifiers and then mixed with a dimethacrylate resin. A few practical advantages of the dual silanization are improved workability of the composite paste, higher filler loadings leading to better modulus values, and nanocomposites with lower polymer shrinkage (Hule and Pochan, 2007; Wilson et al., 2005).

The design of cardiovascular interfaces requires the combination of amphiphilicity and antithrombogenicity. Amphiphilicity determines optimal endothelial cell response at the vascular interface, and thrombogenicity also refers to blood clot formation and can lead to early graft occlusion. Polyhedral oligomeric silsesquioxanes (POSS), which could act as an amphiphile at the water–air interface, have been studied for the possibility of using at the vascular interface. The robust framework having shorter bond length formed via the strong intermolecular interactions between constituent molecules and neighbors makes the POSS nanocomposites more resistant against degradation. Accordingly, the cytologically compatible POSS nanocomposites have a high potential for tissue engineering (Gürses et al., 2014; Hule and Pochan, 2007; Kannan et al., 2005).

In another study, Lee et al. (2003) have used silicate clay nanoparticles (montmorillonite, MMT) to improve the mechanical properties of model scaffolds made by using PLLA. The resulting polymer nanocomposite scaffolds exhibited an increase of 40% for the tensile modulus, compared to the virgin PLLA scaffolds. Because the addition of MMT nanoparticles decreased the polymer crystallinity, the resulting nanocomposite scaffolds showed faster biodegradation than that of virgin polymer. The authors concluded that the mechanical strength and biodegradation of the PLLA nanocomposites could be tailored by the addition of layered silicate nanoparticles. In many related studies, the biodegradation morphology of silicate–PLLA nanocomposites was investigated in more detail. It was reported that the scaffolds obtained by fiber-spinning had improved structural integrity during the biodegradation of polymer. The fiber scaffolds have a high potential for creating further enhancements for bone–cartilage and bone–ligament interfacial applications. This is an expected result of the porous structure of fiber scaffolds, which could promote cellular infiltration, along with strong mechanical properties of fibrous materials. The fact that the presence of silicate nanoparticles can encourage the formation and repair of bone also means that PLLA–clay nanocomposites may have a great biomedical potential (Wu et al., 2010).

Ozkoc et al. (2010) have synthesized porous PLA–MMT nanocomposites by using microcompounding and polymer/particle leaching and also reported that the compression properties of polymer nanocomposites improved as those of the cancellous bone, after the addition of MMT. Because the hydrophilicity of polymer nanocomposite surfaces is directly correlated with the cell adhesion, the addition of MMT can lead to the lower static contact angle and hence to the formation of more hydrophilic PLA surface as result of the decrease in the interfacial tension.

Krikorian et al. (2003) reported significant improvement in mechanical properties of PLLA due to the addition of MMT. Higher amounts of MMT and exfoliated structures gave rise to stiffer materials compared to microphase-separated or intercalated composites. Exfoliated nanoparticles suppressed polymer crystallization due to enhanced surface interactions. Moreover, an increase in silicate concentration and exfoliation resulted in stiffer and transparent PLLA–MMT nanocomposites.

Zhuang et al. (2007) showed that the intercalated structure of MMT–gelatin–chitosan composite has a lower degradation rate compared to a gelatin–chitosan scaffold and that the degradation rate can be altered by changing the MMT ratio. Enhanced cell adhesion and proliferation with the MMT–gelatin–chitosan nanocomposite film have been observed. Chitosan may also be utilized in the production of glycosaminoglycans, which may aid in cartilage integration.

The nanocomposites prepared by using ethylene vinyl acetate and natural Cloisite clay have been investigated by Lewkowitz-Shpuntoff et al. (2009). The clay-dependent mechanical properties and the adhesion and growth of human dermal fibroblasts on the polymer nanocomposite surfaces are reported. The mechanical testing data show that materials with the highest modulus were produced with 10% clay ratio. In a similar way, the highest cell growth was found on the surfaces of polymer nanocomposites containing 10% clay. It was also observed that the cultured fibroblast cells on the substrates with higher clay content had poor growth curves. By adsorption of iron onto Cloisite clay, the resulting polymer nanocomposites became magnetic, which enhanced proliferation of MC3T3 osteoblast cells on the materials surface. Osteoblast cell proliferation was maximized by culture on electrospun aligned fibers in a constant magnetic field. Such an ability to align different cell types provides for the further development of this system for tendon or bone repair (Wu et al., 2010).

5.3.1 Drug Delivery Applications

One of the problems concerned with the biomaterials used for drug delivery that needs to be overcome is the burst release of encapsulated or entrapped drugs. By controlling the release kinetics of drugs, one can not only optimize the therapeutic effects of the drug but also influence their biological activity (Salcedo et al., 2012). The silicate-based polymer nanocomposites, with the tortuous diffusion pathway, demonstrate good barrier properties (Yano et al., 2000). This property can be used for the development of sustained drug-release applications. The nanocomposites synthesized with various combinations of biomedical polymers and clays have been loaded with the model drugs. For example, organically modified silicate nanoparticles (Cloisite clay) have been added to poly(ethylene-co-vinyl acetate) polymer for the investigation of the release kinetics of dexamethasone (Cypes et al., 2003; Wang et al., 2008). It has been shown that the increased silicate nanoparticle ratio provided higher mechanical strength of the polymer nanocomposite and a sustained release of dexamethasone. It has also been suggested that the drug-release kinetics is dependent on the aspect ratio and dispersion degree of the silicate nanoparticles (Sairam et al., 2007). The injectable delivery of drugs can also be achieved with the polymer nanocomposites that can be synthesized by using polymers sensitive to external stimuli such as temperature. Wu et al. (2009) have reported on the temperature-dependent sol-gel transitions in the thermosensitive nanocomposite hydrogels made from laponite nanoparticles and Pluronic-type polymers or poloxamers. Pluronics are thermosensitive tri-block copolymers composed of poly(ethyleneoxide)-poly(propyleneoxide)-poly (ethyleneoxide) (PEO_x-PPO_y-PEO_z) and are important in injectable applications. The fast dissolution properties of pure Pluronic hydrogels hinder their use for long-term drug-release applications. However, the addition of silicate nanoparticles was found to shift the Pluronic phase transition temperature and to enhance the dissolution resistant properties of hydrogels. As a consequence, the release of a macromolecular model drug, albumin, could be significantly lengthened (Wu et al., 2010).

A study by Lee and Chen (2004) described the delivery of model drugs from hydrogels made of acrylic acid–poly(ethylene glycol) (PEG) methyl ether acrylate and natural bentonite clay nanoparticles. It was found that the elution kinetics is depended strongly on the interactions between the charged clay surface and the drug. Vitamin B12 (zwitter ionic), vitamin B2 (uncharged), crystal violet (cationic), and phenol red (anionic) were used as model drugs. The attraction interactions between the negatively charged silicate surfaces and the drug have resulted in slower release rates, while the drug elution rate is increased due to the repulsion interactions between the two. Also, it has been highlighted that the mucoadhesive properties of hydrogels increased the efficiency of drug delivery. A similar study was made to determine the ability of delivery with PEO-polyamide block copolymer-laponite nanocomposite of an uncharged hydrophobic model drug (pyrene) by Takahashi et al. (2005).

Zhou et al. (2010) have prepared poly(vinyl alcohol) (PVA) and PVA nonanocomposite/halloysite films for their use in the tissue engineering industry. It was found that the films exhibited behavior similar to those of the osteoblast-type cell MC3T3-E1 and fibroblast-type cells NIH 3T3, indicating a great use potential for drug-release systems. It was also observed that films prepared with reticulated laponite clay in PEO were biocompatible and feasible for use in bioactive scaffolds (Gaharwar et al., 2011; 2014). Chakraborti et al. (2011) has developed the film that includes poly(lactic acid-co-glycolic acid), alendronate sodium lactate dehydrogenase (LDH), and tetracycline, forming an adequate scaffold for proliferation and differentiation of osteoblasts, with the matrix containing tetracycline and sodium alendronate having the characteristics of controlled release.

Cypes et al. (2003) developed nanocomposites containing polymer-organosilicate (Cloisite20A-dimethyl-dihydrogenated tallow ammonium MMT) in order to control the release of dexamethasone. Takahashi et al. (2005), in another work, developed a PEG-clay carrier to use in the sustained release of hydrophobic substances. On the other hand, Campbell et al. (2008) showed that the nanocomposites containing PEG and Cloisite20A retarded the drug diffusion and affected the dissolution behavior of paracetamol molecules.

Shaikh et al. (2007) prepared transdermal adhesives with an organoclay (MMT and octadecylamine)/polymer nanocomposite, where they obtained improvements in the mechanical properties of the adhesives and in the release of drugs. Other nanocomposites with natural polymers and clays were prepared for the drug-controlled release, including chitosan derivate, where chitosan (CS) enhanced the drug encapsulation efficiency of the nanoparticles and slowed the drug release from the nanocomposites. Salcedo et al. (2012) has demonstrated that the MMT–CS nanocomposite has good biocompatibility, stimulates cell proliferation, and in addition shows mucoadhesive properties combined with low solubility in an acid medium, beyond retarding the drug release in gastric environments and releasing the drug in the intestinal environments in a controlled manner. It is also reported that the intercalation of 5-fluorouracil (5-FU), an anticancer drug, in the interlayer gallery of MMT with the assistance of biopolymer chitosan forms the pattern of controlled release, with the reduced DNA damage and drug toxicity (Kevadiya et al., 2010). Therefore, drug–clay hybrids and composites can be of considerable value in chemotherapy of cancer with reduced side effects (Kevadiya et al., 2012).

Feng et al. (2009) developed poly(lactide)–vitamin E derivative/MMT nanoparticle formulations for the oral delivery of Docetaxel, where an oral dosage administration of these formulations could carry out a sustained chemotherapy for as long as three weeks in comparison to 22-hour intravenous administration of Taxotere® (Standard).

Dong and Feng (2005) prepared bioadhesive nanoparticles of poly(D,L-lactic-co-glycolic acid)/MMT for the release of the antitumoral drug paclitaxel and to obtain a biphasic release pattern, starting with an explosion, followed by the slow and continuous release that was not affected by the presence of MMT. However, the cell capture and absorption of the nanoparticles depend upon the quantity of MMT and the concentration of particles in the incubation.

Especially poly(methyl methylacrylate) (PMMA) particulate carriers are being extensively investigated for the delivery of different classes of drugs via various routes of administration as oral and topical (Bettencourt and Almeida, 2012; Dhana Lekshmi et al., 2010). Poly(methyl methacrylate) (PMMA) has potential use as an effective means to stabilize chemically unstable drugs like antioxidants as well as to solubilize poorly soluble drugs (Sivakumar and Rao, 2002). Several studies address the entrapment of different types of antibiotics, such as gentamicin, cefadroxil, ampicillin, minocycline, and tigecycline (Matos et al., 2010a, b). Potential applications such as oral drug delivery, local drug delivery for biofilm-associated infections, wound dressings, and tissue engineering are envisaged from these formulations to treat many types of bacterial infections (Changerath et al., 2009). Another appealing biomedical application for PMMA could be represented by the local treatment of bone tumors. Mestiri et al. (1993) have proposed the use of PMMA microspheres containing cisplatin, a cytostatic agent whose effectiveness in bone tumors is well-known for local chemotherapy in bone tumors. Paiphansiri et al. (2006) suggested the topical application of antiseptic-loaded PMMA nanocapsules that would provide controlled delivery of the drug to skin-associated bacteria. The antiseptic chlorhexidine digluconate was selected due to its popularity as a hydrophilic model drug. Particles loaded with antidiabetics for the treatment of diabetes mellitus have recently been developed. Liu et al. (2011) assayed the delivery of insulin from PMMA/core-shell particles (CSP). Repaglinide, a lipophilic meglitinide analogue, was loaded into PMMA nanoparticles to achieve prolonged drug release and thereby to reduce the side effects of administration of conventional dosage form of the drug. Other drugs successfully encapsulated in PMMA include antihypertensives such as verapamil HCl and diltiazem HCl, histamine antagonists such as chlorpheniramine (CPM), and bronchodilators such as theophylline (Streubel et al., 2002, 2003).

5.3.2 Tissue Engineering and Repair

Numerous studies suggest that degradable polymers modified with bioactive nanostructured materials containing silicon dioxide (silicate, Bioglass®, wollastonite, silicon-doped calcium phosphate) exhibited excellent bioactivity and improved apatite formation, in vitro and in vivo (Misra et al., 2006). Early studies showed that silicon normally present in vivo was essential in the formation of cartilage and bone by participating in cell metabolism (Carlisle and Alpenfels, 1980; Cava et al., 2006; Hench and Paschall, 1973; Schwarz, 1972, 1973). Schwarz et al. (1972) suggested that silica acted as a crosslinking agent in connective tissue. Hensch et al. (1973) investigated the interactions of living tissue with the bioactive glasses that can be used to repair bone defects. Vogel et al. (2001) have investigated in detail how the solubility of bioactive glass affects the interactions between bone and implants. As a result, it is clear that intensive studies were done in order to investigate the properties of bioactive glass ceramics (Kokubo, 1991; Hensch, 2009). Many studies on implants containing bioactive silicate exert that the implants induce bone formation and can activate osteogenic proliferation and bone-related gene expression (Valerio et al., 2004; Xynos et al., 2001). A desired feature of the implants is to promote bone tissue formation at their surface and bind to the surrounding tissue by providing the implant fixation to be in accordance with hard-tissue engineering applications (Jones et al., 2006; Mansur and Costa, 2008). Many polymer-Bioglass® composite materials and poly(D,L-lactic acid) (PDLLA)-Bioglass® composites were found to be suitable for both hard- and soft-tissue engineering (Blaker et al., 2007). Meanwhile, Verrier et al. (2004) showed that the porous foams from PDLLA-Bioglass® facilitated the growth and cell behavior of human osteosarcoma cells and human lung carcinoma cells and thus opened new avenues to lung tissue engineering applications. Also, Day et al. (2004) have studied the morphology of fibroblasts, as well as the neovascularization at the Bioglass®-coated PGA meshes, in addition to assessment of polyglycolic acid mesh and bioactive glass, for soft-tissue engineering scaffolds.

The combination of natural polymers such as chitin and chitosan with bioactive silicate nanoparticles allows for designing biocompatible, degradable, and cost-efficient biomaterials (Peter et al., 2009, 2010). Recently, nanocomposite scaffolds were prepared by lyophilization of a chitosan mix containing bioactive glass. Scaffold properties such as swelling, degradation, and bioactivity could be modulated, using a chitosan solution or a chitosan–gelatin blend instead of a chitin gel. Addition of gelatin to the chitosan may produce Bioglass® materials that may control cell attachment, cell growth, migration, and differentiation. It has been pointed out that these polymer nanocomposite scaffolds possess the prerequisites needed for further developed tissue engineering applications (Peter et al., 2010).

In another study, Lee et al. (2005) used a sol–gel method to fabricate chitosan–silicate nanocomposite membranes for bone regeneration. The resulting silicate xerogel was found to be uniformly dispersed in the chitosan matrix on the nanoscale. As expected, the addition of silicate resulted in improved mechanical properties of the nanocomposite when compared to pure chitosan. When immersed in simulated body fluid, the chitosan–silicate nanocomposites induced deposition of calcium phosphate minerals, suggesting in vitro bioactivity. A significant increase in osteoblast adhesion, proliferation, and alkaline phosphatase activity was enhanced by the presence of silicate. Histological results of polymer nanocomposite implantation in a rat calvarium model showed a significant increase in bone regeneration when compared to pure chitosan. Overall, improvements in mechanical and biological properties have been attributed to the dispersion of bioactive silica particles in nanoscale into the chitosan matrix. Although virgin PVA shows limited biodegradability compared to modified PEO and poly(lactic-co-glycolic acid) (PLGA), PVA is used frequently in drug delivery, dialysis membranes, wound dressings, and other biomedical devices. Thus PVA and bioactive glass or a combination of PVA, chitosan, and bioactive glass can be used to prepare scaffold materials that include the mechanical strength of synthetic PVA, the hemocompatibility, biodegradability, and bactericidal properties of natural chitosan polymers, and the bioactive properties of silicate (Mansur and Costa, 2008). Heinemann et al. (2009) have also attempted a different strategy to mimic the natural biosilicification processes to fabricate xerogels from silica and collagen under ambient conditions. Sol–gel techniques were used to generate new nanocomposite materials by varying the ratios of collagen and silicate. Addition of calcium phosphate cements further improved bioactivity of the material, accelerated the formation of bone-like apatite, and led to the differentiation of human monocytes into osteoclast-like cells. A similar study reported on silica–chitin-based bionanocomposites fabricated from biological glass sponges. It was found that the chitinous organic matrix could provide a template for the biodirected deposition of silicate mineral phase, and also the resulting structures could be used for tissue engineering applications of bone and cartilage (Ehrlich et al., 2008). Self-assembled nanocomposites from fibrous proteins, such as silk, have been investigated by Foo et al. (2006) who combined silica nanoparticles with chimeric silk proteins. Besides, films and fibers were produced using spider silk by deposition of silica nanoparticles onto the polymer, over silicification reactions. The morphology and structure of the silica nanoparticles were controlled by formulation, and the mechanical properties such as toughness and strength were improved. Silk proteins were also combined with wollastonite to prepare mechanically strong and bioactive scaffolds that support cell growth (Wu et al., 2010; Zhu et al., 2010).

5.3.3 Hydrogels with Biomedical Potential

Hydrogels are accepted as biomaterials possessing the potential to mimic physical, chemical, electrical, and biological properties of most biological tissues (Fisher et al., 2010; Kloxin et al., 2010; Slaughter et al., 2009). Hydrogels possess a 3D highly hydrated polymeric network and can hold up to 20- to 40-fold more water compared to their dry weight. Because of their superior physical properties, these networks can be processed in various sizes and shapes (Gaharwar et al., 2014).

Over the last decade, new applications of hydrogel networks have emerged, particularly in stem cell engineering, immunomodulation, cellular and molecular therapies, and cancer research (Discher et al., 2009; Tibbitt et al., 2009). Most of these applications demand multiple functionalities of the hydrogel network and dynamic interactions between the surrounding matrices and cells (Tibbitt et al., 2009). The developments in polymer chemistry, micro- and nanofabrication technologies, and biomolecular engineering have expanded the limits for designing composite hydrogel networks possessing specific functionality (Annabi et al., 2014). Recent researches also indicate the significant and growing interest in developing nanocomposite hydrogels for various biomedical applications. Nanocomposite hydrogels, also known as hybrid hydrogels, can be defined as hydrated polymeric networks, either physically or covalently crosslinked with each other and/or with nanoparticles or nanostructures.

Besides bulk characteristics, silicate nanoparticles or layered clays can be used to significantly improve the mechanical properties of polymer hydrogels (Yuan et al., 2010). Although hydrogels with similarity to soft tissues have drawn great attention in biomedical engineering, their low mechanical strength often limits their practical applications. In order to improve the mechanical strength of hydrogels and increase their biomedical potential, silicate nanoparticles (or layered clays) in the polymer network have been used as fillers or crosslinkers (de Sousa Rodrigues et al., 2013; Haraguchi et al., 2002, 2006; Schmidt et al., 2000). Stimuli-sensitive poly(*N*-isopropylacrylamide) (PNIPAM) hydrogels are known as attractive biomaterials for drug delivery, bioseparation devices, and culture dishes for cell-sheet engineering. With lower critical solution temperature and thermosensitive coil-globule transition behavior, PNIPAM is available for many biomedical applications (Haraguchi et al., 2006).

Haraguchi et al. (2002) first developed the PNIPAM–laponite nanocomposite hydrogels. They then investigated the cell cultivation and cell-sheet detachment of hydrogel surfaces. Cell adhesion and proliferation of human hepatoma cells, dermal fibroblasts, and umbilical vein endothelial cells were found to be strongly dependent on the silicate concentration. The stimuli-responsiveness of PNIPAM nanocomposites to temperature was used to spontaneously separate confluent cell sheets from hydrogel surfaces (Haraguchi et al., 2006). These studies have provided new strategies for cell-sheet engineering (Wu et al., 2010).

5.4 Wastewater Pretreatment

Some agricultural effluents from olive mills, wineries, soy or coffee bean industry are unsuitable for discharge due to having large amounts of organic and colloidal matters. For example, the annual production of black and highly toxic olive mill wastewater in the countries growing olive in the Mediterranean is estimated at about 10–30 million m^3. The discharge of such effluents to environmental waters without any treatment is known to cause serious environmental problems (Brito et al., 2007; Jail et al., 2010; Paraskeva et al., 2007).

Tendency to clog filtering devices of colloidal particles is an essential problem with effluents. In most cases, colloidal stability of the effluents is due to the following factors (Evangelou, 1998; Lerche, 2002; Myers, 1999; Rytwo et al., 2011):

- Very small particles exhibit very low sinking tendency because the sedimentation rate of any particle in a given liquid is proportional to the particle's diameter squared. Furthermore, the frictional force against the particle's movement (Stokes' law) increases with the actual diameter, and hence small particles possess very low sedimentation rate.
- The weight of the particle also depends on its density. The density difference between the particle and the fluid affects the sedimentation rate. Most of the suspended solids in organic effluents are formed from organic matters having a density close to that of water, hindering separation from the liquid.
- Colloidal particles usually bear an electrical charge. In mineral colloids, the source of these electrical charges may be the imperfections in the lattice of the minerals, while the source for organic matters might be charged functional groups such as amine, carboxyl, phosphate, and hydroxide. The electrostatic repulsion forces between the charged particles can stabilize the colloidal dispersion by preventing their aggregation.

Although the terms "coagulation" and "flocculation" are sometimes used interchangeably, they can be ascribed to two different processes. Coagulation corresponds to the process in which colloidal particles and very fine solid suspensions are destabilized, depending on the agglomeration under suitable conditions. Flocculation refers to the process in which the destabilized particles may conglomerate into larger aggregates, and so they can be separated from the wastewater (Armenante, 1999).

In order to control these effects, some industrial treatments are used. In most cases, it is applied to two separate processes: neutralization, in which the electrostatic repulsion forces between the charged particles were eliminated to achieve coagulation, and bridging, in which several relatively small particles form larger aggregates to provide flocculation. Simultaneous addition of agents such as coagulant and flocculent is usually not an effective treatment due to the interactions between the additives. In this case, unstable colloidal dispersion may emerge, followed by flocculation of the suspended matter with large amounts, which may lead to lower values of total suspended solids (TSS), turbidity, and even higher chemical oxygen demand (COD). However, this can provide improved efficiency for water treatment, thereby reducing environmental hazards (Anastaiou et al., 2009; Coskun et al., 2010; Rytwo, 2010).

Clays and organoclays have been widely used for the pre-treatment of effluents. Many studies have used cationic or anionic polyelectrolytes, combinations of coagulant and polyelectrolytes, or even clay minerals and quaternary ammonium ions. In all cases, considerable changes in the colloidal properties were observed (Amuda and Amoo, 2007; Beall, 2003; Ginos et al., 2006; Mousavi et al., 2006; Rytwo et al., 2007; Sarika et al., 2005). It has also been described as a very effective pretreatment with the two stages based on combinations of organoclay nanoparticles and raw clay and is useable for many industrial effluents, where organoclay and raw clay were used. This pretreatment similar to the process that uses simultaneously coagulation (with cationic polymers and aluminum sulfate or other inorganic polycations) and flocculation (with cationic or anionic polyacrylamide (PAA) derivatives) can provide improved colloidal stability and reduced TSS and turbidity (Rizzo et al., 2010; Rytwo, 2011; Rytwo et al., 2011). Rytwo et al. (2011) have shown that nanocomposites prepared by using the clay mineral sepiolite and the cationic polymer polydiallyldimethylammonium chloride (polyDADMAC) with the same effluent loading in a single step could provide a very efficient pretreatment to reduce both TSS and COD values. However, the general idea behind the use of nanocomposites is to combine neutralization of colloids (coagulation), achieved by the polymer's charged sites, and bridging of the neutralized particles (flocculation), achieved by the fact that the polymer chains are connected to denser and heavier anchoring particles; this might yield very efficient pretreatment of wastewater and the reduction of TSS in one single treatment step (Rytwo, 2012).

5.5 Electricals/Electronics, Optoelectronics, Sensors, and Automobiles

Nanotechnology is an important part of the design of advanced devices for electronic and optoelectronic applications. Nowadays, the dimension scale for advanced electronic devices is in the nano range. The utility of polymer-based nanocomposites in these areas is quite diverse involving many potential applications as well as types of nanocomposites. One specific nanocomposite type receiving considerable interest involves conjugated polymers and carbon nanotubes. In this context, the potential applications of nanotechnology include photovoltaic (PV) cells and photodiodes, supercapacitors, sensors, printable conductors, light-emitting diodes (LEDs), and field-effect transistors (Baibarac et al., 2006). The electrical conductivity of carbon nanotubes in insulating polymers has also been a topic of considerable interest. The potential applications also include electromagnetic interference shielding, transparent conductive coatings, electrostatic dissipation, supercapacitors, electromechanical actuators, and various electrode applications (Baughman et al., 2002). The percolation threshold for electrical conductivity of epoxy composites containing multiwall carbon nanotubes (MWCNTs) was found to be 0.0025 wt.% MWCNT, considerably lower than nanoscale-dispersed carbon black particles (Sandler et al., 2003). The threshold conductivity of single-walled carbon nanotubes (SWCNTs) in epoxy composites was noted to be a function of the SWCNT type, with values as low as 0.00005 vol. fraction (Bryning et al., 2005). Water-dispersed carbon black particles (42 nm) added to acrylic emulsions yielded electrical conductivity percolation levels as low as 1.5 vol.% in dried films. In this system, the carbon black particles concentrate at the interface between the emulsion particles during drying, yielding a percolation network. The modulus of the emulsion system chosen as the matrix was noted to be an important variable with a higher modulus (higher T_g) yielding lower threshold percolation values (Kim et al., 2008). Polymer-based nanocomposites are most relevant materials for the anode, hole injection layer (HIL), light-emitting layer (LEL), and light-harvesting layer (LHL). In the case where flexible transparent substrates are desired, nanocomposite or nanolayered combinations of polymer and barrier nanomaterials can be employed to provide barrier properties to oxygen and water permeation. The basic construction of LED and PV devices is presented in Fig. 5.3 (Paul and Robeson, 2008).

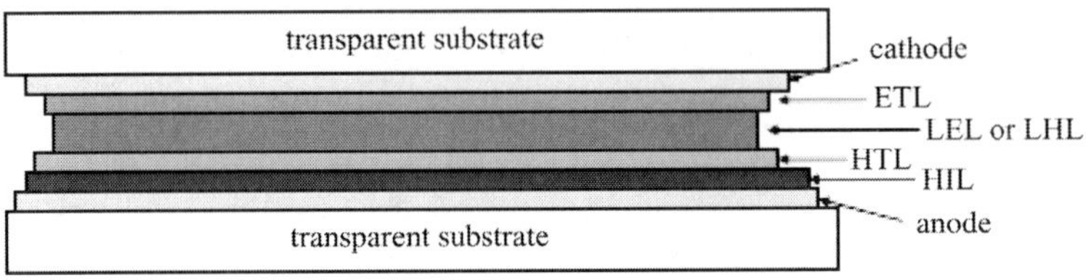

Figure 5.3 Generalized diagram of LED and PV device construction. Reprinted from Paul and Robeson, Copyright 2008, with permission from Elsevier.

The addition of clay into conducting polymers such as polypyrrole, polythiophene, polyaniline sensibly increases conductivity and the impedance effect with improved mechanical properties such as elongation, impact, and scratch resistance and results in a longer life of the product. The polymer nanocomposites (PCNs) are involved in many applications such as solar cells, windmills, electronic circuit boards, battery manufacturing, microchips, and transistors. Nanocomposites consisting of spatially confined liquid crystals are of great interest depending on the possibilities of application in optoelectronic devices, photonic crystals, depolarizers, scattering displays, information storage and recording devices (Anandhan and Bandyopadhyay, 2011; Merhari, 2009). The applications of thermoset/clay/nanocomposites in electronics as well as in the automotive sector are extremely important. It has been suggested that the main function of nanoclay is reducing solvent transmission throughout polymers such as special elastomers, polyimide, and polyurethane. Also, significant reductions have been observed in fuel transmission via polyamide 6/66 reinforced with nanoclay. The major driving force for the usage of polymer nanocomposites (PCNs) in tire fabrication is reduction of weight and processing costs. Clay, which is naturally available and has very low density, can facilitate reduction in weight. The clay-incorporated tires exhibit excellent mechanical properties compared to ordinary tires, besides improved gas barrier performance for tubeless tire applications. Generally, styrene butadiene and natural rubber-based clay nanocomposites are mostly preferred for automobile tire manufacturing and butyl rubber for tubes. It is due to improved abrasion resistance and thermal properties of the tires combined with their longer life (Koo, 2006).

Earliest successes in the automotive industry are based on the sonication process, which has been developed to provide better dispersion of clay into the polymer. Then, various automotive supplies made of nanocomposites based on clay and a thermoplastic olefin have been developed. These materials are not only lighter in weight and stiffer but also more robust at low temperatures and more recyclable. These nanocomposites are also used in the production of lateral parts of automobiles (Sanchez et al., 2005).

Nanocomposite materials based on thermoplastic olefin (TPO) and polypropylene (PP) are mainly used for the production of automotive parts. A great increase is expected in the use of nano PP in the automotive industry as a serious alternative to the existing PP applications. Polypropylene-elastomer (TPO) nanocomposites provide strength equivalent to the talc-filled PP with a 7% to 8% weight saving. Furthermore, TPO nanocomposites improve low-temperature impact strength and surface quality. TPO–clay nanocomposite materials have already been used by many automotive manufacturers for applications such as step-assist, bodywork, and body side molding (see Figs. 5.4 and 5.5) (Kang, 2010; Leaversuch, 2001; Sherman, 2004).

Figure 5.4 An automobile with side moldings from polymer–nanoclay composite material.

Figure 5.5 A bodywork from polymer–nanoclay composite material.

A growing commercialization potential for new applications of hybrid nanoclay materials is nowadays envisaged by automotive manufacturers. Some prominent applications include PP–clays for bodywork with anti-scratch properties; acetal–clays for ceiling lights; PP–clays for panes of doors; consoles and interior decoration due to esthetics and recyclable and lightness properties; nylon–clays for bumpers with enhanced mechanical and lightness properties; nylon–clays for fuel reservoirs with airtightness properties. Besides these properties, the weight advantage could present a significant impact on environmental concerns. It is estimated that the widespread use of PNCs could save millions of liters of gasoline in one year of vehicle production and reduce related carbon dioxide emissions in huge proportions. In relation to environmental applications, organically modified nanoclays are used as effective reinforcement agents in the so-called green nanocomposites (Mohanty et al., 2003). These renewable-resource-based biodegradable nanocomposites are based on cellulosic plastic (plastic made from wood), poly(lactic acid) (corn-derived plastic), or polyhydroxyalkanoate (bacterial polyesters), and they are attractive as a substitute for petroleum feedstock in making plastic in the commercial market (Sanchez et al., 2005; Wilkinson, 2001).

5.6 Other Uses

5.6.1 Engineering Plastics

Most nanocomposites are involved in various applications in engineering fields. The applications include engineering plastics, reinforced materials, retardant materials, functional materials of electronic industry, chemical fibers, molding parts.

5.6.2 Polyester Cloth or Film

Although poly(methyl methacrylate) (PMMA) itself exhibit low hardness and high abrasive property, its polymer matrix properties can be significantly improved by adding nanometer-sized SiO_2 (3.0 wt.%). PMMA-silica nanocomposites can also show very good resistance to abrasion, besides the ideal transparency compared to virgin PMMA matrix. PMMA-silica nanocomposites that are usually safe for food can be used to prepare cloth or films for various food-packaging materials. Silica treated in hexane under its boiling temperature, with coupling agents such as methacryloyloxypropyltrimethoxysilane, KH570, is used to prepare nanocomposites safely for food packaging (Duncan, 2011).

Unsaturated polyester is often used to make a gel film or gel cloth that can protect food. To improve functional properties such as barrier to gas and absorption of harmful light, a polyester composite film covered onto SiO_2 nanoparticles has been prepared.

5.6.3 Reinforcing Plastics

Conventional plastics can offer wide applications with low cost. However, the properties of resin and engineering plastics cannot fulfill most of the requirements of engineering materials. For example, nanoparticles of SiO_2 are used to modify PP; the properties of these nanocomposites are of the level of nylon 6, while their cost is only one-third of the latter. Addition of nanoparticles to plastic matrixes can improve the plastic's toughness. For example, on mixing nanoparticles of SiC/Si_3N_4 mixed particles treated by C10-acid with low-density polyethylene (LDPE) (with a particle weight of 5%), the impact strength of the nanocomposite reaches the level of 55.7 kJ m^{-2}, which is two times that of pure LDPE. Such a nanocomposite material does not break on stretching to 625% of the original size. The extension value at breaking is higher five times than that of virgin LDPE. Similarly, 25% (by weight) of nanometer-sized $CaCO_3$/high density polyethylene (HDPE) nanocomposites has the maximum impact strength, which is 1.7 times higher than that of virgin HDPE, while the composite has an extension value at breaking of 660%. When the $CaCO_3$ load reaches 16% (by weight), all the mechanical properties of the nanocomposite are much better than that of virgin HDPE.

5.6.4 Functionalized Plastics and Fibers

Functionalized nanoparticles are mixed with plastics to prepare nanocomposites with functional properties such as bacteria-killing and antielectrostatic behavior. For example, nanometer-sized ZnO carried on MMT particles acts as antielectrostatic reagents for polymers such as polietilentereftalat (PET), polycarbonate (PC), polypropylene (PP), and saturated polyesters. Functionalized chemicals mixed with polymer fibers provide nanocomposites with improved performance for far-infrared reflection, bacteria killing, and a capability of sequestering odors. For example, nanometer-sized ZnO or nanometer-sized ZnO carried on silica is added to polyester fibers used for insulation against ultraviolet light and for eliminating toxicity and odors. Further, nanocomposite fibers provide bacteria-killing properties if fabricated from Ag carried on SiO_2 or MMT particles by precursors such as tetraethoxysilane (TEOS).

These composites are used in the manufacture of home appliances. Also, mixed composites of nanoparticles of ZnO and SiO_2 on a polyaniline matrix have been prepared to produce fibers for clothes with antibacterial properties and antielectrostatic shielding. The technologies used for the preparation of these materials show close similarities, including melt injection, in situ intercalative polymerization, solution intercalation, and extruding techniques (Barani, 2014; Ibănescu et al., 2014).

5.6.5 UHMWPE–MMT Nanocomposite Plastics

Although ultrahigh molar weight polyethylene (UHMWPE) possesses high abrasion resistance, high impact strength, self-lubrication, and impact energy absorbing property, it is a difficult processable material because it has extremely high viscosity. Nanocomposite technology can easily solve this problem via the melt-intercalation of UHMWPE with organoclay. By mixing UHMWPE with layered silicates, exfoliated organoclay platelets that have a lower frictional coefficient, as well as a lower entanglement of UHMWPE chains, can be obtained. The reduced entanglement itself plays a good self-lubricating role in melt processing. In fact, the mixture of UHMWPE–organic clay can be processed by using common extrusion technology and equipment. For example, the mixture can continuously produce pipes and different forms of bulk materials with better rheological properties than virgin resin. The UHMWPE–MMT nanocomposites enhance the properties of pipes made from virgin resin. The nanocomposites have a better performance–cost ratio than iron, aluminum, plastic, or other common pipes.

5.6.6 Polypropylene-Layered Silicate Nanocomposite Films

Polypropylene (PP) resins, which are blown or biaxially oriented into thin films, are usually used in electronics, packaging, and food storage. Biaxially oriented PP films (BOPP) can be used to create various products such as electronic films, printing films, magnetic films, decoration films, heat-sealing films, money paper films, and metal electronic deposition films.

Traditional PP resins have disadvantages such as easy degradation, poor melt rheological behavior, defect morphology, and a weak controllable isotacticity index. These may lead to many problems such as heterogeneous film transparency, low barrier to gas (O_2, CO_2, and H_2O), and separation of residuals or additives from PP matrix.

For the preparation of a nanocomposite film, first the PP oligomers are grafted with maleic acid to form the composite precursor with layered silicate. The PP-layered silicate nanocomposites can be formed into a film. This can also be done by mixing PP resin with the nanoparticles or other polymers having good barrier performance, such as polyvinylidene chloride (PVDC), polietilen tereftalat (PET), and polyamide 6 (PA6). PA6 is often used to melt-mix with PP or PP-based nanocomposites, where the nanoparticles could be either layered silicate or a silica precursor. PA6–MMT nanocomposites are also used to prepare films for food packaging (Kawasumi, 1997).

5.6.7 PP–Silica Nanoprecursor Films

Silica is used as an antiadhesive for PP films and is usually added indirectly into the PP. For this, superfine nanoparticles of SiO_2 are mixed with PP resins using the nanoprecursor made from SiO_2 particles with surface modification. In application, the nano-sized SiO_2 can be reacted with water-soluble poly(acrylic acid) (PAA) and mixed with polyester oligomers. In this method, the silica nanoprecursor has 20% (by weight) of silica load, which can then be directly melt-blended with PP resins to make pellet products by using a twin screw extruder at a temperature of 220–240°C. The pellets are then hot-blown into films having a thickness of several millimeters.

The polymer–MMT nanocomposite films have unstable colors that affect or limit their applications as films. The silica particles exhibit the highest shielding effect on visible light and hence their nanocomposites possess a lower transparency (Kee and Stroeve, 2005).

References

Adame, D., Beall, G. W. (2009). Direct measurement of the constrained polymer region in polyamide/clay nanocomposites and the implications for gas diffusion. *Appl. Clay Sci.*, **42**, pp. 545–552.

Amuda, O. S., Amoo, I. A. (2007). Coagulation/flocculation process and sludge conditioning in beverage industrial wastewater treatment. *J. Hazard. Mater.*, **141**, pp. 778–783.

Anandhan, S., Bandyopadhyay, S. (2011). Nanocomposites and polymer analytical methods. In *Polymer Nanocomposites: From Synthesis to Applications*, Cuppoletti, J. (Ed.). Croatia: InTech Press, pp. 3–28. ISBN: 978-953-307-352-1.

Anastasiou, N., Monou, M., Mantzavinos, D., Kassinos, D. (2009). Monitoring of the quality of winery influents/effluents and polishing of partially treated winery flows by homogeneous Fe (II) photooxidation. *Desalination*, **248**, pp. 836–842.

Annabi, N., Tamayol, A., Uquillas, J. A., Akbari, M., Bertassoni, L. E., Cha, C., Camci-Unal, G., Dokmeci, M. R., Peppas, N. A., Khademhosseini, A. (2014). 25th anniversary article. Rational design and applications of hydrogels in regenerative medicine. *Adv. Mater.,* **26**, pp. 85–124.

Armenante, P. M. (1999). Coagulation and flocculation. Industrial waste control, physical and chemical treatment. Che 865 course, New Jersey Institute of Technology website, http://Cpe.Njit.Edu/Dlnotes/Che685/Cls07-1.Pdf.

Azizi Samir, M. A. S., Alloin, F., Dufresne, A. (2005). Review of recent research into cellulosic whiskers, their properties and their application in nanocomposite field. *Biomacromolecules*, **6**, pp. 612–626.

Baibarac, M., Gomez-Romero, P. J. (2006). Electrosynthesis of the poly(N-vinyl carbazole)/carbon nanotubes composite for applications in the supercapacitors field. *Nanosci. Nanotechnol.,* **6**, pp. 1–14.

Barani, H. (2014). Preparation of antibacterial coating based on in situ synthesis of ZnO/SiO_2 hybrid nanocomposite on cotton fabric. *Appl. Surf. Sci.*, **320**, pp. 429–434.

Baughman, R. H., Zakhidov, A. A., De Heer, W. A. (2002). Carbon nanotubes: The route toward applications. *Science*, **297**, pp. 787–792.

Beall, G. W. (2000). New conceptual model for interpreting nanocomposite behavior. In *Polymer–Clay Nanocomposites*, Pinnavaia,T. J., Beall, G. W. (Eds.). New York, USA: Wiley, pp. 267–279.

Beall, G. W. (2003). The use of organo-clays in water treatment. *Appl. Clay Sci.*, **24**, pp. 11–20.

Berry, C. C. (2005). Possible exploitation of magnetic nanoparticle–cell interaction for biomedical applications. *J. Mater. Chem.*, **15**, pp. 543–547.

Bettencourt, A., Almeida, A. J. (2012). Poly(methyl methacrylate) particulate carriers in drug delivery. *J. Microencapsulation*, **29**, pp. 357–367.

Black, J. (1992). *Biological Performance of Materials: Fundamentals of Biocompatibility.* New York, USA: Marcel Dekker.

Blaker, J. J., Gough, J. E., Maquet, V., Notingher, I., Boccaccini, A. R. (2003). In vitro evaluation of novel bioactive composites based on Bioglass®-filled polylactide foams for bone tissue engineering scaffolds. *J. Biomed. Mater. Res. A*, **67A**, pp. 1401–1411.

Brito, A. G., Peixoto, J., Oliveira, J. M., Oliveira, J. A., Costa, C., Nogueira, R., Rodrigues, A. (2007). Brewery and winery wastewater treatment: Some focal points of design and operation. In *Utilization of By-Products and Treatment of Waste in the Food Industry*, Oreopoulou, V., Russ, W. (Eds.). New York, USA: Springer, pp. 109–131. ISBN: 9780387335117.

Brody, A. L. (2003). "Nano, nano" food packaging technology. *Food Technol.*, **57**, pp. 52–54.

Brody, A. L. (2006). Nano and food packaging technologies converge. *Food Technol.*, **60**, pp. 92–94.

Brody, A. L. (2007). Nanocomposite technology in food packaging. *Food Technol.*, **61**, pp. 80–83.

Bryning, M. B., Islam, M. F., Kikkawa, J. M., Yodh, A. G. (2005). Very low conductivity threshold in bulk isotropic single-walled carbon nanotube-epoxy composites. *Adv. Mater.*, **17**, pp. 1186–1191.

Cabedo, L., Gimenez, E., Lagaron, J. M., Gavara, R., Saura, J. J. (2004). Development of EVOH–kaolinite nanocomposites. *Polymer*, **45(15)**, pp. 5233–5238.

Campbell, K., Craig, D. Q. M., McNally, T. (2008). Poly(ethylene glycol) layered silicate nanocomposites for retarded drug release prepared by hot-melt extrusion. *Int. J. Pharm.*, **363**, pp. 126–131.

Carlisle, E. M., Alpenfels, W. F. (1980). Silicon requirement for normal growth of cartilage in culture. *Fed. Proc.*, **39**, p. 787.

Cava, D., Gimenez, E., Gavara, R., Lagaron, J. M. (2006). Comparative performance and barrier properties of biodegradable thermoplastics and nanobiocomposites versus PET for food packaging applications. *J. Plast. Film Sheeting*, **22**, pp. 265–274.

Chakraborti, M., Jackson, J. K., Plackett, D., Brunette, M., Burt, H. M. (2011). Drug intercalation in layered double hydroxide clay: Application in the development of a nanocomposite film for guided tissue regeneration. *Int. J. Pharm.*, **416**, pp. 305–313.

Changerath, R., Nair, P. D., Mathew, S., Nair, C. P. (2009). Poly(methyl methacrylate)-grafted chitosan microspheres for controlled release of ampicillin. *J. Biomed. Mater. Res. B*, **89**, pp. 65–76.

Coskun, T., Debik, E., Demir, N. M. (2010). Treatment of olive mill wastewaters by nanofiltration and reverse osmosis membranes. *Desalination*, **259**, pp. 65–70.

Cypes, S. H., Saltzman, W. M., Giannelis, E. P. (2003). Organosilicate polymer drug delivery systems, controlled release and enhanced mechanical properties. *J. Controlled Release*, **90**, pp. 163–169.

Cyras, V. P., Manfredi, L. B., Ton-That, M. T., Vazquez, A. (2008). Physical and mechanical properties of thermoplastic starch/montmorillonite nanocomposite films. *Carbohydr. Polym.*, **73**, pp. 55–63.

Dalmas, F., Cavaillé, J. Y., Gauthier, C., Chazeau, L., Dendievel, R. (2007). Viscoelastic behavior and electrical properties of flexible nanofiber filled polymer nanocomposites. Influence of processing conditions. *Compos. Sci. Technol.*, **67**, pp. 829–839.

Day, R. M., Boccaccini, A. R., Shurey, S., Roether, J. A., Forbes, A., Hench, L. L., Gabe, S. M. (2004). Assessment of polyglycolic acid mesh and bioactive glass for soft-tissue engineering scaffolds. *Biomaterials*, **25**, pp. 5857–5866.

de Azeredo, H. M. C. (2009). Nanocomposites for food packing applications. *Food Res. Int.*, **42**, pp. 1240–1253.

de Azeredo, H. M. C., Mattoso, L. H. C., McHugh, T. H. (2011). Nanocomposites in food packaging: A review. In *Advances in Diverse Industrial Applications of Nanocomposites*, Reddy, B. (Ed.). Crotia: InTech.

de Sousa Rodrigues, L. A., Figueiras, A., Veiga, F., de Freitas, R. M., Nunes, L. C. C., da Silva Filho, E. C., da Silva Leite, C. M. (2013). The systems containing clays and clay minerals from modified drug release: A review. *Colloids Surf.: Biointerfaces*, **103**, pp. 642–651.

Dean, K., Yu, L., Wu, D. Y. (2007). Preparation and characterization of melt extruded thermoplastic starch/clay nanocomposites. *Compos. Sci. Technol.*, **67**, pp. 413–421.

Dhana Lekshmi, U. M., Poovi, G., Kishore, N., Reddy, P. N. (2010). In vitro characterization and in vivo toxicity study of repaglinide loaded poly(methyl methacrylate) nanoparticles. *Int. J. Pharm.*, **396**, pp. 194–203.

Discher, D. E., Mooney, D. J., Zandstra, P. W. (2009). Growth factors, matrices, and forces combine and control stem cells. *Science*, **324**, pp. 1673–1677.

Dong, Y., Feng, S. S. (2005). Poly(d,l-lactide-co-glycolide)/montmorillonite nanoparticles for oral delivery of anticancer drugs. *Biomaterials*, **26**, pp. 6068–6076.

Duncan, T. V. (2011). Applications of nanotechnology in food packaging and food safety: Barrier materials, antimicrobials and sensors. *J. Colloid Interf. Sci.*, **363**, pp. 1–24.

Edlund, U., Albertsson, A.-C. (2002). Degradable polymer microspheres for controlled drug delivery in degradable aliphatic polyesters. *Adv. Polym. Sci.*, **157**, p. 67–112.

Ehrlich, H., Janussen, D., Simon, P., Bazhenov, V. V., Shapkin, N. P., Erler, C., Mertig, M., Born, R., Heinemann, S., Hanke, T., Worch, H., Vournakis, J. N. (2008). Nanostructural organization of naturally occurring composites: part II; silica-chitin-based biocomposites. *J. Nanomater.*, Article ID 670235, pp. 1–8.

Evangelou, V. P. (1998). *Environmental Soil and Water Chemistry*. New York, USA: John Wiley and Sons.

Feng, S.-S., Mei, L., Anitha, P., Gan, C. W., Zhou, W. (2009). Poly(lactide)-vitamin E derivative/montmorillonite nanoparticle formulations for the oral delivery of docetaxel. *Biomaterials*, **30**, pp. 3297–3306.

Fisher, O. Z., Khademhosseini, A., Langer, R., Peppas, N. A. (2010). Bioinspired materials for controlling stem cell fate. *Acc. Chem. Res.*, **43**, pp. 419–428.

Foo, W. P., Patwardhan, S. V., Belton, D. J., Kitchel, B., Anastasiades, D., Huang, J., Naik, R. R., Perry, C. C., Kaplan, D. L. (2006). Novel nanocomposites from spider silk-silica fusion (chimeric) proteins. *Proc. Natl. Acad. Sci.*, **103**, pp. 9428–9433.

Gaharwar, A. K., Peppas, N. A., Khademhosseini, A. (2014). Nanocomposite hydrogels for biomedical applications. Tissue engineering and delivery systems. *Biotechnol. Bioeng.*, **111**, pp. 441–453.

Gaharwar, A. K., Schexnailder, P. J., Kline, B. P., Schmidt, G. (2011). Assessment of using laponite cross-linked poly(ethylene oxide) for controlled cell adhesion and mineralization. *Acta Biomaterials*, **7**, pp. 568–577.

Gilman, J. W., Jackson, C. L., Morgan, A. B., Harris, R., Manias, E., Giannelis, E. P., Wuthenow, M., Hilton, D., Phillips S. H. (2000). Flammability properties of polymer layered silicate nanocomposites. Propylene and polystyrene nanocomposites. *Chem. Mater.*, **12**, pp. 1866–1873.

Ginos, A., Manios, T., Mantzavinos, D. (2006). Treatment of olive mill effluents by coagulation-flocculation-hydrogen peroxide oxidation and effect on phytotoxicity. *J. Hazard. Mater.*, **133**, pp. 135–142.

Gupta, A. K., Gupta, M. (2005). Synthesis and surface engineering of iron oxide nanoparticles for biomedical applications. *Biomaterials*, **26**, pp. 3995–4021.

Gürses, A., Ejder-Korucu, M., Doğar, Ç. (2015). Polymer-silicate nanocomposites for biomedical applications. In *Encyclopedia of Biomedical Polymers and Polymeric Biomaterials*, Mishra, M. (Ed.), 11 Volume Set. USA: CRC Press. ISBN 9781439898796.

Haraguchi, K., Li, H. J. (2006). Mechanical properties and structure of polymer–clay nanocomposite gels with high clay content. *Macromolecules*, **39**, pp. 1898–1905.

Haraguchi, K., Takehisa, T. (2002). Nanocomposite hydrogels: A unique organic-inorganic network structure with extraordinary mechanical, optical, and swelling/de-swelling properties. *Adv. Mater.* **14**, pp. 1120–1124.

Haraguchi, K., Takehisa, T., Ebato, M. (2006). Control of cell cultivation and cell sheet detachment on the surface of polymer/clay nanocomposite hydrogels. *Biomacromolecules*, **7**, pp. 3267–3275.

Heinemann, S., Heinemann, C., Bernhardt, R., Reinstorf, A., Nies, B., Meyer, M., Worch, H., Hanke, T. (2009). Bioactive silica-collagen composite xerogels modified by calcium phosphate phases with adjustable mechanical properties for bone replacement. *Acta Biomater.*, **5**, pp. 1979–1990.

Heinemann, S., Heinemann, C., Ehrlich, H., Meyer, M., Baltzer, H., Worch, H., Hanke, T. (2007). A novel biomimetic hybrid material made of silicified collagen: Perspectives for bone replacement. *Adv. Eng. Mater.*, **9**, pp. 1061–1068.

Hench, L. L. (2009). Genetic design of bioactive glass. *J. Eur. Ceram. Soc.*, **29**, pp. 1257–1265.

Hench, L. L., Paschall, H. A. (1973). Direct chemical bond of bioactive glass-ceramic materials to bone and muscle. *J. Biomed. Mater. Res.*, **7**, pp. 25–42.

Huang, J., Lin, Y. W., Fu, X. W., Best, S. M., Brooks, R. A., Rushton, N., Bonfield, W. (2007). Development of nano-sized hydroxyapatite reinforced composites for tissue engineering scaffolds. *J. Mater. Sci. Mater.*, **18 (11)**, pp. 2151–2157.

Hule, R. A., Pochan, D. J. (2007). Polymer nanocomposites for biomedical application. *MRS Bull.*, **32**, pp. 354–358.

Ibănescu (Buşilă), M., Muşat, V., Textor, T., Badilita, V., Mahltig, B. (2014). Photocatalytic and antimicrobial Ag/ZnO nanocomposites for functionalization of textile fabrics. *J. Alloy Compd.*, **610**, pp. 244–249.

Jail, A. F., Boukhoubza, A., Nejmeddine, S., Sayadi, Hassani, L. (2010). Co-treatment of olive-mill and urban wastewaters by experimental stabilization ponds. *J. Hazard. Mater.*, **176**(1–3), pp. 893–900.

Jones, J. R., Ehrenfried, L. M., Hench, L. L. (2006). Optimising bioactive glass scaffolds for bone tissue engineering. *Biomaterials*, **27**, pp. 964–973.

Jordan, J., Jacob, K. I., Tannenbaum, R., Sharaf, M. A., Jasiuk, I. (2005). Experimental trends in polymer nanocomposites: A review. *Mater. Sci. Eng. A*, **393**(1–2), pp. 1–11, ISSN 0921-5093.

Kang, H.-Y. (2010). *A Review of the Emerging Nanotechnology Industry: Materials, Fabrications, and Applications.* Department of Toxic Substances Control Pollution Prevention and Green Technology.

Kannan, R. Y., Salacinski, H. J., Butler, P. E., Seifalian, A. M. (2005). Polyhedral oligomeric silsesquioxane nanocomposites: The next generation material for biomedical applications. *Acc. Chem. Res.*, **38**(11), p. 879.

Kawasumi, M., Hasegawa, N., Kato, M., Usuki, A., Okada, A. (1997). Preparation and mechanical properties of polypropylene-clay hybrids. *Macromolecules,* **30**, pp. 6333–6338.

Ke, Y. C., Stroeve, P. (2005). *Background on Polymer-Layered Silicate and Silica Nanocomposites*, 1st Ed. Amsterdam, The Netherlands: Elsevier.

Kevadiya, B. D., Joshi, G.V., Bajaj, H. C. (2010). Layered bionanocomposites as carrier for procainamide. *Int. J. Pharm.*, **388**, pp. 280–286.

Kevadiya, B. D., Patel, T. A., Jhala, D. D., Thumbar, R. P., Brambhatt, H., Pandya, M. P., Rajkumar, S., Jena, P. K., Joshi, G. V., Gadhia, P. K., Tripathi, C. B., Bajaj, H. C. (2012). Layered inorganic nanocomposites: A promising carrier for 5-fluorouracil (5-FU). *Eur. J. Pharm. Biopharm.*, **81**, pp. 91–101.

Kilcast, D., Subramaniam, P. (2000). Introduction. In *The Stability and Shelf-Life of Food*, Kilcast, D., Subramaniam, P. (Eds.). Cambridge, UK: Woodhead Publishing.

Kim, H. W., Kim, H. E., Salih, V. (2005). Stimulation of osteoblast responses to biomimetic nanocomposites of gelatin-hydroxyapatite for tissue engineering scaffolds. *Biomaterials*, **26**, pp. 5221–5230.

Kim, J., Hu, C., Woo, R., Sham, M. (2005). Moisture barrier characteristics of organoclay-epoxy nanocomposites. *Compos. Sci. Technol.*, **65**, pp. 805–813.

Kim, Y. S., Wright, J. B., Grunlan, J. C. (2008). Influence of polymer modulus on the percolation threshold of latex-based composites. *Polymer*, **49**, pp. 570–578.

Kloxin, A. M., Kloxin, C. J., Bowman, C. N., Anseth, K. S. (2010). Mechanical properties of cellularly responsive hydrogels and their experimental determination. *Adv. Mater.,* **22**(31), pp. 3484–3494.

Koh, H. C., Park, J. S., Jeong, M. A., Hwang, H. Y., Hong, Y. T., Ha, S. Y., Nam, S. Y. (2008). Preparation and gas permeation properties of biodegradable polymer/layered silicate nanocomposite membranes. *Desalination,* **233**, pp. 201–209.

Kojima, Y., Usuki, A., Kawasumi, M., Okada, A., Kurauchi, T., Kamigaito, O. (1993). One-pot synthesis of nylon 6–clay hybrid. *J. Polym. Sci. A*, **31**, pp. 1755–1758.

Kokubo, T. (1991). Bioactive glass ceramics, properties and applications. *Biomaterials*, **12**, pp. 155–163.

Koo, J. H. (2006). *Polymer Nanocomposites: Processing, Characterization and Applications*. New York, USA: McGraw-Hill, pp. 235–261.

Krikorian, V., Pochan, D. J. (2003). Poly(L-lactic acid)/layered silicate nanocomposite: Fabrication, characterization and properties. *Chem. Mater.*, **15**, pp. 4317–4324.

Kronenthal, R. L. (1975). *Polymer Medicine and Surgery*. NewYork, USA: Plenum Press.

Kulshrestha, A. S., Mahapatro, A. (2008). *Polymers for Biomedical Applications*. ACS Symposium Series. Washington, DC: American Chemical Society.

Lange, J., Wyser, Y. (2003). Recent innovations in barrier technologies for plastic packaging: A review. *Packaging Technol. Sci.*, **16**, pp. 149–158.

Leaversuch, R. (2001). Nanocomposites: Broaden roles in automotive, barrier packaging. *Plastic Technol.*, **47**(10), pp. 64–69.

Lee, E. J., Shin, D. S., Kim, H. E., Kim, H. W., Koh, Y. H., Jang, J. H. (2009). Membrane of hybrid chitosan-silica xerogel for guided bone regeneration. *Biomaterials*, **30**, pp. 743–750.

Lee, J. H., Park, T. G., Park, H. S., Lee, D. S., Lee, Y. K., Yoon, S. C., Nam, J. D. (2003). Thermal and mechanical characteristics of poly(L-lactic acid) nanocomposite scaffold. *Biomaterials*, **24**, pp. 2773–2778.

Lee, W. F., Chen, Y. C. (2004). Effect of bentonite on the physical properties and drug-release behavior of poly(AA-co-PEGMEA)/bentonite nanocomposite hydrogels for mucoadhesive. *J. Appl. Polym. Sci.*, **91**, pp. 2934–2941.

Lee, Y. H., Lee, J. H., An, I. G., Kim, C., Lee, D. S., Lee, Y. K., Nam, J. D. (2005). Electrospun dual-porosity structure and biodegradation morphology of montmorillonite reinforced PLLA nanocomposite scaffolds. *Biomaterials*, **26**, pp. 3165–3172.

Lerche, D. (2002). Dispersion stability and particle characterization by sedimentation kinetics in a centrifugal field. *J. Dispersion Sci. Technol.*, **23**, pp. 699–709.

Lewkowitz-Shpuntoff, H. M., Wen, M. C., Singh, A., Brenner, N., Gambino, R., Pernodet, N., Isseroff, R., Rafailovich, M., Sokolov, J. (2009). The effect of organo clay and adsorbed FeO_3 nanoparticles on cells cultured on ethylene vinyl acetate substrates and fibers. *Biomaterials*, **30**, pp. 8–18.

Liu, Z., Zhao, G., Yu, J., Zhang, J., Ma, X., Han, G. (2011). Preparation and properties of chitosan-graft-poly(methyl methacrylate) nanoparticles using potassium diperiodatocuprate (III) as an initiator. *J. Appl. Polym. Sci.*, **120**, pp. 2707–2715.

Lotti, C., Isaac, C. S., Branciforti, M. C., Alves, R. M. V., Liberman, S., Bretas, R. E. S. (2008). Rheological, mechanical and transport properties of blown films of high density polyethylene nanocomposites. *Eur. Polym. J.*, **44**, pp. 1346–1357.

Ludueña, L. N., Alvarez, V. A., Vasquez, A. (2007). Processing and microstructure of PCL/clay nanocomposites. *Mater. Sci. Eng. A*, **460–461**, pp. 121–129.

Mangiacapra, P., Gorrasi, G., Sorrentino, A., Vittoria, V. (2006). Biodegradable nanocomposites obtained by ball milling of pectin and montmorillonites. *Carbohydr. Polym.*, **64**, pp. 516–523.

Mansur, H. S., Costa, H. S. (2008). Nanostructured poly(vinyl alcohol)/bioactive glass and poly(vinyl alcohol)/chitosan/bioactive glass hybrid scaffolds for biomedical applications. *Chem. Eng. J.*, **137**, pp. 72–83.

Marras, S. I., Kladi, K. P., Tsivintzelis, I., Zuburtikudis, I., Panayiotou, C. (2008). Biodegradable polymer nanocomposites: The role of nanoclays on the thermomechanical characteristics and the electrospun fibrous structure. *Acta Biomater.*, **4**(3), pp. 756–765.

Marsh, K., Bugusu, B. (2007). Food packaging: Roles, materials, and environmental issues. *J. Food Sci.*, **72**(3), pp. 38–55.

Matos, A., Castro, M., Almeida, A. J., Bettencourt, A. (2010a). Incorporation of minocycline into poly(methylmethacrylate) drug delivery systems. *Proceedings of the 7th World Meeting on Pharmaceutics, Biopharmaceutics and Pharmaceutical Technology*, March 8–11, 2010, Valletta, Malta.

Matos, A., Padrela, L., Rodrigues, M. A., Matos, H. A., Castro, M., Duarte, A., Almeida, A. J., Bettencourt, A. (2010b). Antibacterial activity of tigecycline micro/nanoparticles under supercritical fluids conditions. *Proceedings of the 3rd International NanoBioConference*, August 24–27, 2010, Zurich, Switzerland.

Meneghetti, P., Qutubuddin, S. (2006). Synthesis, thermal properties and applications of polymer-clay nanocomposites. *Thermochim. Acta*, **442**, pp. 74–77.

Merhari, L. (Ed.) (2009). *Hybrid Nanocomposites for Nanotechnology: Electronic, Optical, Magnetic and Biomedical Applications.* (pp. 231–287). Berlin, Germany: Springer, pp. 231–287. ISBN 978-0-387-30428-1.

Mestiri, M., Puisieux, F., Benoit, J. P. (1993). Preparation and characterization of cisplatin-loaded polymethyl methacrylate microspheres. *Int. J. Pharm.*, **89**, pp. 229–234.

Misra, S. K., Valappil, S. P., Roy, I., Boccaccini, A. R. (2006). Polyhydroxyalkanoate (PHA)/inorganic phase composites for tissue engineering applications. *Biomacromolecules*, **7**, pp. 2249–2258.

Mohanty, A. K., Drzal, L. T., Misra, M. (2003). Nano reinforcements of bio-based polymers: The hope and the reality. *Polym. Mater. Sci. Eng.*, **88**, pp. 60–61.

Moniruzzaman, M., Winey, K. I. (2006). Polymer nanocomposites containing carbon nanotubes. *Macromolecules*, **39**, pp. 5194–5205.

Moore, S. (1999). Nanocomposites achieves exceptional barrier in films. *Mod. Plast.*, **76**, pp. 31–32.

Moraru, C. I., Panchapakesan, C. P., Huang, Q., Takhistov, P., Liu, S., Kokini, J. L. (2003). Nanotechnology, a new frontier in food science. *Food Technol.*, **57**, pp. 24–29.

Morgan, A. B., Chu, L. L., Harris, J. D. (2005). A flammability performance comparison between synthetic and natural clays in polystyrene nanocomposites. *J. Fire Mater.*, **29**, pp. 213–229.

Mousavi, S. M., Alemzadeh, I., Vossoughi, M. (2006). Use of modified bentonite for phenolic adsorption in treatment of olive oil mill wastewater. *Iranian J. Sci. Technol. B*, **30**, pp. 613–619.

Myers, D. (1999). *Surfaces, Interfaces, Colloids*. New York, USA: John Wiley and Sons. ISBN: 0-471-22111-2.

Nielsen, L. E. (1967). Models for the permeability of filled polymer systems. *J. Macromol. Sci., A*, **1**(5), pp. 929–942.

Oksman, K., Mathew, A. P., Bondeson, D., Kvien, I. (2006). Manufacturing process of cellulose whiskers/polylactic acid nanocomposites. *Compos. Sci. Technol.*, **66**(15), pp. 2776–2784.

Ozkoc, G., Kemaloglu, S., Quaedflieg, M. (2010). Production of poly(lactic acid)/organoclay nanocomposite scaffolds by microcompounding and polymer/particle leaching. *Polym. Compos.*, **31**, pp. 674–683.

Paiphansiri, U., Tangboriboonrat, P., Landfester, K. (2006). Polymeric nanocapsules containing an antiseptic agent obtained by controlled nanoprecipitation onto water-in-oil miniemulsion droplets. *Macromol. Biosci.*, **6**, pp. 33–40.

Paraskeva, C. A., Papadakis, V. G., Tsarouchi, E., Kanellopoulou, D. G., Koutsoukos, P. G. (2007). Membrane processing for olive mill wastewater fractionation. *Desalination*, **213**, pp. 218–229.

Park, H. M., Lee, W. K., Park, C. Y., Cho, W. J., Ha, C. S. (2003). Environmentally friendly polymer hybrids, part I. Mechanical, thermal, and barrier properties of the thermoplastic starch/clay nanocomposites. *J. Mater. Sci.*, **38**, pp. 909–915.

Park, J. B. (1984). *Biomaterials Science and Engineering*. New York, USA: Plenum Press.

Paul, D. R., Robeson, L. M. (2008). Polymer nanotechnology: Nanocomposites. *Polymer,* **49**, pp. 3187–3204.

Paul, M. A., Alexandre, M., Degee, P., Henrist, C., Rulmont, A., Dubois, P. (2003). New nanocomposite materials based on plasticized poly(L-lactide) and organomodified montmorillonites, thermal and morphological study. *Polymer*, **44**, pp. 443–450.

Pendhari, S. S., Kant, T., Desai, Y. M. (2008). Application of polymer composites in civil construction: A general review. *Compos. Struct.*, **84**, pp. 114–124.

Peter, M., Binulal, N. S., Nair, S. V., Selvamurugan, N., Tamura, H., Jayakumar, R. (2010). Novel biodegradable chitosan–gelatin/nanobioactive glass ceramic composite scaffolds for alveolar bone tissue engineering. *Chem. Eng. J.*, **158**, pp. 353–361.

Peter, M., Binulal, N. S., Soumya, S., Nair, S. V., Furuike, T., Tamura, H., Jayakumar, R. (2010). Nanocomposite scaffolds of bioactive glass ceramic nanoparticles disseminated chitosan matrix for tissue engineering applications. *Carbohydr. Polym.*, **79**, pp. 284–289.

Peter, M., Kumar, P. T. S., Binulal, N. S., Nair, S. V., Tamura, H., Jayakumar, R. (2009). Development of novel [alpha]-chitin/nanobioactive glass ceramic composite scaffolds for tissue engineering applications. *Carbohyd. Polym.* **78**, pp. 926–931.

Petersson, L., Oksman, K. (2006). Biopolymer based nanocomposites: Comparing layered silicates and microcrystalline cellulose as nanoreinforcement. *Compos. Sci. Technol.,* **66**, 2187–2196.

Qiao, R., Brinson, L. C. (2009). Simulation of interphase percolation and gradients in polymer nanocomposites. *Compos. Sci. Technol.*, **69**(3–4), pp. 491–499.

Ramakrishna, S., Mayer, J., Weintermantel, E., Leong, K. W. (2001). Biomedical applications of polymer-composite materials: A review. *Compos. Sci. Technol.,* **61**, pp. 1189–1224.

Ray, S. S., Yamada, K., Okamoto, M., Ueda, K. (2002). Polylactide-layered silicate nanocomposite: A novel biodegradable material. *Nano Lett.*, **2**, pp. 1093–1096.

Ray, S., Easteal, A., Quek, S. Y., Chen, X. D. (2006). The potential use of polymer-clay nanocomposites in food packing. *Int. J. Food Eng.*, **2**, Article 5.

Rhim, J.-W., Park, H.-M., Ha, C.-S. (2013). Bio-nanocomposites for food packaging applications. *Prog. Polym. Sci.*, **38**, pp. 1629–1652.

Rizzo, L., Lofrano, G., Belgiorno, V. (2010). Olive mill and winery wastewaters pre-treatment by coagulation with chitosan. *Sep. Sci. Technol.*, **45**, pp. 2447–2452.

Rolland, J. P., Maynor, B. W., Euliss, L. E., Exner, A. E., Denison, G. M., Desimone, J. M. (2005). Direct fabrication and harvesting of monodisperse, shape-specific nanobiomaterials. *J. Am. Chem. Soc.*, **127**, pp. 10096–10100.

Rutnakornpituk, M., Baranauskas, V. V., Riffle, J. S., Connolly, J., Pierre, St. T. G., Dailey, J. P. (2002). Polysiloxane fluid dispersions of cobalt nanoparticles in silica spheres for use in ophthalmic applications. *Eur. Cells Mater.* **3**, pp. 102–105.

Rytwo, G. (2011). Method for treatment of industrial organic effluents with nanocomposites of a polymer adsorbed on anchoring particles, Provisional US patent application 61/500, 296.

Rytwo, G. (2012). The use of clay-polymer nanocomposites in wastewater pretreatment. *Sci. World J.*, Article ID 498503, pp. 1–7.

Rytwo, G., Kohavi, Y., Botnick, I., Gonen, Y. (2007). Use of CV- and TPP montmorillonite for the removal of priority pollutants from water. *Appl. Clay Sci.*, **36**, pp. 182–190.

Rytwo, G., Rettig, A., Gonen, Y. (2011). Organo-sepiolite particles for efficient pretreatment of organic wastewater: Application. *Appl. Clay Sci.*, **51**, pp. 390–394.

Rytwo, G., Varman, H., Bluvshtein, N., König, T. N., Mendelovits, A., Sandler, A. (2011). Adsorption of berberine on commercial minerals. *Appl. Clay Sci.*, **51**, pp. 43–50.

Sairam, M., Ramesh Babu, V., Krishna Rao, K. S. V., Aminabhavi, T. M. (2007). Poly(methylmethacrylate)-poly(vinyl pyrrolidone) microspheres as drug delivery systems, indomethacin/cefadroxil loading and in vitro release study. *J. Appl. Polym. Sci.*, **104**, pp. 1860–1865.

Salcedo, I., Aguzzi, C., Sandri, G., Bonferoni, M. C., Mori, M., Cerezo, P., Sanchez, R., Viseras, C., Caramella, C. (2012). In vitro biocompatibility and mucoadhesion of montmorillonite chitosan nanocomposite: A new drug delivery. *Appl. Clay Sci.*, **55**, pp. 131–137.

Sanchez, C., Julian, B., Belleville, P., Popall, M. (2005). Applications of hybrid organic–inorganic nanocomposites. *J. Mater. Chem.*, **15**, pp. 3559–3592.

Sandler, J. K. W., Kirk, J. E., Kinloch, I. A., Shaffer, M. S. P., Windle, A. H. (2003). Ultra-low electrical percolation threshold in carbon-nanotube epoxy composites. *Polymer*, **44**, pp. 5893–5899.

Sarika, R., Kalogerakis, N., Mantzavinos, D. (2005). Treatment of olive mill effluents, part II. Complete removal of solids by direct flocculation with poly-electrolytes. *Environ. Int.*, **31**, pp. 297–304.

Schmidt, G., Nakatani, A. I., Butler, P. D., Karim, A., Han, C. C. (2000). Shear orientation of viscoelastic polymer-clay solutions probed by flow birefringence and SANS. *Macromolecules*, **33**, pp. 7219–7222.

Schwarz, K. (1973). A bound form of silicon in glycosaminoglycans and polyuronides. *Proc. Natl. Acad. Sci.*, **70**, pp. 1608–1612.

Schwarz, K., Milne, D. B. (1972). Growth-promoting effects of silicon in rats. *Nature*, **239**, pp. 333–334.

Shaikh, S., Birdi, A., Qutubuddin, S., Lakatosh, E., Baskaran, H. (2007). Controlled release in transdermal pressure sensitive adhesives using organosilicate nanocomposites. *Ann. Biomed. Eng.*, **35**, pp. 2130–2137.

Sherman, L. M. (2004). Chasing nanocomposites. *Plastic Technol.*, **50**(11), pp. 56–61.

Singh, R. P., Anderson, B. A. (2004). The major types of food spoilage: An overview. In *Understanding and Measuring the Shelf-Life of Food*, Steele, R. (Ed.). Cambridge, UK: Woodhead, pp. 3–23.

Sivakumar, M., Rao, K. P. (2002). In vitro release of ibuprofen and gentamicin from PMMA functional microspheres. *J. Biomater. Sci. Polym. Ed.*, **13**, pp. 111–126.

Slaughter, B. V., Khurshid, S. S., Fisher, O. Z., Khademhosseini, A., Peppas, N. A. (2009). Hydrogels in regenerative medicine. *Adv. Mater.*, **21**(32–33), pp. 3307–3329.

Streubel, J., Siepmann, J., Bodmeier, R. (2002). Floating microparticles based on low density foam powder. *Int. J. Pharm.*, **241**, pp. 279–292.

Streubel, J., Siepmann, J., Bodmeier, R. (2003). Multiple unit gastroretentive drug delivery systems: A new preparation method for low density microparticles. *J. Microencapsulation*, **20**, pp. 329–347.

Takahashi, T., Yamada, Y., Kataoka, K., Nagasaki, Y. (2005). Preparation of a novel PEG-clay hybrid as a DDS material, dispersion stability and sustained release profiles. *J. Controlled Release*, **107**, pp. 408–416.

Tibbitt, M. W., Anseth, K. S. (2009). Hydrogels as extracellular matrix mimics for 3D cell culture. *Biotechnol. Bioeng.*, **103**(4), pp. 655–663.

Toprak, M. S., McKenna, B. J., Waite, J. H., Stucky, G. D. (2007). *Chem. Mater.*, **19**, pp. 4263–4269.

Tsivintzelis, I., Marras, S. I., Zuburtikudis, I., Panayiotou, C. (2007). Porous poly(L-lactic acid) nanocomposite scaffolds prepared by phase inversion using supercritical CO_2 as antisolvent. *Polymer*, **48**, pp. 6311–6318.

Tsuji, H., Ikada, Y. (2000). Properties and morphology of poly(l-lactide) 4. Effects of structural parameters on long-term hydrolysis of poly(l-lactide) in phosphate-buffered solution. *Polym. Degrad. Stab.* **67**, pp. 179–189.

Vadala, M. L., Rutnakornpituk, M., Zalich, M. A., Pierre, St. T. G., Riffle, J. S. (2004). Block copolysiloxanes and their complexation with cobalt nanoparticles. *Polymer*, **45**, pp. 7449–7461.

Valerio, P., Pereira, M. M., Goes, A. M., Leite, M. F. (2004). The effect of ionic products from bioactive glass dissolution on osteoblast proliferation and collagen production. *Biomaterials*, **25**, pp. 2941–2948.

Verrier, S., Blaker, J. J., Maquet, V., Hench, L. L., Boccaccini, A. R. (2004). PDLLA/ Bioglass® composites for soft-tissue and hard-tissue engineering: An in vitro cell biology assessment. *Biomaterials*, **25**, pp. 3013–3021.

Vogel, M., Voigt, C., Gross, U. M., Müller-Mai, C. M. (2001). In vivo comparison of bioactive glass particles in rabbits. *Biomaterials*, **22**, pp. 357–362.

Wang, X., Du, Y., Luo, J. (2008). Biopolymer/montmorillonite nanocomposite: Preparation, drug-controlled release property and cytotoxicity. *Nanotechnology*, **19**, pp. 1–7.

Weiss, J., Takhistov, P., McClements, D. J. (2006). Functional materials in food nanotechnology. *J. Food Sci.*, **71**(9), R107–R116.

Williams, D. F. (1979). *Cunningham Journal of Materials in Clinical Dentistry*. Oxford, UK: Oxford University Press.

Wilson, K. S., Zhang, K., Antonucci, J. M. (2005). Systematic variation of interfacial phase reactivity in dental nanocomposites. *Biomaterials*, **26**(25), p. 5095.

Wu, C. J., Gaharwar, A. K., Schexnailder, P. J., Gudrun, S. (2010). Development of biomedical polymer-silicate nanocomposites: A materials science perspective. *Materials*, **3**, pp. 2986–3005.

Wu, C. J., Schmidt, G. (2009). Thermosensitive and dissolution properties in nanocomposite polymer hydrogels. *Macromol. Rapid Commun.* **30**, pp. 1492–1497.

Xu, R., Manias, E., Snyder, A. J., Runt, J. (2001). New biomedical poly(urethane urea)-layered silicate nanocomposites. *Macromolecules,* **34**, pp. 337–339.

Xu, Y., Ren, X., Hanna, M. A. (2006). Chitosan/clay nanocomposite film preparation and characterization. *J. Appl. Polymer Sci.,* **99**(4), pp. 1684–1691.

Xynos, I. D., Edgar, A. J., Buttery, L. D. K., Hench, L. L., Polak, J. M. (2001). Gene-expression profiling of human osteoblasts following treatment with the ionic products of Bioglass® 45S5 dissolution. *J. Biomed. Mater. Res. A*, **55**, pp. 151–157.

Yano, K., Usuki, A., Okada, A. (2000). Synthesis and properties of polyimide-clay hybrid films. *J. Polym. Sci. A*, **35**, pp. 2289–2294.

Yoo, Y., Kim, S. S., Won, J. C., Choi, K. Y., Lee, J. H. (2004). Enhancement of the thermal stability, mechanical properties and morphologies of recycled PVC/clay nanocomposites. *J. Polym. Bull.*, **52**, pp. 373–380.

Yuan, Q., Shah, J., Hein, S., Misra, R. D. K. (2010). Controlled and extended drug release behavior of chitosan-based nanoparticle carrier. *Acta Biomater.*, **6**, pp. 1140–1148.

Zhang, Q., Zha, L., Ma, J., Liang, B. (2007). A novel route to the preparation of poly(*n*-isopropylacrylamide) microgels by using inorganic clay as a cross-linker. *Macromol. Rapid Commun.*, **28**, pp. 116–120.

Zhou, W. Y., Guo, B., Liu, M., Liao, R., Rabie, A. B., Jia, D. (2010). Poly(vinyl alcohol)/halloysite nanotubes bionanocomposite films: Properties and in vitro osteoblasts and fibroblasts response. *J. Biomed. Mater. Res. A*, **93**, pp. 1574–1587.

Zhu, H. L., Shen, J. Y., Feng, X. X., Zhang, H. P., Guo, Y. H., Chen, J. Y. (2010). Fabrication and characterization of bioactive silk fibroin/wollastonite composite scaffolds. *Mater. Sci. Eng. C*, **30**, pp. 132–140.

Zhuang, H., Zheng, J. P., Gao, H., De Yao, K. (2007). In vitro biodegradation and biocompatibility of gelatin/montmorillonite-chitosan intercalated nanocomposite. *J. Mater. Sci. Mater. Med.*, **18**, pp. 951–957.

Index

ABS, *see* acrylonitrile-butadiene-styrene
acrylonitrile-butadiene-styrene (ABS) 10
adhesives 12, 16, 20, 22–23, 55, 309
admicelles 116–119, 130
adsorption
 counterion 107–108
 physical 115
adsorption mechanisms 115, 122, 129, 161–162
AFM, *see* atomic force microscopy
aliphatic polyesters 28, 303
alkylammonium ions 88, 114, 124–125, 145, 151, 153, 155, 172, 192, 196, 200
 pseudotrimolecular layers of 154, 196
alkylammonium salt 132, 196–197
alkylammonium salts, quaternary 158, 200
aluminum 89–90, 94, 111–112, 195, 323
amino acids 25–26, 31, 137, 250
ammonium cations 134, 199, 254–255
ammonium groups 126, 153, 155
ammonium salt 134, 201, 204, 212, 214
 polymeric quaternary 133, 198, 277
anilinium cations 136–137, 150–151
anionic polymerization 51–53, 59
aromatic diamines 32–33
aromatic dianhydrides 32, 36
aromatic polyamides 34
aromatic polyimides 32, 34
atomic force microscopy (AFM) 106, 256, 261

Bakelite 1, 3, 5, 16–17
ball clays 94
barrier properties 271–272, 278, 294, 307, 318
basal spacings 101, 113, 123, 126, 133–137, 144, 146, 148–152, 154–156, 158–160, 196, 199, 201–203, 242, 251
bentonites 96, 112, 134, 198–199, 203, 205, 215–216, 223, 297
benzene 10, 194, 222–223, 247
berthierine 91–92
bioactive glass 311–312
biodegradation 301, 305
biomaterials 300, 307, 313–314
 polymer nanocomposite 301
biomedical polymer–silicate nanocomposites 300
biomolecules 137
Bragg's law 101–102, 256, 258
butadiene 45–46, 54

carbanions 51–52
carbon atoms 10, 14–15, 25–26, 155–156, 170, 254
carbon dioxide 52, 265–266, 294–295, 297
carbon monoxide 265–266
carbon nanotubes 239, 317
cation exchange 115, 135, 144, 191, 195–197, 200–209, 213, 215–216, 248, 254

cation exchange capacity (CEC) 111, 113, 125, 127, 129, 131–133, 148–151, 159, 171, 196, 201–202, 225, 250, 254–255, 297
cationic surfactants 117, 125, 132, 134, 144, 148, 158, 164, 196, 200, 222, 225, 276
  long-chain 131, 152
cations, alkylphosphonium 122, 169, 196
CC, *see* cone calorimeter
CEC, *see* cation exchange capacity
chain reaction polymerization 47
chemical oxygen demand (COD) 316
chlorites 100, 106
chlorobenzene 222–223
clay dispersion 215, 259, 263, 278, 319
clay layer separation 243
clay minerals
  dried 158, 199
  mixed-layer 100
  modification of 137, 158, 200
  modified 156
  properties of 98, 144
  structure 89
  surface modification of 87, 115, 203
clay nanocomposites 202, 259, 267, 269–271, 297
clays
  adsorptive 144–145
  flint 94
  inorganic 241
  layered 241, 314
  organic 147, 323
  organophilic 144–145, 223
  pillared 102, 111
coagulation 23, 315–316
COD, *see* chemical oxygen demand
composites, polymer–clay 277
condensation polymerization 54
cone calorimeter (CC) 256, 264–265
controlled release 302–303, 308–309
coordination polymerization 50–51, 53–54

dexamethasone 307–308
dicarboxylic acid 14, 28, 31
differential scanning calorimetry (DSC) 71, 103, 146, 169, 247, 256, 261, 276–278
differential thermal analysis (DTA) 71, 103, 146, 193
DMA, *see* dynamic mechanical analysis
DMTA, *see* dynamic mechanical thermal analysis
DSC, *see* differential scanning calorimetry
DTA, *see* differential thermal analysis
dynamic mechanical analysis (DMA) 73–74, 269, 274
dynamic mechanical thermal analysis (DMTA) 73, 256

elastomers 4, 7, 55, 66, 277
electrostatic interactions 119, 137, 151
epichlorohydrin 16, 54, 56
epoxy resins 15–16, 56, 138, 271, 273, 297

fireclay 94
flame retardancy 71, 240, 270–271, 294
flame retardants 270–271
flocculation 315–316
formaldehyde 3, 16–18, 36–38

gas permeability 275–276, 299
glass transition temperature 9, 60, 66–68, 71, 269

heat-release-rate (HRR) 264–266, 299
hemi-micelles 117–119, 164
hexamethylenediamine 14–15, 31–32
HIL, *see* hole injection layer
hole injection layer (HIL) 317
HRR, *see* heat-release-rate
humic substances 224
hydrated polymeric networks 313–314
hydrocarbon chains 118–119, 126, 130–131, 147, 159
hydrophobic bonding 122, 129–131, 162–163
hydrophobic groups 115, 124, 130
hydrophobic interactions 117–119, 158, 164, 225
hydrophobic surface 107, 111, 116–117

illite 97–98, 105–106, 225
interlayer cations 100, 103, 114, 123, 126, 192, 195, 199, 223
interlayer packing density 146, 155
interlayer spacing 101, 122–123, 151, 169, 196, 251, 258
ion adsorption 107–108
ion exchange 87, 114, 122–125, 127, 129, 131–132, 135, 144–145, 150, 161, 164, 192, 195–196, 205–206
ion exchange capacity 112, 131, 134
ion pairing 122, 125, 129, 161
ion–dipole interactions 143, 148, 195, 199–200
ionic polymerizations 50–51

kaolin 93, 110–111
kaolinite 87, 92–94, 98, 105, 193–194

L-lactic acid 302–303, 311
lauryl acrylate 134, 212, 214
LEL, *see* light-emitting layer
LHL, *see* light-harvesting layer
light-emitting layer (LEL) 317
light-harvesting layer (LHL) 317
living polymerization 53

macromolecules 1–2, 5, 22–23, 25, 27, 31, 59, 66, 68, 114
magnesium 89–90, 112, 195
melamine 18, 37–38
melamine formaldehyde (MF) 18, 37
melt intercalation 166, 244, 252, 263, 275–276
methanol 9, 30, 52, 247, 299
MF, *see* melamine formaldehyde
micelles 50, 116
microspheres 49–50, 302
MMT, *see* montmorillonite
  calcium 96, 111–112
  modified 134–136, 141, 159, 167, 248, 250
MMT-based organoclays 130, 170
MMT modification 144, 165
montmorillonite (MMT) 96–98, 129, 135–137, 149–150, 158–159, 169–170, 193–194, 198–200, 202–203, 213–215, 247–252, 274, 276, 305–306, 309

nacrite 92–93
nanocomposite composition 262, 267
nanocomposite films 278, 323
  polymer–MMT 324
nanocomposite formation, polypropylene–MMT 169
nanocomposite hydrogels 313
nanocomposite polyethylene 262
nanocomposite scaffolds 305, 311

nanocomposites
epoxy 136, 251
exfoliated 242, 255
hybrid 246
intercalated 242, 258
nitrile rubber–organoclay 299
organoclay–polyethylene 135
polyamide-6/MMT 277
polyamide–clay 269
polyethylene terephthalate–MMT 135
polymer-based 317
polymer–organoclay 200
polyurethane 279
thermoplastic 250, 254
natural rubber 2, 23–24
NMR, *see* nuclear magnetic resonance
nuclear magnetic resonance (NMR) 69, 102–103, 138, 147, 169, 256, 263

octadecyl-ammonium chloride 204, 206, 209
octahedral sheets 90, 93–94, 96, 99, 110–111, 138, 195
odinite 92–93
OMLS, *see* organically modified layered silicate
organically modified layered silicate (OMLS) 87, 131, 252, 304
organobentonites 133–134, 198–199, 277–278
organoclay dispersion 253, 259
organoclays, surfactant-intercalated 202

palygorskite 87, 92, 99, 112, 193
PF, *see* phenol formaldehyde
phenol formaldehyde (PF) 16, 37–38
phosgene 28, 30–31
phyllosilicates 94, 97, 110, 112, 132
PLS, *see* polymer-layered silicate
PLS nanocomposites 240, 242, 246, 250, 256, 263
PMMA, *see* poly(methyl methacrylate)
PMMA-silica nanocomposites 321
poly-3-hydroxybutyrate 26
poly(methyl methacrylate) (PMMA) 9, 44, 47, 245, 310, 321
polyamides 4, 14, 28, 30–31, 34, 56, 127, 158, 200, 243, 277–278, 318, 324
polybenzimidazoles 34–36
polybenzoxazoles 34–35
polycaprolactam 56, 58
polycaprolactone 29, 245, 247, 303
polycarbonates 30
polycondensation 29
polyesters 4–5, 10, 16, 18, 27–29, 303, 321
polyethers 34, 39–40
polyethylene 3, 5, 7–8, 10, 19, 47, 68, 158, 200, 275, 302, 322
polyimides 28, 32, 34, 248, 279, 318
polyisoprene 24
polymer-based nanocomposite systems 245, 249
polymer characterization 69–73
polymer-clay nanocomposites, exfoliated 244
polymer coatings 273
polymer composites 74, 293
polymer crystallinity 72, 74, 305
polymer intercalation 248, 254
polymer-layered silicate (PLS) 97, 169, 240, 293
polymer nanocomposites, silicate-based 307
polymer–clay nanocomposites, exfoliated 263

polymeric quaternary ammonium salts (PQAS) 133, 198, 277
polymerization
  cationic 51–52
  chain-growth 27, 40
  diene 45
  emulsion 49–50
  free-radical 41, 44, 51–52, 165, 304
  intercalative 244, 252, 322
  interparticle 114, 192
  low-temperature 58–59
polymerization of ethylene 7–8
polymerization of monomers 133, 169, 196
polymerization of propylene oxide 55–56
polymers
  aromatic ladder 35–36
  biodegradable 240, 302
  cationic 52, 316
  ladder 36
  linear 6, 22, 277
  living 53
  natural 1–2, 5, 22, 26, 301–302, 309, 311
  polyurethane 19, 21
  step-growth 27–29
  synthetic 3, 5, 12, 14, 16, 239
  thermoset 7, 36, 61
  virgin 240, 269–270, 294, 299, 305
polyols 19–20, 28
polypropylene 8, 63, 68, 158, 200, 263, 277, 319, 322–323
polystyrene 1, 3, 5, 9–10, 44, 47–49, 270, 273, 299
polysulfides 34, 40
polysulfone 28, 34, 40, 273
polytetrafluoroethylene 19
polyurethane 5, 19–21, 28, 318
polyvinyl alcohol 13
PQAS, *see* polymeric quaternary ammonium salts
propylene oxide 54–55

QACs, *see* quaternary ammonium cations
quaternary ammonium cations (QACs) 127, 144, 164, 201

ring-opening polymerization (ROP) 28, 54–55, 167, 250, 303
ROP, *see* ring-opening polymerization
rubber nanocomposites 274

scanning electron microscopy (SEM) 104–105, 145, 148, 256, 258, 260
SEM, *see* scanning electron microscopy
sepiolite 92, 99, 112, 135, 193, 206
silica 111, 116, 118, 312, 321–322, 324
silica nanoparticles 304, 313
silicones 21–22
smectite clays 112, 143
smectites 95–98, 103, 111–113, 127, 137, 164, 166, 192–193, 201, 223
smectitic clay minerals 114, 195
sodium montmorillonite 96, 111, 134–135, 158–159, 204–206, 209, 254
sodium montmorillonite cloisite 206–207, 214–215
sodium montmorillonite kunipia 204, 209
sol–gel techniques 245–246, 312
step-growth polymerization 27–29

substitution
  isomorphic 95–96, 107, 110
  isomorphous 94, 96, 98, 100, 133
sulfur 4, 24
surfactants
  intercalated 146–147, 197
  nonionic 116–117, 134, 196
  organic 125, 131, 145, 159, 195

TEM, *see* transmission electron microscopy
terephthalic acid 29, 34–35
TGA, *see* thermal gravimetric analysis
thermal gravimetric analysis (TGA) 70–71, 103, 141, 145–147, 247, 256, 261, 269–270, 278
thermoplastic polyurethanes 20
transmission electron microscopy (TEM) 105, 145, 148, 172, 246–247, 256, 258–260, 263, 277, 297
trichloroethylene 222, 224

urea formaldehyde 3, 18, 37, 259–260

vermiculites 92, 99, 112, 123, 132, 153, 158, 195, 199, 299
vinylbenzyl chloride 134, 212, 214
vulcanization 2, 4, 24

wastewater pretreatment 314–315
wastewaters 224, 315–316

X-ray diffraction (XRD) 71–72, 101, 103, 145–146, 148, 150, 199, 202, 247–248, 256, 258, 263, 276, 278, 297
XRD, *see* X-ray diffraction

zeta potential 160–162